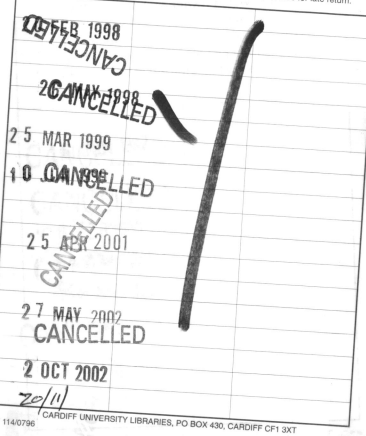

THE QUATERNARY HISTORY OF IRELAND

THE
QUATERNARY HISTORY
OF IRELAND

Edited by

KEVIN J. EDWARDS

Department of Geography
University of Birmingham
Birmingham, England

WILLIAM P. WARREN

Quaternary and Geotechnical Section
Geological Survey of Ireland
Dublin, Ireland

1985

ACADEMIC PRESS
Harcourt Brace Jovanovich, Publishers

London   Orlando   San Diego   New York
Austin   Montreal   Sydney   Tokyo   Toronto

ACADEMIC PRESS INC. (LONDON) LTD.
24–28 Oval Road
LONDON NW1 7DX

United States Edition published by
ACADEMIC PRESS, INC.
Orlando, Florida 32887

British Library Cataloguing in Publication Data

The Quaternary history of Ireland.
    1. Geology, Stratigraphic-----Quaternary
    2. Geology-----Ireland
    I. Edwards, Kevin J.   II. Warren, William P.
    551.7'9'09415   QE696

Library of Congress Cataloging in Publication Data

Main entry under title:

The Quaternary history of Ireland.

    Includes index.
    1. Geology, Stratigraphic--Quaternary.   2. Geology--
Irish-Republic.   3. Geology--Northern Ireland.
I. Edwards, Kevin J.   II. Warren, William P.
QE696.Q327   1985   551.7'9'09415   84-24554
ISBN 0-12-232730-6 (alk. paper)

PRINTED IN THE UNITED STATES OF AMERICA

85 86 87 88   9 8 7 6 5 4 3 2 1

To the memory of
FRANCIS MILLINGTON SYNGE
(1923–1983)
Scientist and Friend

PREFACE

Our record of Irish Quaternary history begins late in the Quaternary Period. The start of this period may have been 2 million years ago; for the time being, we remain largely ignorant of events for perhaps 90% of the Irish Quaternary sequence. Nevertheless, the visible landscape provides many tantalizing glimpses of its origins. Apart from selected papers in *Irish Geographical Studies in Honour of E. Estyn Evans* (1970, edited by N. Stephens and R. E. Glascock), G. F. Mitchell's excellent treatment of *The Irish Landscape* (1976), and sections of *Ireland* (1978, by G. L. Herries Davis and N. Stephens, with contributions from F. M. Synge) there has been no extended treatment of the Quaternary history of Ireland in this century, although the special Quaternary issue of the *Irish Naturalist's Journal* of 1934 should be noted.

In preparing this book our aim has been to provide a text at the senior undergraduate and postgraduate levels and for all who require a knowledge of the Quaternary history of Ireland, whether for advanced study or as a starting point for research. Each chapter reflects an approach to Quaternary studies from a particular specialist viewpoint. No attempt has been made to harmonize opinions as this would run counter to the intention of producing a work that will reflect the many interpretations of a constantly developing field of study. The value in including differences of opinion lies in the fact that the strengths and weaknesses of various approaches can be assessed by the reader. We hope this will also stimulate further research aimed at resolving the many outstanding problems. Although we have attempted to provide a comprehensive treatment of Irish Quaternary history it is unfortunate that, as yet, there is insufficient material to warrant specific chapters on such topics as palaeoclimates, fossil insects, diatoms, or radiochronometric methods other than radiocarbon dating. However, many of these topics receive mention in the text.

It is a pleasure to thank the referees for their advice and to record the great debt we owe to the many typists and cartographers. The support of Konrad Guettler of Academic Press has also been much appreciated. Both of us have benefited from the encouragement of fellow committee members of a fledgling Irish Association for Quaternary Studies (IQUA), in particular Ned Culleton and the late Francis Synge. Needless to say, the whole venture would not have been possible without the ensuing interest, expertise, and forebearance of the contributors. Finally, very special thanks are due to Leonie Warren for her assistance and patience and Carol for her understanding.

July 1984   KEVIN J. EDWARDS
                                                    WILLIAM P. WARREN

CONTRIBUTORS

Numbers in parentheses indicate the pages on which the authors' contributions begin.

MICHAEL G. L. BAILLIE (294), Department of Archaeology and Palaeo-ecology Centre, The Queen's University of Belfast, Belfast BT7 1NN, Northern Ireland

EDWARD B. CULLETON[1] (133, 318), An Foras Talúntais, Dublin 5, Ireland

EUGENE P. DALY (331), Groundwater Section, Geological Survey of Ireland, Dublin 4, Ireland

KEVIN J. EDWARDS (1, 187, 280, 302), Department of Geography, University of Birmingham, Birmingham B15 2TT, England

MICHAEL GARDINER (133, 318), An Foras Talúntais, Dublin 5, Ireland

KENNETH R. HIRONS (302), Department of Geography and Palaeo-ecology Centre, The Queen's University of Belfast, Belfast BT7 1NN, Northern Ireland

COLIN A. LEWIS[2] (95), Department of Geography, University College, Dublin, Ireland

A. MARSHALL McCABE (67), School of Environmental Sciences, University of Ulster at Jordanstown, Newtonabbey BT37 0QB, Northern Ireland

G. FRANK MITCHELL (17), Quaternary Studies, Trinity College, Dublin 2, Ireland

MICHAEL O'MEARA (309), Geological Survey of Ireland, Dublin 4, Ireland

JONATHAN R. PILCHER (294), Department of Botany and Palaeo-ecology Centre, The Queen's University of Belfast, Belfast BT7 1NN, Northern Ireland

[1]Present address: Commission of the European Communities, 1040 Brussels, Belgium.
[2]Present address: Geography Department, University of Transkei, Umtata, Republic of Transkei, via South Africa.

ANTHONY J. STUART (222), Department of Zoology, University of Cambridge, Cambridge CB2 3EJ, England

FRANCIS M. SYNGE† (115), Quaternary Section, Geological Survey of Ireland, Dublin 4, Ireland

ROY THOMPSON (302), Department of Geophysics, University of Edinburgh, Edinburgh EH9 3JZ, Scotland

WILLIAM P. WARREN (1, 39, 343), Quaternary and Geotechnical Section, Geological Survey of Ireland, Dublin 4, Ireland

WILLIAM A. WATTS (155), Trinity College, Dublin 2, Ireland

LOUISE H. VAN WIJNGAARDEN-BAKKER (233), Albert Egges van Giffen Instituut voor Prae- en Protohistorie, University of Amsterdam, 1012 WP Amsterdam, The Netherlands

PETER C. WOODMAN (251), Department of Archaeology, University College, Cork, Ireland

†Deceased.

CONTENTS

4.   GLACIAL GEOMORPHOLOGY
A. Marshall McCabe

5.   PERIGLACIAL FEATURES
Colin A. Lewis

12.   CHRONOLOGY

Kevin J. Edwards, Michael G. L. Baillie, Jonathan R. Pilcher,
Kenneth R. Hirons, and Roy Thompson

13.   ECONOMIC ASPECTS OF THE QUATERNARY

William P. Warren, Michael O'Meara, Eugene P. Daly,
Michael J. Gardiner, and Edward B. Culleton

LIST OF FIGURES

LIST OF TABLES

CHAPTER 1

QUATERNARY STUDIES IN IRELAND

Kevin J. Edwards
Department of Geography
University of Birmingham
Birmingham, England

William P. Warren
Quaternary and Geotechnical Section
Geological Survey of Ireland
Dublin, Ireland

THE TRADITION

Louis Agassiz, the Swiss glacial theorist, visited Ireland in the autumn of 1840 and demonstrated evidence of glaciation to a sceptical and generally incredulous geological establishment. Within 2 years, however, C. W. Hamilton (1842) delivered a lecture on the topic of glacial scratches and ice-moulded surfaces in Counties Cork and Kerry to the Geological Society of Dublin. This paper was not published in the Society's journal, perhaps because it was regarded as too radical. Certainly, Oldham, a pillar of the geological establishment, was trenchant in his refusal to countenance the idea of glaciation in Glanmalure, Co. Wicklow, and described the origins of distinct moraines there as "referable to the ordinary action of water"; he further commented, "There is no necessity for resorting to the idea of glacial action to account for such deposits" (Oldham, 1846, p. 198).

Despite observations such as those of Ball (1849) on cirques and those of Rowan (1852), who, on the subject of erratic blocks, questioned whether glacial ice would transport debris uphill, it took many decades to establish

THE QUATERNARY HISTORY OF IRELAND
1

the glacial theory and the concept of an "Ice Age" in Ireland. It is without question that Maxwell Close's (1867) clear and incisive account of the glaciation of Ireland, which introduced the Gaelic terms droimnín (drumlin) and eiscir (esker) to the scientific literature, played a crucial role in convincing the Irish geological community of the fact of glaciation. His paper may be regarded as a turning point in the geological sciences in Ireland and indeed is a classic in the world geological literature. By the end of the 1860s, geologists of the Geological Survey were recording glacial features as a matter of routine and investigation had begun into the stratigraphy of the Quaternary deposits (Harkness, 1869).

By 1894 the value of Close's work was already recognized. In the introduction to H. Carvill Lewis's posthumous book *The Glacial Geology of Great Britain and Ireland,* it is remarked: "One of the most valuable aids in Professor Lewis's work in Ireland were the detailed observations of the Rev. Maxwell H. Close M.R.I.A. & c., every one of which was found to be absolutely correct" (Lewis, 1894, p. xxxiv). This evaluation has been re-echoed in succeeding generations (Wright, 1927; Warren, 1978), and this remarkable man's work is as valid to day as it was 120 years ago.

H. Carvill Lewis visited Ireland from North America in 1885–1886 and again in 1887. It is clear that during his first visit he recognized the feature since named the Southern Irish End-Moraine (Charlesworth, 1928) as an important terminal moraine (Lewis, 1886, 1894). He regarded it as the southern limit of general glaciation. Charlesworth (1928), who grossly underestimated the importance of Lewis's observations, particularly with reference to the moraine and the features associated with the glaciation of the southwest, described this moraine in detail and, following Geikie (1914), regarded it as the limit of the Younger Drift. Thus began the morphostratigraphic school in Irish Quaternary studies. Wright and Muff (1904) had earlier laid the foundations for a stratigraphic framework in their description of the Quaternary sediments of the south coast. Their work was later incorporated into the morphostratigraphic framework (see Mitchell *et al.,* 1973).

The origin of eskers puzzled geologists from the 1860s to the end of the nineteenth century. As this was a period during which the theory of glacial submergence was popular, it is not surprising that Kinahan (1878) interpreted them as submarine forms and even categorized them on the basis of specific current conditions. It was Sollas and Praeger (see Sollas, 1896) who finally put an end to Kinahan's marine theory and established eskers as ice-contact glaciofluvial deposits. Francis M. Synge (1950) demonstrated the depositional mechanism for eskers using examples near Trim, Co. Meath, and his model is broadly accepted today.

In another seminal paper on Quaternary geology by an Irish geologist, G.

H. Kinahan (1894) first described and accurately interpreted what have since come to be termed protalus ramparts. These features would more aptly and accurately fit the name by which he called them, clocha sneachta (not knowing Gaelic well he spelled it cloghsnatty), meaning snow stones, having heard the term from sheep farmers in Connemara.

Farrington's detailed studies in Co. Wicklow, particularly in the glacio-fluvial gravels at Enniskerry (1944) and Brittas (1942), are the first systematic stratigraphic interpretation of such deposits through the medium of provenance in Ireland. Synge (1966) extended systematic stone-type counts and drew the first published Quaternary maps which show both the lithology and petrological composition of Quaternary sediments. Synge was also the first Irish geologist to study systematically the raised shorelines in the north and northeast, using techniques learned in Finland.

The soils of Ireland began to receive attention in the mid nineteenth century when the Griffith Valuation requested the valuators to take account of the nature of the soil and subsoils (Anonymous, 1853). Robert Kane's project for a soil survey of Ireland included chemical analyses of soils. His coloured soil maps were lodged with the Museum of Irish Industry about 1862 but later disappeared (Simington and Wheeler, 1945). The Geological Survey of the then United Kingdom was instituted in 1845, and the examination of soils initially formed an important part of its regular work. Soil examination was subsequently abandoned in favour of a more rapid mapping programme and was not resumed until 1901. *A Description of the Soil-Geology of Ireland, Based upon Geological Survey Maps and Records, with Notes on Climate* appeared in 1907 (Kilroe). Although various research papers on pedogenesis appeared in the first half of this century (e.g., Gallagher, 1942; Gallagher and Walsh, 1942), a formalized National Soil Survey of Ireland was not set up until 1968. There is no soil survey for Northern Ireland, but pedological work has been carried out by Cruickshank (see review, 1982).

Most early commentaries on the Irish Quaternary flora were the result of observations in peat bogs (reviewed in Jessen, 1949). Thus, Kinahan (1878) noted the distinct layers of oak and pine stumps, separated by and eventually covered by peat in Irish bogs—a view of the stratigraphy which was in general agreement with Geikie's (1881) system for Scottish bogs. A corresponding pattern observed in Norway by Blytt and in Sweden by Sernander, produced the theory of climatic change with terms such as "Boreal" and "Atlantic", which themselves came to be applied to Irish peat profiles (Erdtman, 1928; Jessen, 1949). Robert Lloyd Praeger (1892) showed that submerged peats in the north-east of Ireland contained remains of pine, hazel, oak, and willow. Forbes (1914) found layers of tree stumps in peat similar to those described by Kinahan, but he questioned the correctness of Kinahan's wood identifications and their stratigraphic positions. Unlike the earlier writer, Forbes also drew climatic inferences from the data, as did Brooks (1921). The question of the survival of the Irish flora during glacial/interglacial times had been extensively discussed by Praeger (1932, 1934a, 1939). Following Forbes (1846), Praeger made a claim for the survival in Ireland of Lusitanian and American elements of the flora during glaciation, rather than their extinction. Jessen (1949) in general agreed with this but stressed the fact that many other elements (especially woodland plants) must have reached Ireland during the postglacial period.

Other contributions of interest to the past flora of Ireland include the *Chara* marl finds of Reid (1895, 1904); the discovery of *Salix herbacea* leaves at Ballybetagh (Stelfox, 1927); the records of *Dryas octopetala* and *Betula nana* in lateglacial deposits in Counties Monaghan and Louth (Mitchell, 1942a,b); and literature on diatoms (e.g., Foged, 1977; Battarbee, 1978).

In 1902, Robert Lloyd Praeger wrote (p. 60), "We can never hope to understand our phytogeography till its problems have been attacked by the historical method." In saying this, Praeger had in mind the work on the macroremains of plants recovered from peat bogs and other deposits. As a result of visits to Ireland in 1924 and 1926, the Swedish worker Gunnar Erdtman first applied the techniques of pollen analysis (palynology) to Irish deposits (Erdtman, 1924, 1927, 1928). This encouraged Praeger to enthuse

> What is wanted is a study of the plant (and animal) remains of all our Postglacial, Interglacial and Early Preglacial deposits of whatever kind—work like that of Clement Reid in England and Bennie in Scotland; work that deals with seeds and fruits, twigs and wood, as well as pollen, and in the animal world not only with bones, but insect elytra and all other faunistic remains [Praeger, 1934b, p.127].

Earlier, an informal meeting had taken place at the Academy House, Dublin, on 25 January 1933. This had led to the formation of the Royal Irish Academy's Committee for Quaternary Research in Ireland with Praeger as chairman and Anthony Farrington as secretary. In spite of financial problems the Committee was able to invite the Danish palaeobotanist Knud Jessen to visit in 1934 and 1935. One of the desires of the Committee had been to "arrange for the instruction of Irish students with the aim of founding permanent centres of quaternary research in Ireland" (Farrington, 1934, p.128). As a consequence, Professor Jessen (1934, p.130) wrote, "My assistant, Mr. H. Jonassen, also participated, as well as Mr. G. F. Mitchell and Mr. T. Maher, who were the Irish students accepted by the Committee to receive instructions in the field methods of the research." Jessen's visit was a great success and resulted in the investigation of 43 sites in 1934 alone. Delays occasioned by war prevented the appearance of Jessen's major study on Ireland's lateglacial and postglacial vegetational history until 1949. Jessen's research had also resulted in the first determination of Irish lateglacial deposits (with the traditional tripartite biolithostratigraphic divisions) at Ballybetagh, Co. Dublin, Ralaghan, Co. Cavan, Frenchpark, Co. Roscommon, and Roundstone, Co. Galway (Jessen and Farrington, 1938). The interglacial stratotype at Gort, Co. Galway, the pivot of much Irish Quaternary stratigraphic debate, was discovered by Kinahan (1878) but was first investigated by Jessen, Andersen, and Farrington (1959).

Palynological studies in archaeological contexts as well as in more conventional anthopogenically orientated peat and lake deposit studies were also published by Erdtman (1928) and Jessen (1936, 1949). In 1945, Mitchell published the results of other investigations by himself and Jessen, followed by research which focussed particularly on man (Mitchell, 1951, 1956). Most Irish pollen studies have come from Trinity College, Dublin, where Frank Mitchell and William Watts have researched all periods of the Quaternary, and The Queen's University of Belfast, where Alan Smith (now of Cardiff), Jonathon Pilcher, and associates have focussed on the impact of early man. The Belfast laboratory was established at The Queen's University after a successful appeal to the Nuffield Foundation in 1954 by a number of University departments. The Nuffield Quaternary Research Unit (later to become the Palaeoecology Centre) was envisaged as an interdisciplinary enterprise with Alan Smith as leader, and Michael Morrison and Bruce Proudfoot became his first assistants (Nuffield Foundation, 1954; Evans, 1975). The Palaeoecology Centre contains the only surviving radiocarbon dating unit in Ireland.

Ireland's fauna has long received attention, and this has extended to a recognition of the restricted nature of the terrestrial mammalian fauna. The absence of snakes and frogs, for example, had been remarked upon by various commentators before the ninth century AD (Praeger, 1950). Later, scientific interest in the Irish fauna led to William Thompson's three volumes on the birds (1849–1851) and a fourth volume, published posthumously, on the mammals (1856). R. F. Sharff was particularly concerned with the former distribution of animal species, and he was also actively involved in the study of bones from cave deposits and archaeological sites (Sharff, 1904, 1906, 1928; Scharff et al., 1906, 1918).

The extinct giant Irish deer (*Megaloceros giganteus*), long known as "Irish elk" (*Cervus giganteus*), was recorded as early as the sixteenth century AD (O'Riordan, 1980). The research at the Ballybetagh bogs (Jessen and Farrington, 1938) had itself been prompted by a desire to discover the environment under which the giant deer had flourished—the bogs had long been famous for their large yields of *Megaloceros* bones.

Robert Lloyd Praeger was able to draw on the Scottish researches of Jamieson (1865) and Robertson (1877) to demonstrate a postglacial warm period from the molluscan remains in the estuarine clays of Belfast Lough

(Praeger, 1888, 1892, 1896). This discovery of the "Climate Optimum" was wrongly attributed to Praeger by W. B. Wright in the dedication of the second edition of his book *The Quaternary Ice Age* (Wright, 1937).

Aspects of the history of archaeology in Ireland have been documented in, for example, Mahr (1934, 1937), Mitchell (1976), Herity and Eogan (1977), and Woodman (1978). The Harvard University Missions directed by Hencken were particularly influential, especially the third under the assistant directorship of Hallam Movius who encouraged the participation of environmental specialists (Movius, 1942). The palaeoenvironmental context of archaeological sites has been investigated most notably by palynologists (Erdtman, 1928; Jessen, 1936; Mitchell, 1945; Pilcher, 1969; Pilcher and Smith, 1979), soil scientists (Gardiner and Walsh, 1966; Edwards *et al.*, 1983), or both (Proudfoot and Simmons, 1958; Case *et al.*, 1969; appendices in O'Kelly, 1982). The recent burgeoning of such papers (e.g., see Reeves-Smyth and Hamond, 1983) is a response to an increased awareness of the value of environmental studies in archaeology. This has not been matched to the same extent by spatioenvironmental studies, although mention might be made of Watson (1945) and Woodman (1974, 1981).

A SELECT DISCUSSION OF CURRENT VIEWS

In the century and a half since Agassiz's visit, our understanding of the Quaternary history of Ireland has moved from the realm of religious myth to that of a science of an increasingly exact nature. Yet, although we seem to have a nominal framework of glacial events, interglacial floras, and postglacial floral and faunal developments, there is much that remains unclear and many questions which seem no closer to resolution now than they were a half century ago.

Although Quaternary deposits cover more than 90% of the land surface of Ireland and are more extensive than the deposits of any other period, the stratigraphic sequence seems very much incomplete. Deposits of the early and middle Quaternary appear to be absent. If they do occur, the palaeontological evidence on which they might be dated has so far eluded researchers. Close examination of infillings in karst depressions, particularly in the southern synclinal valleys, has revealed no evidence of early Quaternary deposition. This may, in some cases, be due to the fact that they have not always been approached from the point of view of possible Quaternary redeposition. It is these areas, however, that offer the greatest possibility of identifying early Quaternary sequences, many of which may, for the most part, take the form of reworked Tertiary sediments and require very close analysis for Quaternary indicators.

Given the nature of the late Quaternary glaciation of the country this hiatus is not surprising, for it must be assumed that considerable amounts of friable surface material, if not removed by preceding periglacial activity, were removed by glacial action (Mitchell, Chap. 2, this volume). Thus, our examination of the Quaternary history of Ireland must move from Mitchell's interpretation of the preglacial landscape directly to a consideration of the stratigraphic sequence of the late Quaternary, which provides a framework to which the more detailed examination of its components can be fitted or counterposed.

The question of the number and extent of glaciations which occurred in Ireland seems no nearer solution than when Wright (1914) discussed the question of the occurrence of "Older Drift" in Ireland. The morphostratigraphic approach which endured for 50 years has been criticized and the consensus achieved by Mitchell et al. (1973) undermined (Bowen, 1973; Warren, 1979). But the strict stratigraphic approach advocated as an alternative does not increase the number of glacial events recognized; the reverse tends to be the case (Table I). This is ironic in view of the dramatic evidence in the deep-sea core record which suggests that the four or five glaciations traditionally recognized in northern Europe may underestimate the number of glacial cycles considerably (see Kukla, 1977).

The question of the number of interglacials is the nub of the matter, for without evidence of interglacials we cannot count glacials. Knowledge of the palynological sequence and inferred climatic conditions of the Gortian Interglacial has not expanded significantly since Watts's review (Watts, 1967; but see Warren, 1979). Mitchell's (1976) phases of that interglacial are not based on any new evidence. Yet there are now at least seven accepted Gortian sites widely spread from Co. Tyrone to Co. Limerick. Still, the question of the number of interglacial events has not been resolved palynologically. The "Hidden Interregnum" of Mitchell (1976) has not emerged within the context in which he placed it.

Watts, in Chapter 8 of this volume, clearly anticipates the kingdom of the last interglacial and seeks a hornbeam-dominated flora in suitable stratigraphic position with which to crown it. This, of course, is not the only way in which this problem might be solved. Warren's approach (Chap. 3) is to treat the "Hidden Interregnum" as a misinterpretation of an obscure or repetitive reference in the royal annals of Quaternary history and to place the Gortian within the realm of the last interglacial.

The disposition of the glacial and glaciofluvial morphological features is well known, in particular the esker networks and the drumlin swarms

Table I Quaternary Stratigraphic Columns for Ireland, Britain, and Northern Europe,[a] Each Compared with the $^{18}O/^{16}O$ Stages of the Deep-Sea Record

	Ireland[b]		Ireland[c]	$^{18}O/^{16}O$ Stages[d]	Britain[c]		N. Europe[d]
	Littletonian Stage		Littletonian Stage	1	Flandrian		Holocene
10,000 bp	Ballybetagh Substage	Fenitian Stage	Late Midlandian	2	Late	Devensian Stage	Weichsel
	Woodgrange Substage						
	Maguiresbridge Substage		—— 26,000 bp ——				
		Midlandian Stage	Middle Midlandian	3	Middle		
30,500 bp	Derryvree Substage		—— 50,000 bp ——				
	Fermanagh Substage		Early Midlandian	4	Early		
	Gortian Stage		Last Interglacial (Hidden Interregnum, Mitchell, 1976)	5	Ipswichian Stage		Eem-Skaerumhede
				— 128,000 bp —			
	? Ballybunnionian Stage		Munsterian Stage	6	Wolstonian Stage		Warthe
			Gortian Stage	7	Hoxnian Stage		Eem-Ehrmgsdorf
				— 225,000 bp —			
			Pre-Gortian Stage	8	Anglian		? Saale-Rehburg

[a] No correlation is implied between the stratigraphic columns of Ireland, Britain, and northern Europe. Each should be related independently to the $^{18}O/^{16}O$ column.
[b] Warren (Chap. 3, this volume).
[c] Mitchell et al. (1973).
[d] Kukla (1977).

(Sollas, 1896; Charlesworth, 1928, 1963, 1973; Synge, 1970). The nature of the sediments and their morphology south of the Southern Irish End-Moraine are not so well known and still await detailed study. The traditional explanation, that the more "subdued" morphology south of the moraine reflects a periglacially degraded glacial landscape that was once significantly pronounced, remains to be tested (see Charlesworth, 1928). McCabe's treatment of the Quaternary geomorphology in Chapter 4 reflects this situation. In concentrating on the evolution of the classic surface expression of the glacial/glaciofluvial landforms, he provides us with new and challenging insights but keeps within the framework of Mitchell *et al.* (1973) by regarding the deposits beyond the Southern Irish End-Moraine as having been greatly altered by periglacial activity.

Lewis (Chap. 5) points out that the known occurrence of periglacial features, in particular ice-wedge casts, correlates very closely with the distribution of areas studied in detail. He notes that these features are common inside the "End-Moraine" and questions whether the same might not prove true of features such as pingos. Data from mapping projects (W. P. Warren, unpublished) indicate that glacial and glaciofluvial features, where they occur south of the moraine (e.g., west of Cork Harbour and Watergrass Hill), are not significantly more degraded than similar features to the north of it. It does seem, however, as if the pattern of deposition, as well as the underlying topography, was quite different on either side of the moraine. There is evidence of a thin cover of till but no evidence that there were ever significant numbers of eskers, moraines, or drumlins to the south.

A perceived difference in the degree of weathering of sediments on either side of the moraine was a common argument used by some writers in support of a morphostratigraphic approach which recognized the moraine as the limit of the last glaciation. But Culleton and Gardiner (Chap. 7) point out that an outstanding feature of Irish soils is their youth. The soils are not highly weathered, and therefore soil types relate closely to parent materials. A corollary of this has been that, since its inception in 1968, the National Soil Survey has, in studying soil parent materials, added much to our knowledge of specific Quaternary sediments. Soil is the interface between the study of Quaternary sediments and the floral, faunal, and human environmental branches of Quaternary studies.

Culleton and Gardiner recognize two phases in pedogenesis separated around the time of the Neolithic. These authors are reluctant to invoke a specifically anthropogenic cause for the expansion of heath vegetation and blanket bog. Edwards (Chaps. 9 and 12) concurs but expresses no reluctance in referring to the depredations of Neolithic people as exemplified, for example, by the evidence from pollen diagrams. He also reports relatively recent evidence for pioneer Neolithic or even late Mesolithic agriculture in Ireland (see Groenman-van Waateringe, 1983; Edwards and Hirons, 1984).

There is considerable information regarding the extent of human colonization during the Mesolithic period. Communities of this time are known to have occupied inland sites and to have been much more than the limited coastal Larnian groups that had so long been regarded as the first inhabitants. Yet the Palaeolithic period is still an enigma. A palaeolithic-style flint flake found in a gravel pit near Drogheda (Mitchell and Sieveking, 1972) suggests human activity in the Irish Sea basin area. And the tantalizing, if controversial, date obtained on the basis of a flourine test would make Kilgreany Man's bones more than 10,000 years old (see Raftery, 1963; but see also Stuart, Chap. 10). As Woodman points out (Chap. 11), the earliest accepted date for human occupation of Ireland has been pushed back to about 9,000 bp.*

The question of the earliest inhabitants of Ireland also raises the topic of a land bridge. Certainly one of the most stimulating suggestions in this regard is Synge's proposal (Chap. 6) of a Celtic Sea land bridge which would have opened Ireland directly to land influences from southern Britain and the Continent. Proof for the existence of a land bridge for a short period in lateglacial time would add a new dimension to our understanding of faunal history, including the introduction of *Megaloceros giganteus* as well as that of man (much of this had been presaged by Jamieson, 1865). Its early closure would also partially explain some of the puzzling deficiencies in our present faunal list (see Stuart and van Wijngaarden-Bakker, Chap. 10), but it would also raise the vexed questions of immigration rates, the possibility of refugia for a wide range of species during glaciation, and time lags in the pollen record. Edwards (Chap. 9) invokes the disappearance of the land bridge as much as the restricted range of habitats for the relative poverty of the flora (some 67% of the total British list). Certainly, an early closure of the bridge might also help to put this in perspective. Again, a southern location of the proposed bridge would have facilitated, as Synge (Chap. 6) points out, the introduction of the Lusitanian element. This, along with the thriving alpine elements in some western localities, might encourage less emphasis being placed on the limited available range of habitats. A conference on the postglacial colonization of Ireland favoured a lateglacial land bridge at the northern end of the Irish Sea (Woodman, 1984).

The chronology of the Irish Quaternary is for the most part relative. The problem of obtaining absolute dates relates largely to the lack of suitable techniques that can be applied to the period earlier than the mid Midlandian (Fenitian of Chap. 3). Radiocarbon techniques, the calibration and dating

*Usage of lowercase bp indicates a radiocarbon date and capital BP, a calibrated date.

Table II Terminology of Irish Lateglacial and Postglacial Stratigraphy

14C Years bp	Blytt-Sernander/ Jessen (1949)	General	Mitchell (1976)	Pollen zones — Jessen (1949)	Pollen zones — Mitchell (1956)	Archaeology	14C Years bc/ad
0	Sub-Atlantic (Ireland only)	Postglacial	Littletonian	VIII	X	Modern	1,000
1,000					IX	Historic	0
2,000					VIIIb	Early Christian Iron Age	1,000
3,000	Sub-Boreal					Bronze Age	2,000
4,000				VIIb	VIIIa	Neolithic	3,000
5,000	Atlantic			VIIa	VII		4,000
6,000				VI	VI	Mesolithic	5,000
7,000	Boreal			V	V		6,000
8,000				IV	IV		7,000
9,000	Pre-Boreal						8,000
10,000	Younger Dryas	Lateglacial	Nahanagan Stadial	III			9,000
11,000	Allerød		Woodgrange Interstadial	II			10,000
12,000	Older Dryas			I			11,000
13,000							

Fig. 1. The counties of Ireland (bold face) and other localities.

afforded by dendrochronological studies, and, for the north of Ireland, palaeomagnetism, have provided excellent dating controls for the late- and postglacial periods (discussed in Chap. 12). Some interstadial deposits have also been dated, but the "infinite" dates for deposits at Fenit and Newton, for example, are of limited use. Palaeomagnetism may hold the key to some earlier dates, provided that suitable undisturbed lacustrine deposits can be found.

Chapter 13, the final chapter of this volume, deals with the economic and practical aspects of Quaternary studies. This chapter was conceived to provide an indication, and no more, of the areas where Quaternary studies have a direct economic application. Its purpose is twofold. First, it is to draw the attention of those working in applied areas to the nature, history, and evolution of the deposits with which they operate and to guide them toward the value of a more complete understanding of the totality of Quaternary deposits. Second, it is designed to create an awareness of the fact that Quaternary studies can have a practical end and to encourage some to engage themselves in the areas outlined in the chapter.

Tables I and II present comparative temporal data relating to Ireland. Within the text and in keeping with current practice, uncalibrated ^{14}C dates are designated by lowercase letters (bp, bc, ad), and corrected radiocarbon dates (or calendar years) are denoted by capitals (BP, BC, AD). Figure 1 depicts the counties of Ireland, to which frequent mention is made in the text.

References

Anonymous. (1853). "Instructions to Valuators and Surveyors Appointed under the Act 15 and 16 Vict." Cap. 63, for the Uniform Valuation of Lands and Tenements in Ireland. Alex Thom, Dublin.

Ball, J. (1849). *J. Geol. Soc. Dublin* **4**, 151–154.

Battarbee, T. E. (1978) *Philos. Trans. R. Soc. London, Ser. B* **281**, 304–420.

Bowen, D. Q. (1973). *Proc. Geol. Assoc.* **84**, 249–272.

Brooks, C. E. P. (1921). *Q. J. R. Meteorol. Soc. London* **47**, 173–194.

Case, H. J., Dimbleby, G. W., Mitchell, G. F., Morrison, M. E. S., and Proudfoot, V. B. (1969). *J. R. Soc. Antiq. Irel.* **99**, 39–53.

Charlesworth, J. K. (1928). *Q. J. Geol. Soc. London* **84**, 293–342.

Charlesworth, J. K. (1963). *Proc. R. Ir. Acad. Sect. B* **62**, 295–322.

Charlesworth, J. K. (1973). *Proc. R. Ir. Acad., Sect. B* **73**, 79–86.

Close, M. H. (1867). *J. R. Geol. Soc. Irel.* **1**, 207–242.

Cruickshank, J. G. (1982). *In* "Northern Ireland Environment and Natural Resources" (J. G. Cruickshank and D. N. Wilcock, eds.), pp. 164–184. The Queen's Univ. of Belfast and The New Univ. of Ulster, Belfast and Coleraine.

Edwards, K. J., and Hirons, K. R. (1984). *J. Archaeol. Sci.* **11**, 71–80.

Edwards, K. J., Hamond, F. W., and Simms, A. (1983). *Proc. R. Ir. Acad., Sect. C* **83**, 351–376.

Erdtman, G. (1924). *Sven. Bot. Tidskr.* **18,** 451–459.

Erdtman, G. (1927). *Ir. Nat. J.* **1,** 1–4.

Erdtman, G. (1928). *Geol. Foeren. Stockholm Forh.* **50,** 123–192.

Evans, E. E. (1975). *In* "The Effect of Man on the Landscape: the Highland Zone" (J. G. Evans, S. Limbrey, and H. Cleere, eds.), pp. 1–5. CBA Res. No. 11, London.

Farrington, A. (1934). *Ir. Nat. J.* **5,** 128–130.

Farrington, A. (1942). *Proc. R. Ir. Acad., Sect. B* **42,** 279–291.

Farrington, A. (1944). *Proc. R. Ir. Acad., Sect. B* **50,** 133–157.

Foged, N. (1977). "Freshwater Diatoms in Ireland." Bibliotheca Phycologica, Band 34, Vaduz, Liechtenstein.

Forbes, A. C. (1914). *Proc. R. Ir. Acad., Sect. B* **31,** 1–32.

Forbes, E. (1846). *Mem. Geol. Surv. G. B.* **1,** 336–432.

Gallagher, P. H. (1942). *Proc. R. Ir. Acad., Sect. B* **48,** 213–229.

Gallagher, P. H., and Walsh, T. (1942). *Proc. R. Ir. Acad., Sect. B* **47,** 205–249.

Gardiner, M. J., and Walsh, T. (1966). *Proc. R. Ir. Acad., Sect. C* **45,** 29–35.

Geikie, J. (1881). "Prehistoric Europe." Stanford, London.

Geikie, J. (1914). "Antiquity of Man in Europe." Oliver & Boyd, Edingburgh.

Groenman-van Waateringe, W. (1983). *In* "Landscape Archaeology in Ireland" (T. Reeves-Smyth and F. Hamond, eds.), pp. 217–232. BAR Brit. Ser. No. 116, Oxford.

Hamilton, C. W. (1842). Unpublished, but see *J. Geol. Soc. Dublin* **2,** 10–13.

Harkness, R. (1869). *Geol. Mag.* **6,** 542–550.

Herity, M., and Eogan, G. (1977). "Ireland in Prehistory." Routledge & Kegan Paul, London.

Jamieson, T. F. (1865). *Q. J. Geol. Soc. London* **21,** 161–203.

Jessen, K. (1934). *Ir. Nat. J.* **5,** 130–134.

Jessen, K. (1936). *Proc. R. Ir. Acad., Sect. C* **43,** 31–37.

Jessen, K. (1949). *Proc. R. Ir. Acad., Sect. B* **52,** 85–290.

Jessen, K., Andersen, S. Th., and Farrington, A. (1959). *Proc. R. Ir. Acad., Sect. B* **60,** 1–77.

Jessen, K., and Farrington, A. (1938). *Proc. R. Ir. Acad., Sect. B,* **44,** 205–260.

Kilroe, J. R. (1907). "A Description of the Soil-Geology of Ireland, Based upon Geological Survey Maps and Records, with Notes on Climate." HM Stationery Office, Dublin.

Kinahan, G. H. (1878). "Manual of the Geology of Ireland." C. Kegan Paul and Co., London.

Kinahan, G. H. (1894). *Ir. Nat.* **3,** 236–240.

Kukla, G. J. (1977). *Earth Sci. Rev.* **13,** 307–374.

Lewis, H. C. (1886). *Rep. Br. Assoc. Adv. Sci.* **1886,** 632–665.

Lewis, H. C. (1894). "The Glacial Geology of Great Britain and Ireland." Longman, Green and Co., London.

Mahr, A. (1934). *Ir. Nat. J.* **5,** 137–144.

Mahr, A. (1937). *Proc. Prehist. Soc.* **3,** 261–436.

Mitchell, G. F. (1942a). *Nature (London)* **149,** 502.

Mitchell, G. F. (1942b). *J. Co. Louth Archaeol. Soc.* **10,** 97–99.

Mitchell, G. F. (1945). *Proc. R. Ir. Acad., Sect. C* **50,** 1–19.

Mitchell, G. F. (1951). *Proc. R. Ir. Acad., Sect. B* **53,** 111–206.

Mitchell, G. F. (1956). *Proc. R. Ir. Acad., Sect. B* **57B,** 85–251.

Mitchell, G. F. (1976). "The Irish Landscape." Collins, Glasgow.

Mitchell, G. F., and Sieveking, G. de G. (1972). *J. R. Soc. Antiq. Irel.* **102,** 174–177.

Mitchell, G. F., Penny, L. F., Shotton, F. W., and West, R. G. (1973). *Spec. Publ. Geol. Soc. London* **4.**

Movius, H. L. (1942). "The Irish Stone Age." Harvard Univ. Press, Cambridge, Massachusetts.

Nuffield Foundation. (1954). "Nuffield Foundation Report for the Year Ended 31 March 1954." Oxford Univ. Press, Oxford.
O'Kelly, M. J. (1982). "Newgrange, Co. Meath." Thames and Hudson, London.
Oldham, T. (1846). *J. Geol. Soc. Dublin* **3,** 197–199.
O'Riordan, C. E. (1980). "Extinct Terrestrial Mammals of Ireland in the National Museum." British Museum (Natural History), London.
Pilcher, J. R. (1969). *Ulster J. Archaeol.* **32,** 73–90.
Pilcher, J. R., and Smith, A. G. (1979). *Philos. Trans. R. Soc. London* **B286,** 345–369.
Praeger, R. Ll. (1888). *Proc. Belfast Nat. Field Club* **2,** Appendix 2, 29–51.
Praeger, R. Ll. (1892). *Proc. R. Ir. Acad. 3rd Ser.,* **2,** 212–289.
Praeger, R. Ll. (1896). *Proc. R. Ir. Acad., Sect. B* **4,** 30–54.
Praeger, R. Ll. (1902). *Proc. R. Ir. Acad., Sect. B* **24,** 1–60.
Praeger, R. Ll.(1932). *Proc. R. Ir. Acad., Sect. B* **41,** 95–142.
Praeger, R. Ll. (1934a). "The Botanist in Ireland." Hodges, Figgis & Co., Dublin.
Praeger, R. Ll. (1934b). *Ir. Nat. J.* **5,** 126–128.
Praeger, R. Ll. (1939). *Proc. Linn. Soc. London* **151,** 192–213.
Praeger, R. Ll. (1950). "Natural History of Ireland: a Sketch of its Flora and Fauna." Collins, London.
Proudfoot, V. B., and Simmons, I. G. (1958). *Ulster J. Archaeol.* **21,** 54–61.
Raftery, J. (1963). *J. R. Soc. Antiq. Irel.* **93,** 101–114.
Reeves-Smyth, T., and Hamond, F. (eds.) (1983). "Landscape Archaeology in Ireland." British Archaeological Reports, Brit. Ser. No. 116, Oxford.
Reid, C. (1895). *Ir. Nat.* **4,** 131.
Reid, C. (1904). *Ir. Nat.* **13,** 162.
Robertson, D. (1877). *Trans. Geol. Soc. Glasgow* **5,** 192–200.
Rowan, A. B. (1852). *J. Geol. Soc. Dublin* **5,** 201–203.
Sharff, R. F. (1904). *Proc. R. Ir. Acad., Sect. C* **25,** 16–19.
Sharff, R. F. (1906). *Proc. R. Ir. Acad., Sect. B* **26,** 1–12.
Sharff, R. F. (1928). *Proc. R. Ir. Acad., Sect. C* **38,** 122–124.
Sharff, R. F., Ussher, R. J., Cole, A. J., Newton, E. T., Dixon, A. F., and Westropp, T. J. (1906). *Trans. R. Ir. Acad., Sect. B* **33,** 1–76.
Sharff, R. F., Seymour, H. J., and Newton, E. T. (1918). *Proc. R. Ir. Acad., Sect. C* **34,** 33–72.
Simington, R. C., and Wheeler, T. S. (1945). *Studies* **34,** 539–551.
Sollas, W. J. (1896). *Trans. R. Dublin Soc.* **5,** 785–822.
Stelfox, A. W. (1927). *Nature (London)* **119,** 781.
Synge, F. M. (1950). *Proc. R. Ir. Acad., Sect. B* **53,** 99–110.
Synge, F. M. (1966). *In* "Soils of County Limerick" Soil Survey Bulletin 16 (T. F. Finch and P. Ryan, eds.), pp. 12–20. An Foras Talúntais, Dublin.
Synge, F. M. (1970). *In* "Irish Geographical Studies in Honour of E. Estyn Evans" (N. Stephens and R. E. Glasscock, eds.), pp. 34–48. The Queen's Univ. of Belfast, Belfast.
Thompson, W. (1849–1856). "Natural History of Ireland," 4 volumes, London.
Warren, W. P. (1978). "The Glacial History of the MacGillycuddy's Reeks and the Adjoining Area." Unpublished Ph.D. thesis, National Univ. of Ireland.
Warren, W. P. (1979). *Geol. Surv. Irel. Bull.* **2,** 315–332.
Watson, E. (1945). *Ulster J. Archaeol.* **8,** 80–99.
Watts, W. A. (1967). *Proc. R. Ir. Acad., Sect. B* **65,** 339–347.
Woodman, P. C. (1974). *Ulster J. Archaeol.* **36–37,** 1–16.
Woodman, P. C. (1978). "The Mesolithic in Ireland: Hunter–Gatherers in an Insular Environment." British Archaeological Reports, Brit. Ser. No. 58, Oxford.

Woodman, P. C. (1981). *In* "Irish Antiquity" (D. Ó. Corráin, ed.), pp. 93–110. Tower Books, Cork.
Woodman, P. C. (1984). Conference report in *Mesolithic Miscellany* **5,** 9.
Wright, W. B. (1914). "The Quaternary Ice Age." MacMillan, London.
Wright, W. B. (1927). "The Geology of Killarney and Kenmare." *Mem. Geol. Surv. Irel.*
Wright, W. B. (1937). "The Quaternary Ice Age" (2nd Ed.). MacMillan, London.
Wright, W. B., and Muff, H. B. (1904). *Sci. Proc. R. Dublin Soc.* **10,** 250–324.

CHAPTER 2

THE PREGLACIAL LANDSCAPE

G. Frank Mitchell

Quaternary Studies
Trinity College
Dublin, Ireland

INTRODUCTION

When we look at the configuration of the modern Irish landscape, we must remember that it is the response to at least four major factors:

1. The lithology of the rocks involved
2. The rock structures (including the times at which those structures were imposed)
3. The varying climatic regimes, and consequently the different types of weathering, to which they have been subjected
4. The varying lengths of time over which weathering types have had opportunity to operate

In total length of time we need not look back beyond the beginning of the Tertiary, about 70 million years ago, as there must be few, if any, parts of Ireland that have not had their topography completely altered within that period of time. With regard to climate, it is beyond the scope of this short chapter to enter into the controversial field of climatic variation in the Tertiary (see Savin, 1977). But it is clear that Tertiary variations in time and climate were considerably greater than those of the Quaternary, restricted to perhaps less than 2 million years.

But even in that short space, varying climate could produce spectacular effects. There was humidity and water flow in warm stages (including the current Littletonian) and frost and ice in cold stages. Particularly with regard to the Littletonian, I use the term "water flow" and not "fluvial

17

erosion" because if we look at the puzzling pattern of waterways in modern Ireland, we are usually not looking at a series of rivers in organized valleys, but rather at a constantly renewed supply of rainwater moving seaward under the influence of gravity over fine-textured glacial materials whose drainage is impeded.

Frost and ice certainly have had opportunity to impose their influences. In the British Pleistocene no fewer than eight temperate and seven cold stages are currently recognized: massive ice sheets were present during the last three cold stages. In Ireland only two cold stages (each with massive ice) have been recognized, and all interglacial deposits so far examined could be assigned to a single warm stage (see Warren, 1979).

The rock topography of Ireland is essentially pre-Pleistocene in age, although its details have been trimmed by the passage of ice or by periglacial action. Watershed breaches, either by ice or by spillway water, have changed the courses of some rivers. Six major topographical areas can perhaps be recognized: (1) the Caledonian uplands of Connemara and Co. Donegal, (2) the Caledonian lowlands of Counties Down, Armagh, Cavan, and Louth, (3) the Caledonian granite and associated structures of southeast Ireland, (4) the Hercynian folds of southwest Ireland, (5) the Carboniferous area of central Ireland, with a low-lying heart of Tournaisian and Visean limestones surrounded by a higher, much-dissected, rim of Namurian shales and sandstones, and (6) the Tertiary Northeast, part plateau, part mountain.

Profound dissection influenced by structure and lithology is clearly apparent. This is particularly marked in Connemara, and Linton (1964) suggested that the high humidity here might have enabled denudation to proceed more rapidly than in less humid Great Britain. But Linton was writing in 'pre–Plate Tectonic days, and the opening of the Atlantic Ocean must have had dramatic effect on the tectonic postures of Ireland and Great Britain. When this effect has been identified more precisely and a time scale has emerged, the question of relative dissection can be returned to. Some small outcrops in Wexford have an age of 2,400 million years, and the "Caledonian grain" that is still so marked in the Irish landscape was first impressed about 400 million years ago. But the specific topography we see today cannot be older than the Tertiary.

Before proceeding further, I must make two disclaimers. First, when considering the courses of Irish rivers today, it is difficult to avoid the uniformitarian feeling that there were always such rivers and so speak of such things as "the preglacial Shannon." But today's Shannon is quite a "young" river, and were all glacial influences, both erosional and depositional, to be removed from what is now its catchment area, the surface waters of that area might find entirely different routes to the sea. When we look at a modern river, we must not expect to find its "preglacial channel" somewhere in the vicinity. We may indeed find a "subglacial valley", but can we say that the overall route of water that flowed in it bore any relationship to the modern rivers of the area? Second, our uniformitarian feelings also prompt us to consider current sea level as immutable and to refer all erosional processes to that base level automatically. But the formation of ice alone can lower that level by at least 150 m, while the effects of geoid deformation are still quite incalculable.

INTERGLACIAL EFFECTS

The Pleistocene period in Ireland must have been, as in other parts of Europe, interrupted by warm stages when there was widespread development of forest. There are in Ireland numerous deposits of muds and peats buried below glacial deposits, but as yet it cannot be unambiguously demonstrated that more than one type of interglacial climax woodland is recorded. This woodland type, the Gortian (Jessen *et al.*, 1959), finds its closest modern analogue in the forests on the southern slopes of the Caucasus. During the warm stages there would also have been weathering of rock, soil development, and fluvial erosion—all with influence on landscape.

PERIGLACIAL EFFECTS

Before Pleistocene freeze–thaw activity began, Ireland would have been largely covered by a deeply weathered mantle that had developed during the Tertiary. We do not find these weathered materials in Irish glacial deposits, which are generally composed of relatively fresh rock debris, and therefore we must assume that the mantle was stripped away by alternating periods of frost and rain during the preglacial Pleistocene.

Solifluction would have carried the materials downslope in the cold phases, and the rivers would have flushed the debris away during the warm phases. There would have been a splendid development of tors on higher ground. In granite, chemical weathering moves downward with a very irregular front, and, when frost action strips away the weathered material, tors appear. Good examples are to be seen both in the Tertiary Mourne Mountains and in the Caledonian Leinster Mountains. Very dramatic tors of volcanic ash stand on the top of Croghan Mountain on the Wicklow/Wexford border. Great inclined fangs of cleaved rock still stand to a height of 6 m, while their bases are surrounded by debris, presumably shattered by frost. Thus, the tors were once still higher, and indicate a truly remarkable thickness of former regolith.

Sometimes, remnants of fossil indurated scree or head suggest a complicated history. Immediately east of the landing point on Great Skellig Island, sheets of indurated head can be seen plastered against the steep rock slope. Such sheets must be the last vestiges of formerly much more extensive deposits which were banked against the rock at a time when wave action cannot have been operating at its present level. Was this volume of material produced by mass wasting, such as we can see today in regions with a high diurnal temperature range, or was it produced by frost action when sea level was low in a cold stage?

GLACIAL EFFECTS

The passage of an ice mass has two major effects, the one erosional, the other depositional. Erosional power is well evidenced in Ireland, whether in the roches moutonnées of Co. Cork or in the glacial breaches of watersheds in Co. Donegal. Glacial deposits of many forms cover wide areas, reaching thicknesses of over 100 m in some places.

But while glacial effects can be most impressive in limited areas, by and large the Irish landscape is rock landscape, blocked out in preglacial times. Nevertheless, important changes in the coastline have been effected. In southeast Co. Wexford, where considerable areas of rock are below modern sea level, glacial deposits have been built up above that level. Today, if we stand on the top of Forth Mountain and look to the southeast, we see farmland at an altitude of about 30 m stretching away for approximately 22 km to the granite headland of Carnsore Point. Had we been on the mountain in preglacial time and had the sea been at its present level, however, we would have looked out on a waste of skerries, running from Greenore Point to Kilmore Quay. We see similar outbuilding of land up along the east coast, in Killiney Bay, Co. Dublin, and from Balbriggan, also in Co. Dublin, to Carlingford Lough in the northeast.

PREGLACIAL TOPOGRAPHY

To George (1967) all Neogene landforms in the north of Ireland are more recent than a mid-Tertiary wave-planed benched plateau, little deformed tectonically, which was superimposed on all rocks and structures in the north and west of Ireland; the summit accords of Britain and Ireland arise from the dissection of this plateau. In contrast, Reffay (1972) argues that there is in the north a fundamental planation surface—a peneplain—cut by fluvial erosion and chemical corrosion, subdivided into two steps of different age, one at 300–391 m, the other at 180–270 m, each one showing some undulations, and both being separated by a clear break of slope. This peneplain was subsequently distorted tectonically and dissected subaerially to give us the landscape of today. In the case of Wales, Battiau-Queney (1980) considers that the mountains of that country were created by final Tertiary and Quaternary tectonism.

The doctrine of Tertiary tectonic stability which allowed Ireland's hypothetical mantle of Cretaceous chalk to be dissipated slowly and so bring about the superimposition of many river systems held good until the 1950s. But in 1957 Watts demonstrated that Oligocene plant fossils occurred in Co. Tipperary at an altitude of 75 m, and in 1960 Walsh identified the white limestone, earlier recorded by the Geological Survey at Ballydeenlea (Irish Grid Ref. V 952 977) in Co. Kerry at an altitude of 90 m, as Senonian chalk. In 1970 Davies published his paper "The Enigma of the Irish Tertiary," and the possibility of widespread Tertiary diastrophism in Ireland could no longer be denied. Mitchell (1980, 1981) has carried this point further.

Of course, to speculate that there has been late tectonic movement is one thing; to prove it is quite another. It can be proved where we have dated deposits, as in northeast Ireland, which show that there has been extensive post-Oligocene faulting (see Mitchell, 1981, Fig. 150). Jones (1981) discusses macroflexuring in southern England during the Oligocene and Neogene, and Keen and Lautridou (1982) suggest that neotectonics have had a strong influence on the geomorphology of Normandy, France.

I shall largely confine my speculations to southeast Co. Wicklow (Fig. 1), an area of particularly puzzling topography, in whose creation recent tectonic movement may have played a part. It is by no means an ideal area, because the Lower Palaeozoic sediments contain lenses of contemporary igneous materials and were later deformed by the Caledonian orogeny and altered during the emplacement of the Leinster granites. There is thus a very strong "grain," both lithological and structural, in the topography, but there is also, I feel, an element of recent tectonic uplift.

Tectonic Influence

Fault Blocks and Interblock Depressions

Annagh/Avoca Block.   River entrenchment in the Avoca region is very striking, and I believe that this could be the consequence of the elevation of a sickle-shaped block, the Annagh/Avoca Block (Fig. 1), whose base runs from Barnbawn (292 m; T 204 919) north-northeast of Rathdrum to Ballymoyle Hill (281 m; T 255 788) north of Arklow, and whose point runs away southwest to Annagh Hill (457 m) and beyond.

Above Rathdrum the Avonmore River, which flows south from Laragh, is in a relatively open valley; at Rathdrum it enters the "block," and its

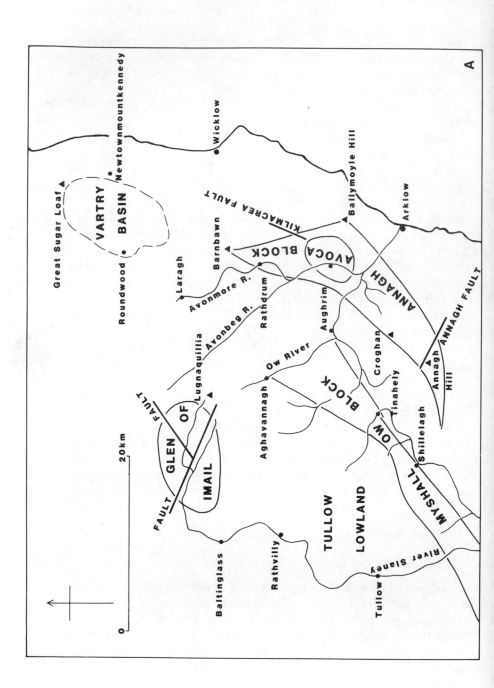

A

Fig. 1. (A) Some erosional and tectonic features in part of southeast Ireland. (B) Satellite image of the same part of southeast Ireland (Landsat-2, 13/11/76). Permission to reproduce this image of Ireland was acquired through the European Space Agency; Earthnet network was obtained through the National Board for Science and Technology for Ireland.

valley becomes a gorge trenched into a "surface" at 120 m. Opposite Avondale the gorge shows an incised meander. Where it enters the stretch known as "The Vale of Avoca" (probably fault influenced) a further complication arises. The village of Avoca lies at the centre of a basin with its floor at 90 m and surrounded by a rock rim ranging between 153 and 262 m; the centripetal drainage focusses on Avoca. The gorge cuts through the basin, its floor at Avoca being at 25 m, and the centripetal streams are entrenched to this level. At Woodenbridge the river collects the Aughrim River (similarly entrenched) as a tributary, and, adopting its line of flow, turns southeast to the sea at Arklow. At Shelton Abbey (T 218 753), a short distance upstream from Arklow, the river crosses the southeast border of the "block," and the gorge ceases.

At Woodenbridge the river also collects as tributaries the Gold Mine River (T 170 760) and the Clanwilliam River (T 180 750) which flow down from the northeast slope of Croghan Mountain, whose northeast shoulder has a striking "surface" at 120 m. Both tributaries are deeply and dramatically incised into the "surface," and the incision here provides some of the most convincing evidence for block elevation.

A point perhaps worth noting is that the lower reaches of the river are not flooded by the tide for a long distance inland, as are those of the rivers of the southeast corner of Ireland, the Slaney, the Barrow, the Nore, and the Suir. The tide does not run above Shelton Abbey, only 4 km from the sea.

On the right margin of the "block," at a short distance northwest of Redcross, a fault, the Kilmacrea Fault (already noted by the Geological Survey), can be traced southward for a distance of about 4 km. The eastern side has been moved laterally northward for a distance of about 1 km, thus creating a line of weakness along which erosion has taken place. It is most obvious in Kilmacrea Pass (T 239 850), a short steep defile followed by the road from Redcross to Rathdrum. North of the pass, as far as Bolagh (T 227 873), the 120 m "surface" southeast of Rathdrum ends abruptly against a steep, straight, west-facing slope which rises to 244 m. South of the pass, as far as Kilmacoo Upper (T 245 823) a steep, straight, east-facing slope runs down to the lower ground around Redcross. Morainic deposits of Midlandian age lie on the "surface" immediately northwest of the pass, and almost certainly meltwater flowed away down the pass, but its genesis is tectonic and not glacial.

The southwest tip of the block curls away through Croghan Mountain (607 m) and Annagh Hill (457 m), and peters out just south of Coolattin. Lithology has certainly influenced the topography here, because the upper part of Croghan Mountain is of indurated ash and the summit of Annagh Hill is of diorite. But a great fault, the Annagh Fault, which can be traced for 8 km, cuts through the sickle blade just east of Annagh Hill; erosion has formed the Wicklow Gap, which carries the road from Gorey to Tinahely.

Before the faulting took place, the headwaters of the River Bann were flowing in a strike-controlled valley southwest between the ridge from Croghan Mountain to Annagh Hill and a subsidiary parallel lower ridge to the southeast. As Annagh Hill was approached the subsidiary ridge turned due west. The fault moved Annagh Hill and the western end of the sickle about 2 km to the southeast. North of the gap, the cut western end of the Croghan Mountain ridge slopes steeply down into the basin filled with soliflucted Munsterian glacial deposits that lies east of Coolboy. In the gap, the eastern triangular fault face of Annagh Hill blocked the valley of the Bann, and the river was forced to turn at right angles to the southeast for a distance of 2.5 km before regaining its strike valley and continuing to the southwest. The truncated northeast end of the transported subsidiary ridge lies immediately northwest of Killinierin; the ridge can be traced southwest for about 4 km. It would seem that movement on the fault was transcurrent rather than vertical.

The southwest end of the sickle has, from the point where it crosses the Aughrim River as far as its tip south of Coolattin, a very pronounced northwest-facing steep slope which cuts across rocks of varying lithology. It certainly was not cut by the existing streams of the area, and it does not seem to be the result of glacial erosion. It is tempting therefore to regard it as a result of the block elevation already postulated. If it is the result of tectonic movement, then that movement must be older than the Annagh Fault, as that fault dislocated the scarp laterally but left no indication of vertical movement.

Myshall/Ow Block.   Myshall lies beyond the southwest limit of Fig. 1. The Myshall/Ow Block, a long, narrow, apparently raised block, is essentially formed by part of the metamorphic aureole of the Tullow Unit of the Leinster Granite. Its morphology has long been a puzzle. Davies and Stephens (1978, p. 108) call the block "the Tinahely Hills" and note that, although the schist and marginal granite which form them rise some 280 m above the level of the Tullow (granite) Unit to the north, along most of their length they do not form a watershed. Five rivers, including the Slaney, cut through the hills in a series of narrow, steep-sided valleys, while near Moylisha (S 922 652) there is a dry gap which seems to have carried a similar southwestward flowing river until comparatively recently. They note Farrington's (1927) view that at the time when these rivers first flowed across the schist aureole the granite of the Tullow Unit must have stood at least as high as the summits of the present Tinahely Hills.

More recent studies (Brück and O'Connor, 1980) suggest that the Tullow (granite) Unit has not been deeply eroded, as features associated with roofing still remain; the picture of the transridge valleys as consequent streams is difficult to sustain. Tectonism may be appealed to for an explanation. Davies (1970) has already suggested late tectonic uplift in the Wicklow massif. He pointed out that although granitic debris was to be found in the Old Red Sandstone of Kilkenny, 300 million years later the summit of Lugnaquillia still carried the batholith's schist roof. Thus, for a very long time the granite must have lain at a low level beneath a protective blanket, which was only stripped away after the granite had been uplifted.

Brück and O'Connor (1980) suggest that the Ow Valley follows a fracture line. We may picture its valley between Aghavannagh and Aughrim as forming the base of this second block, which stretches away southwestward to Myshall. The block is tilted as well as uplifted and shows a scarp face to the northwest and a steady slope to the southeast into the Aughrim/Kildavin corridor of Davies and Stephens (1978), here interpreted as an interblock depression.

Two of the through valleys may be briefly referred to. The valley leading to Tinahely holds an unusual deposit on its southwest side at Curravanish (T 018 736). There is a slight hollow in the schist hillside at 230 m, and resting on this is an indurated coarse deposit (with clasts about 10 cm in diameter) composed of rounded, highly rotted cobbles of granite and fragments of schist set in a reddish sandy matrix. Synge (1979) has described this as a Tertiary breccia, but the granite erratics suggest a glacial origin. It is perhaps an outwash gravel of Munsterian age. Both Munsterian and Midlandian ice will have occupied the valley, but it has not been created by glacial action.

Immediately east of Moyne (which lies at 183 m at the head of a through valley) a small rock platform at 244 m protrudes north from the scarp; from this a small stream drains south through the scarp. If the uplift was intermittent, this platform may be a survivor from a former level of planation on the Tullow Unit.

Aughrim/Kildavin Interblock Depression. Between the two elevated blocks that we have been discussing, there is a long, flat-floored valley, running parallel with the Caledonian strike. The valley extends for 32 km from northeast of Aughrim, across the River Slaney at Kildavin beyond the southwest limit of Fig. 1, to end in the northeast slopes of the Mount Leinster massif. It has already been noted by Davies and Stephens (1978) as the Aughrim/Kildavin Corridor, a long strike valley. It can perhaps be regarded as an interblock depression, associated with the tectonism we have been postulating.

If the depression received only the water that falls on its flanks, then the amount of water to be got rid of would be small. But, as we have already noted, several extraneous streams breach its northern flank and discharge large quantities of water into it. Some of this water flows northeast as the Derry Water to join the Aughrim River, and some flows southwest as the Derry River to join the Slaney at Kildavin. But water movement in the middle of the depression is very sluggish, and its floor is choked with outwash gravels, soliflucted glacial materials, alluvium, and bog; pingos are also to be seen. Neither fluvial nor glacial erosion can have brought about the present form of the depression, and its origin is probably tectonic.

Fault Basins.   Ireland's most dramatic fault basin is that at Kingscourt, Co. Cavan, where Permo–Triassic and Carboniferous beds are thrown down against a fault scarp of Lower Palaeozoic rocks. Lough Neagh lies in a great depression caused by the downwarping of Tertiary basalts; the lowering was accompanied by large-scale faulting which persisted until at least Oligocene time.

These basins both hold relatively young deposits. But what of other basins in Ireland now without deposits of Mesozoic or Tertiary age, though possibly robbed of them by glacial scouring? Are these also of tectonic origin? Bloom (1978) discussed two types of dissection which follow on tectonic movement: in the first, a consequent river initiated along a fault line is capable of carrying away not only its own debris but also that provided by its smaller dissecting tributaries, so creating a basin. In the second, the downfaulting of a block creates a basin with a higher rim (the basin–range type of fault-block landscape), and debris from the rim tends to accumulate on the floor of the basin.

In west Co. Wicklow we have the Glen of Imail, a splendid basin with centripetal drainage, though now rather choked by Midlandian glacial deposits. Brück and O'Connor (1980) show a fault running from the west into the centre of the glen; the main fault line continues on up the slopes of Lugnaquillia, passing just south of the summit and on down the line of the Ow River valley. From the centre a branch fault runs northeast to Lough Ouler. The modern River Slaney essentially follows the line of the main fault. Has the glen been created by river erosion exploiting zones of weakness arising from tectonic activity?

Can it be that we have also intermontane basins of the basin–range type with downfaulted centres in Ireland? On the east side of the Wicklow Mountains we have the Vartry Basin (Fig. 1), with a relatively flat floor surrounded by a rim of hills. Freeman (1960) described it as the Vartry peneplain, 6.5 km wide and 16 km long, lying between 200 and 330 m. It is clearly bounded by the quartzite Great Sugar Loaf (599 m) on the north, by knoblike hills of quartzite reaching 330–400 m on the east, and by the hillfoot of Djouce and other mica-schist hills on the west. The drainage in the basin is centripetal. If its eastern rim had not been breached by a deep cut, the Devil's Glen, the basin would today hold a natural lake; it does hold an artificial one, held in by the dam of a waterworks. To Farrington (1934) the glen was a glacial meltwater channel, but it is of dramatically large size, and faulting may be involved. Today the basin floor is encumbered by young glacial deposits (see Farrington, 1934), but it appears to be composed largely of fresh rock (see Davies, 1966). A marked break in slope separates the floor from the rim; at Djouce the break in slope is at 335 m. Is the lower slope area a pediment?

Pediments

What is a pediment? Bryan (1922) used the fuller title "mountain pediment" for a plain cut in rock which lies at the foot of mountains in an arid region or in headwater basins within a mountain mass. A well-marked break of slope created by slope retreat separated mountain from pediment. Wood (1942) initiated further extensive discussion of the retreat of slopes.

In my 1980 paper, I referred to Smerwick Harbour in Kerry as a dismembered valley: further study compels me to regard it as a pedimented basin, with a gently sloping rock floor and a steeply sloping rim. The rim is now breached by the sea at the southwest between Sybil Head and Clogher Head (Q 304 026) and at the north between Carrigbream (Q 351 089) and Ballydavid Head. The break in slope between floor and rim comes at about 40 m; at Ardamore (Q 385 082) the floor lies at about 30 m. At several points round the shores of the harbour, rock buried by glacial deposits can be seen sinking below modern sea level. The centripetal streams that now flow across the floor of the basin cannot have excavated it, though in pre-Pleistocene times they probably carried alluvium down into it. But, if ever there, the alluvium has long since disappeared, because the north breach must be older than the "preglacial" beach which is well seen at Ballynagall (Q 372 020; Bryant, 1966); solifluxted glacial deposits, presumably of Munsterian age, are to be seen at sea level at Feohanagh (Q 388 095; Lewis, 1974) and Clogher (Q 315 033; King and Gage, 1961). That the area may have previously experienced a different climatic regime is suggested by the richly coloured weathered rock that can be seen in the bed of the stream that flows southwest from Brandon Mountain down to Ballybrack (Q 442 100).

The pediment in Smerwick Harbour is not unique, as similar landforms are well developed on both sides of the west end of Dingle Bay. These pediments are under study (Mitchell et al., 1983).

Influence of Lithology

Calcareous Rocks

By far the greater part of central Ireland is underlain by calcareous Carboniferous rocks, differing in facies, but ready to yield to the attack of dissolved carbonic acid. In preglacial time karstification was widespread, probably almost universal. The surface showed a wide development of hums and dolines, together with vertical pipes and horizontal tubes.

Vertical solution pipes are widely distributed, and depths of up to 150 m are recorded. Pipes were common on the site of the large industrial development at Aughinish Island, Co. Limerick, and one of them was explored in detail (Clark et al., 1981). A borehole in the centre of the pipe was abandoned at 62 m without reaching bedrock (Fig. 2). The fill was of mottled grey and orange sandy clay and grey and black carbonaceous silty clay with some gravel; there were some lignitic horizons, one of which yielded Tertiary pollen and spores. Tongues of "living rock" (i.e., still in direct continuity with bedrock) projected horizontally from the walls towards the centre, with fill above and below. The same feature was noted in the opencast workings at Tynagh, Co. Tipperary, where weathered ore had accumulated on karstic limestone; tongues of living rock had ore above and below. Fossiliferous fillings are extremely rare, but an Oligocene age is recorded at Ballymacadam, Co. Tipperary (Watts, 1957), and early Pliocene at Hollymount, Co. Carlow (Boulter, 1980).

Walsh (Mitchell et al., 1980, p. 341) has suggested that the pipes may have been created as sinkholes during periods of falling sea level. During the Oligocene the shoreline in the southwestern approaches withdrew to the outer shelf (Evans and Hughes, 1984). After an early Miocene transgression, there was a worldwide fall in sea level in the late Miocene (Vail and Hardenbol, 1979). Thus there may be two generations of pipes, an earlier containing Oligocene fossiliferous deposits, and a later with Pliocene material, both at Hollymount and at Brassington in Derbyshire, England (Walsh et al., 1972).

Many valleys and quarries exhibit several "stories" of almost horizontal phreatic tubes; "Dermot and Grania's Cave," a phreatic tube truncated by a cliff, is 30 m wide and 15 m high, and stands, at an elevation of 550 m, 300 m above the floor of Gleniff in Co. Sligo. Stalactitic deposits are widespread, and some may be of very considerable age.

At several localities where there are overlying glacial deposits—for example, the limestone quarry at Ballyellin, the Mullinahone drainage tunnel, beneath the fluvioglacial deposits of the Curragh in Kildare—the surface of

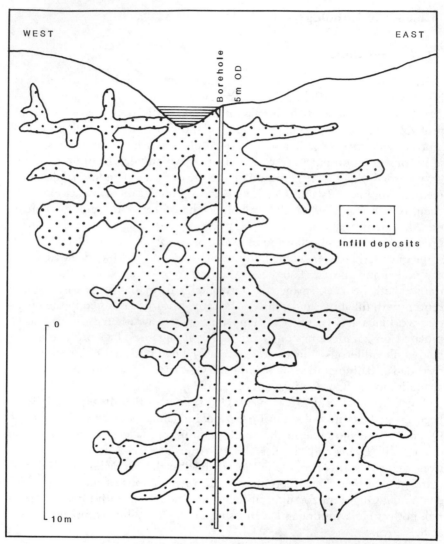

Fig. 2. Schematic section through karst feature at Aughinish Island, Co. Limerick (after Clark *et al.*, 1981).

the limestone is dissected by karstic solution. The process of karstification in Ireland may have a very long history. Evans and Hughes (1984) record Eocene limestone with an irregular surface of rugged relief of tens of metres at a very considerable depth in the southwestern approaches and suggest it is probably a palaeokarst surface.

The masking of karstic topography by glacial deposits can be well seen in the lower valley of the Boyne. As the river approaches Navan, it crosses an open area of undulating glacial deposits (Synge, 1950). At Navan the Boyne meets the Blackwater, and makes a right-angle turn to the northeast toward Slane. Higher limestone immediately closes in on the river from either side, rising in some places to over 20 m above the river. The limestone is markedly karstic, with irregular knolls and large dolines. Through this country the river makes its way, sometimes constricted in a rock-walled gorge, sometimes crossing a doline. Some dolines are empty, save for the floodplain and fragments of outwash terraces; sometimes (as at Broadboyne Bridge, N 916 714), a doline has been loaded with coarse but well-bedded outwash gravel. At Slane Castle the river enters a large doline set at an angle to the previous course and makes another right-angle bend, this time to the southeast. I think it fair to regard the Boyne as traversing preexisting gullies and dolines more or less at random, as opposed to having cut a valley of its own across the terrain. Also, it would not be unreasonable to regard the "preglacial channel" of the Liffey (see Naylor, 1965) not as a water-excavated valley but as a series of karstic features buried beneath glacial deposits.

In picturing the areas of calcareous rock in preglacial Ireland, I would see the higher ground dissected into pinnacles, which sometimes rose into towering hums, interspersed with extensive low-lying dolines. It is impossible to say whether the lower dolines held standing water and were interconnected by sluggish streams, or whether all drainage was essentially underground. Davies and Whittow (1975) discussed this problem, when developing their theory on the origin of the drainage pattern in the south of Ireland. Where slopes were gentle, there was a deep mantle of acid soil, composed of the nonsoluble residues.

There is evidence in Ireland that some calcareous sediments, as they lay on the floor of the Carboniferous Sea, had not a horizontal surface but rose into reefs and mudbanks, forming knolls in which bedding planes would not have been prominent. If buried by argillaceous sediments, such knolls would be entombed and could, after elevation and the passage of many millions of years, reemerge when the overlying shale was stripped away by erosion. Knolls of poorly bedded limestone do rise in the Irish landscape, and these may well be exhumed primary knolls. But there are also knolls of well-bedded rock which cannot have had a preexistence on the sea floor and must be hums of karstic origin.

Fine-Grained Rocks

In general, the Namurian shales and sandstones stand at an altitude of more than 100 m. It is perhaps risky to describe them as a dissected rim surrounding the older limestones which generally lie below 100 m. Their greatest extent is in the southwest, on both sides of the lower Shannon Estuary, which cuts across them in a rather surprising fashion. A small patch appears again near Castlebar, Co. Mayo, and in the north they lie on both sides of Lough Allen. In the east, there is a downfaulted, but still topographically high, patch near Kingscourt, Co. Cavan, as well as small outliers in the vicinity of Navan. On the southeast we have the high ground running from Timahoe to Killenaule. Where substantial areas of these rocks occur, they are deeply dissected by fluvial action.

When we come to the fine-grained argillaceous Lower Palaeozoic rocks, usually influenced by metamorphic action, which cover extensive areas in Ireland, we encounter real difficulty. These rocks have a relatively uniform texture, and here one would think that fluvial processes of erosion would have dominated. But instead of integrated valleys, we find small, isolated rock knolls, now usually rounded and oval in shape. Their current shape is probably influenced by the erosive action of ice, and, in the area of Counties Cavan, Monaghan, and Down where drumlin forms are abundant, these may be of entirely glacial material, of a combination of glacial material and rock, or of rock alone.

Wherever this type of rock is present, this landform appears: the knolls are to be found in the peninsula of Corca Dhuibhne, Co. Kerry, in the Slieve Bernagh and the Arra Mountains, in Connemara, as well as in low-lying Wexford (see Mitchell, 1980, Fig. 4). If the areas in which they appear were relatively horizontal, then we could toy with the idea that they represent some type of inselberg and that they once had a much more abrupt form and rose above wash plains rather than organized valleys. But they are not restricted to flat ground and occur on the south slopes of the Slieve Bernagh and Arra Mountains—although the possibility of late tilting cannot be ruled out. Here we can also see their relationship to lithology. Crudely speaking, the north slopes of these mountains are coarse in lithology with conglomerates; the south slopes are in finer-grained rock. We have rugged outlines on the north slopes; on the south slopes, typical rounded oval hills (which, except for their large size, could be easily mistaken at first glance for drumlins) appear (e.g., west of Killaloe at R 685 735, and north-east of the same town at R 728 753).

Siliceous Rocks

When we try to picture what Ireland's outcrops of siliceous rocks would have looked like in the late Tertiary, we have to take into consideration their degree of cementation. Where they were fully silicified, as in true quartzites, they were almost immune to chemical attack and will have tended to break down into screes. Peaks like Errigal, Co. Donegal, and the Great Sugar Loaf, Co. Wicklow, will probably have looked much as they do today, though the screes that now surround them may largely have developed during Midlandian freeze–thaw activity.

Just as large quartzite masses developed screes, the lesser ones would also have done so. Where these lay at low levels and were overrun by Midlandian ice, their screes were swept away and their profiles rounded. If we look at the crest of Bray Head, Co. Wicklow, today, we see a series of rounded quartzite knobs; if we had looked at it in pre-Munsterian days, we would have seen a row of mini-sugarloaves; Carriggollan (O 230 202) would have had a more impressive cone.

Where we had cemented sandstones, considerable disintegration probably took place, and a mantle of sand with scattered sarsen stones would have formed. When freeze–thaw set in, the loose sand would have been washed or blown away, and prominent tors would have appeared. The outcrop of Carrickleck Sandstone near Ardagh, Co. Meath (N 828 950), still carries occasional tors, but as this area was heavily glaciated in the Midlandian cold stage, the tors may have been rejuvenated in Nahanagan time.

Granitic Rocks

With regard to granite, we are perhaps on safer ground. Almost certainly the outcrops would have been covered with irregularly heaped core stones, which concealed and, in part, protected deeply weathered rock. By the end of the Tertiary, weathering may have proceeded to depths of more than 30 m, as such depths are known today in Hong Kong and in the Snowy Mountains of Australia.

When ice advanced over such terrain, it carried off the core stones and rotted rock and left fresh rock exposed. In Donard, west of the Wicklow granite, the ground was so closely littered with transported core stones that it was mapped as granite even though different solid rock lay underneath. If periglacial conditions obtained, the core stones and the rock debris moved downslope, leaving tors crowning the fresh rock exposed.

How long does granite take to weather? Carnsore Point, Co. Wexford, was apparently overrun by Midlandian ice, and one crop of its transported core stones litter the south Wexford coast as far as Cullenstown and the Saltee Islands. But today core stones lie on the flanks of the point, and in places we can see further core stones apparently in the process of formation. Does each warm stage produce its own crop of core stones?

On the other hand, in striking contrast, some tors can be of very great antiquity. Near Christianstad in south Sweden, erosion is stripping chalk off a high point in gneissic granite. The revealed granite surface stands up in small tors. Where the tor surfaces show undercut features, residual patches of chalk can be found in the undercut, showing that the formation of tors preceded the deposition of the Cretaceous chalk (Magnusson and Lidmar-Bergstrom, 1983).

River Valleys

Simple Sections

There are in many parts of Ireland simple valleys apparently cut in rock by fluvial action, and these are *prima facie* of very considerable antiquity. The southeast flanks of the Slieve Bernagh Hills, in Co. Clare below Glennagalliagh, are drained by the Ardclooney River which enters the Shannon above O'Briensbridge, and the southwest slopes of the Arra Mountains are drained by the Grange River which enters the Shannon below Killaloe. In each case one has the impression of simple fluvial excavation, although the volume of the modern rivers seems small compared to the size of the valleys they occupy.

Sectors of Some Apparent Antiquity

In considering the present courses of many Irish rivers, it is necessary to take into consideration the possibility of their having been created in part by some glacial catastrophe, perhaps of short duration, such as the overspilling of a glacial lake. But some stretches of river courses show incised meanders, and it is difficult to see these being produced by a rapid glacial event. The finest incised meanders are on the Nore between Inistioge and Rathsnagadan (S 667 335), Co. Kilkenny; immediately northwest of Rathsnagadan the meander cliff rises 150 m above river level. The Barrow between Graiguenamanagh, Co. Kilkenny, and Ballynacoolagh (S 730 343), Co. Wexford, also shows splendid incision into a "surface" at about 60 m. Both these gorges cut through a Caledonian ridge, and the combination of rock topography and late ice movements suggest that glacial events could have been the causative factor. But the well-developed meanders indicate that these river stretches must have a longer history. The Avonmore between Rathdrum and the Meeting of the Waters (T 188 832) is incised into a "surface" at 120 m, and it has a splendid meander scar at T 203 836; here again this stretch of the river must be relatively old.

Buried Channels

East of Lough Foyle, south down the Irish Sea coast and round Carnsore Point as far as Youghal, Co. Cork, there is evidence of river valleys having formerly been related to base levels much lower than those of the present; several of these buried valleys are filled by glacial deposits. At the head of Lough Foyle, rock head is at −30 m OD; in Belfast it is −60 m; an incised valley for the Boyne is not fully proved; in Dublin it is −40 m; in Wexford −22 to −45 m; in Dungarvan, Co. Waterford, −30 in Youghal −40 m (Davies and Stephens, 1978). The valleys also exist on the east side of the Irish Sea (e.g., the Mersey).

The Lee at Cork is also often cited as evidence of former river erosion to a great depth, but the position here is very puzzling. The synclinal valleys of the Blackwater, the Lee, and the Bandon may have been excavated a very long time ago, as among other weathered debris to be found in them the so-called Colbond Clays are probably of Jurassic age. Also, more recent surveys have failed to find any buried channel at the Lee exit at Roche's Point, where Lamplugh *et al.* (1905) postulated a channel depth of −78 m.

The size of the buried channels and the depth of their incision suggest a long period of development, probably pre-Quaternary. In a cold stage, however, sea level would be low and river volumes would probably be small. In a warm stage, sea level would not be low and the period of time involved probably too short.

CONCLUSIONS

Arthur Holmes once wrote that theories must have more than plausibility to commend them and that no heaping of speculation on speculation can lead to defensible conclusions. One can only say that when one surveys the scene, certain impressions keep reappearing.

The strongest is that an enormous quantity of material has been removed from Ireland by erosion (see Mitchell, 1980). The Carboniferous area of central Ireland was left as a chaotic region of limestone hums and dolines, mantled by acid residual soils, with little organized surface drainage. Vestiges of a former rim of higher Namurian rocks have been deeply dissected by fluvial action. In the south, we have the Hercynian fold country, and here the hilly ridges must be of great antiquity because the residual clays in the Cork syncline suggest that the etching away of the limestone must have been going on for a long time. To the northwest the Caledonian trend of Connemara and Co. Donegal, established in rocks of varying lithology, must long have been evident in the topography, although high outlying patches of basal Carboniferous rock show us that either much has disappeared here also or that there has been recent uplift—possibly both. The Tertiary basalt plateau of the northeast corner tells a different story and warns that great quantities of Mesozoic rock have gone as well. The penecontemporaneous Mourne granites also indicate vast erosion, a degree of erosion that surprises us when we see the much older Leinster batholith still part roofed. Between lies the knobby lowland of Counties Down, Armagh, and Cavan; why the Lower Palaeozoic rocks should take this form in Ireland, when to the east across the Irish Sea they rise into the Scottish Southern Uplands, remains a mystery. Granite everywhere would have been capped by core stones, and many quartzite masses would have been surrounded by screes.

The second impression is that when we consider the low-lying water-logged area that stretches from Upper Lough Erne to Lough Ree, and picture the glacial deposits removed, we are almost compelled to envisage a very large lake. If we regard sea level as remaining constant at its present level and Ireland as tectonically stable, it is almost impossible to picture how such a lake basin could have been excavated. Has there been in Tertiary time a sinking of north-central Ireland and possibly a balanced rise of the coastal rim in the northwest?

Unlike the estuaries of southeast Ireland, up which the tides penetrate very deeply, several northwestern estuaries have waterfalls at their heads. At Ballyshannon, Co. Donegal, we have the Cathleen Falls on the Erne, there is a high waterfall at Ballysadare, Co. Sligo, there are the Aasleagh Falls at the head of Killary Harbour, while the Corrib tumbles down a series of rapids into Galway Bay.

The scree-surrounded peaks of Errigal and Muckish in north Co. Donegal are well-known. Slieve League and the other quartzite highs in southwest Donegal do not show such features; have they been raised relatively recently?

At present we can only see preglacial Ireland through a glass darkly. Let us hope for future enlightenment.

References

Battiau-Queney, Y. (1980). "Contribution a l'étude geomorphologique du Massif Gallois." Lille.
Bloom, A. L. (1978). "Geomorphology." Prentice-Hall, Englewood Cliffs, New Jersey.
Boulter, M. C. (1980). *J. Earth Sci. (Dublin)* **3**, 1–12.
Brück, P. M., and O'Connor, P. J. (1980). *Geol. Surv. Irel. Bull.* **2**, 349–370.
Bryan, K. (1922). *Geol. Surv. Bull. U.S.* **730B**, 19–90.
Bryant, R. H. (1966). *Ir. Geogr.* **5**, 188–203.
Clark, R. G., Gutmanis, J. C., Furley, A. E., and Jordan, P. G. (1981). *Q. J. Eng. Geol.* **14**, 231–239.
Davies, G. L. (1966). *Ir. Geogr.* **5**, 150–160.
Davies, G. L. (1970). *In* "Irish Geographical Studies" (N. Stephens and R. E. Glasscock, eds.), pp. 1–16. The Queen's Univ. of Belfast, Belfast.
Davies, G. L. Herries, and Stephens, N. (1978). "Ireland." Methuen, London.

Davies, G. L., and Whittow, J. B. (1975). *Ir. Geogr.* **8**, 24–41.

Evans, C. D. R., and Hughes, M. J. (1984). *J. Geol. Soc. London* **141**, 315–326.

Farrington, A. (1927). *Proc. R. Ir. Acad., Sect. B* **37**, 181–192.

Farrington, A. (1934). *Proc. R. Ir. Acad., Sect. B* **42**, 173–209.

Freeman, T. W. (1960). "Ireland" (2nd ed.). Methuen, London.

George, T. N. (1967). *Scott. J. Geol.* **3**, 414–448.

Jessen, K., Andersen, S. Th., and Farrington, A. (1959). *Proc. R. Ir. Acad., Sect. B* **60**, 1–77.

Jones, D. K. C. (1981). "The Geomorphology of the British Isles: Southeast and Southern England." Methuen, London.

Keen, D. H., and Lautridou, J. P. (1982). *Quat. Newsl.* **38**, 32–38.

King, C. A. M., and Gage, M. (1961). *Ir. Geogr.* **4**, 202–207.

Lamplugh, G. W., Kilroe, J. R., M'Henry, A., Seymour, H. J., Wright, W. B., and Muff, H. B. (1905). "The geology of the country around Cork and Cork Harbour." *Mem. Geol. Surv. Irel.*

Lewis, C. A. (1974). *Sci. Proc. R. Dublin Soc.* **5A**, 207–235.

Linton, D. L. (1964). *In* "The British Isles: a Systematic Geography" (J. W. Watson and J. B. Sissons, eds.), pp. 110–130. Nelson, Edinburgh.

Magnusson, S. E., and Lidmar-Bergstrom, K. (1983). *Sven. Geogr. Arsb.* **59**, 124–137.

Mitchell, G. F. (1980). *J. Earth Sci. (Dublin)* **3**, 13–34.

Mitchell, G. F. (1981) *In* "The Geology of Ireland" (C. H. Holland, ed.), pp. 231–272. Scottish Academic Press, Edinburgh.

Mitchell, G. F., Coxon, P., and Price, A. (1983). *IQUA Field Guide No. 6.*

Mitchell, G. F., Scannell, M. J. P., Walsh, P. T., and Watts, W. A. (1980). *Proc. R. Ir. Acad. Sect. B* **80**, 335–342.

Naylor, D. (1965). *Sci. Proc. R. Dublin Soc.* **2A**, 175–188.

Reffay, A. (1972). "Les Montagnes de l'Irlande." Grenoble, Imp. Allier.

Savin, S. M. (1977). *Annu. Rev. Earth Planet. Sci.* **5**, 319–355.

Synge, F. M. (1950). *Proc. R. Ir. Acad., Sect. B* **53**, 99–110.

Synge, F. M. (1979). *Quat. Newsl.* **28**, 1–18.

Vail, P. R., and Hardenbol, J. (1979). *Oceanus* **22**(3), 71–79.

Walsh, P. T. (1960). *Proc. Geol. Soc. London* No. 1581, 112–113.

Walsh, P. T., Boulter, M. C., Ijtaba, M., and Urbani, D. M. (1972). *J. Geol. Soc. London* **128**, 519–559.

Warren, W. P. (1979). *Geol. Surv. Irel. Bull.* **2**, 315–332.

Watts, W. A. (1957). *Sci. Proc. R. Dublin Soc.* **27**, 309–311.

Wood, A. (1942). *Proc. Geol. Assoc* **53**, 128–140.

CHAPTER 3

STRATIGRAPHY

William P. Warren

Quaternary/Geotechnical Section
Geological Survey of Ireland
Dublin, Ireland

INTRODUCTION

There has for many years been a tendency to treat the stratigraphy of Quaternary sediments very differently from that of any other period in geology. This is reflected in the exclusive application of climate units (glacials and interglacials) as distinct from the normal chronostratigraphic units of solid geology in interpreting the sediments. The recognition that Quaternary sediments need separate treatment and interpretation has probably caused as many problems as it has solved. The Irish situation illustrates this clearly: there is a broadly accepted sequence of climate units but litho- and biostratigraphic units have not been adequately defined or described, and in many cases there are no recognized stratotypes. There are many reasons for this, but the most outstanding one seems to have been the reluctance on the part of Quaternary geologists to alter the interpretative model in the changing light of evolving field evidence. This, coupled with a reluctance to establish a climate stratigraphic sequence based on Irish field data without prior reference to, or correlation with, sequences established elsewhere, has produced an inflexible model incapable of adjustment to new data and unable to withstand close scrutiny based on the field data (see Warren, 1979).

A tripartite subdivision of Irish Quaternary sediments has persisted since Harkness (1869) identified the sequence: upper boulder clay, middle sands and gravels, lower boulder clay. This sequence, which originally identified two glacial events separated by an interglacial has, with the passage of time, translated into a sequence indicating an interglacial followed by two glacials which are separated by a presumed interglacial, the "hidden interregnum" of Mitchell (1976), for which no evidence has yet come to light. The sequence has now four parts, one of which is hidden.

BACKGROUND

Following Harkness's paper, Hull (1871, 1878) embraced the threefold sequence with enthusiasm, and Kinahan (1872), although more cautious, eventually accepted a modified form of it (1878). The case was argued for and against the older till and the newer till separated by (interglacial) middle sands. However, by 1914, W. B. Wright, although he recognized an Older and Newer Drift in Britain, suggested that all of the surface tills in Ireland relate to the Newer Drift and that the manurial "Middle" sands and gravels of Wexford are glacial. At this time too Wright cautioned strongly against interpreting the age of deposits on the basis of topographic expression.

By the time Wright revised his book for the second edition in 1937, both Charlesworth (1928) and Farrington (1934) had contributed what were to be very influential papers on Quaternary stratigraphic thinking in Ireland. In his second edition Wright not only accepted the Southern Irish End-Moraine as separating Newer and Older Drift in Ireland but suggested that the Newer Drift correlated with a penultimate European glaciation. Moreover, Wright (1937, p. 91–92) accepted a subdivision based entirely upon differences in topographic expression, distinguishing between a "monotonous sheet of boulder clay with but little associated gravel, and scarcely any original diversity of surface" south of the moraine, and "an ever-varying range of types" to the north. There was, at this stage, no reference to interglacial sediments.

In 1928 Charlesworth had drawn attention to the "End Moraine" that had previously been described by Carvill Lewis (1894) and suggested that it represented the southern limit of the equivalent of the Newer Drift of England. This distinction between a Newer and an Older Drift was based largely on topographical expression of the sediments and followed Geikie (1914).

By 1948 Mitchell had provided a stratigraphic sequence which separated what were then called the Eastern General and Midland General glaciations on the basis of evidence of interglacial pollen in a deposit at Ardcavan, Co. Wexford. Farrington (1949) had hinted that he regarded these glaciations as equivalent to the older and younger drifts and later (1954) clearly stated this.

Mitchell's (1948) scheme found such general acceptance that even when it was demonstrated that the Ardcavan deposit was in fact postglacial (Colhoun and Mitchell, 1971) the stratigraphic interpretation was not altered, rather the topographical argument was once more resorted to. However, the correlation of the interglacial lake muds at Gort (Jessen et al., 1959) with the Hoxne of Britain and Holstein of Europe seemed to confirm the established sequence, and after the Ardcavan interglacial had been rejected it was thought to be only a matter of time before a suitable alternative would be found. Synge (1970a,b) used the degree of weathering of limestone till clasts, the occurrence of frost erected clasts, and topographic characteristics to distinguish between tills of the Older Drifts and the Younger Drifts. Mitchell in 1972, for reasons unstated, retained the Southern Irish End-Moraine as the southern limit of last glaciation ice, and, although this interpretation has since been challenged (Bowen, 1973; Warren, 1979) and there is much support for the alternative views, there remains a strong attachment to the principle of subdividing deposits on the basis of differences in topographic expression (see McCabe, Chap. 4, and Watts, Chap. 8, this volume). Thus the established sequence at the time of this writing may be said to be that outlined by Mitchell (1976):

Littletonian (postglacial)
Midlandian Glaciation
"Hidden Interregnum"
Munsterian Glaciation
Gortian Interglacial
Pre-Gortian cold stage

Within the Midlandian, Mitchell (1977) recognized two interstadials (Woodgrange and Castlepook) and three stadials (Nahanagan, an unnamed stadial, and Saltmills).

Bowen (1973) pointed out that there was no stratigraphic evidence that the tills south of the End–Moraine were older than the last glaciation. Also, he interpreted the Courtmacsherry Raised Beach as most recent interglacial in age. Warren (1979) agreed with Bowen and, following the logic of the stratigraphic approach, interpreted the Gortian Interglacial as last interglacial in age. The differences between the two schools can be reduced to, on one hand, an approach that is partly stratigraphic and partly morphological and, on the other, the standard geological stratigraphic approach. There is also the classic conflict that often arises between the lithostratigraphic and biostratigraphic approaches. The stratigraphic model that is presented below is entirely independent of the constraints of the traditional approach, is primarily lithostratigraphic (for reasons outlined below), and is based on the principles of the International Stratigraphic Guide (Hedberg, 1976). This allows for a fresh look at the evidence.

LITHOSTRATIGRAPHY

Because geological processes in the Quaternary Period have been characterized by extensive erosion as well as by deposition, a continuous onshore record is unlikely in formerly glaciated areas. This is particularly true of Ireland as there are very few, if any, parts of the country which were not glaciated. A stratigraphic sequence, if it is to be capable of widespread application to an area the size of Ireland, must be based on an isochronous unit of common occurrence, that is, a chronostratigraphic horizon. Such a unit of sediment will represent the same specific unit or point of time no matter where it occurs. Given stable tectonic conditions, a marine margin should provide such a unit, and the raised beach sediments of the south and southwest coasts may possibly be considered as such. The major objection to using beach sediments is that they may represent a number of different periods when sea level achieved approximately the same level. However, given the very small altitudinal range within which the beach materials are found (Synge, 1977a) and given that they are nowhere seen to be separated by another unit, the assumption that the beach deposits (Courtmacsherry Raised Beach of Wright and Muff, 1904) represent an isochronous unit can be made, subject to adjustment of the stratigraphy if it is later demonstrated not to be so. The Courtmacsherry Raised Beach of Wright and Muff or the 4- to 6-m raised beach of Synge (1977a) will therefore be used as a chronostratigraphic horizon and referred to as the Courtmacsherry Formation. Its stratotype is the section along the northern shore of Courtmacsherry Bay as described by Wright and Muff (1904).

The Courtmacsherry Formation

This deposit was first identified by Wright and Muff (1904) as a discontinuous feature occurring along the exposed coastal sections on the south coast between west Cork and Wexford. Owing to its position relative to modern sea level it is normally interpreted as an interglacial deposit (see Mitchell *et al.*, 1973). This interpretation is accepted here. It is invariably overlain by head which in turn is usually overlain by till.

Correlation with the beach deposits at the same level at Ross Behy (the Ross Behy Formation), Baile na nGall, Fenit, and Ballybunnion in Co. Kerry is at first sight problematic, as it is possible that some if not all were influenced by isostatic or tectonic activity. However, the coincidence in level of these deposits and their varied positions, from Ross Behy deep in Dingle Bay to Dingle and Baile na nGall in the western end of the peninsula of Corca Dhuibhne, to Fenit in Tralee Bay, and to Ballybunnion on the relatively unindented north coast of Co. Kerry, tends to confirm rather than question their contemporaneity and the lack of isostatic influence. For, any interpretation involving land displacement would call for a number of different sea levels and very complex adjustment patterns.

Given that the Ballybunnion beach sediments (Warren, 1981) represent a relatively warm period, the same period as is represented by the Courtmacsherry Formation, then the sequence at Ballybunnion

3. Geliflucted till
2. Beach
1. Till

represents a cold–warm–cold sequence there. The normal south coast stratigraphy

3. Till
2. Head
1. Beach

reflects the warm–cold part of the sequence. Thus, a climate stratigraphic sequence, glacial–interglacial–glacial, can be independently interpreted from the lithostratigraphy of the Quaternary sediments of the south coast of Ireland.

Deposits Below the Courtmacsherry Formation

Quaternary deposits underlying the Courtmacsherry Formation are rare, but they do occur, notably at Ballybunnion in north Kerry (Warren, 1981) and possibly at Clogga, Co. Wicklow (Synge, 1977a). As there is some doubt as to whether or not the deposit at Clogga is a beach (Huddart, 1977) more attention will be paid to the deposit at Ballybunnion. The beach sediments at Ballybunnion rest on a level platform cut in a compact till containing Carboniferous limestone and shale and are overlain by a soliflucted till unit. The platform rises slightly from north to south. The beach deposit narrows in the same direction and pinches out at 7.8 m OD. This probably reflects a diagonal cut across the former coastline rather than a tilt in the platform, as a horizontal beach occurs at the same general level 4 km away, on the other side of the Cashen estuary.

The till in which the platform is cut is likely to represent a glacial event predating the interglacial represented by the raised beach. The till is here termed the Ballybunnion Formation. This till is composed of Carboniferous sandstone, shale, and limestone in a stiff grey calcareous matrix. No far-travelled erratics have yet been identified in it.

Deposits Overlying the Courtmacsherry Formation

The Fenit Formation

Without exception, where either the Courtmacsherry Formation or its equivalents have been recognized the immediately superjacent sediment is a head deposit: the Lower Head of Wright and Muff (1904) or the Stookaniller Member of Warren (1978). This deposit, which varies in thickness from 1.5 m at Courtmacsherry to 30 m at Ross Behy (see Warren, 1981), is normally interpreted as a product of gelifluction (see Lewis, Chap. 5, this volume). It is a stony diamicton of usually angular clasts and is frequently, but not always, erratic free. It often exhibits steeply dipping pseudobedding structures. Along the south coast and as far north as Ross Behy on the southwest coast, this deposit is usually overlain by a till unit. Occasionally the till unit is in turn overlain by a second head unit, the Upper Head of Wright and Muff (1904), but this unit rarely exceeds 2 m in thickness.

The till can be subdivided into three main units of differing petrology and provenance. These are the Ballycroneen till of Irish Sea basin provenance, the Bannow till (and equivalents) of inland, northern provenance, and the Garryvoe till of inland Kerry/Cork provenance. Each of these units has been described by Wright and Muff (1904). They are each given formation status here. North of Ross Behy the till is usually missing. This situation is best expressed at Fenit and Ballybunnion. At Fenit the head contains erratics, and at Ballybunnion it is clearly a soliflucted facies of the till which underlies the raised beach.

The stratotype of this formation extends from Fenit to Spa on the north coast of Tralee Bay, Co. Kerry. The Quaternary sequence was first described by Mitchell (1970) who has since accepted a revised description by Warren (1981). The revised sequence identified a raised beach at approximately 6 m MSL overlain by polleniferous peat reflecting an open woodland vegetation. This in turn is overlain by grey silts with interbedded organic horizons. The silt interdigitates with an overlying head deposit into which it passes. The head unit is composed mainly of Carboniferous shale but contains significant amounts of Devonian sandstone erratics. A concentration of large Devonian erratics about two-thirds of the way up the sequence may represent a change in the dynamics of the slope process. The sequence from the base of the grey silts to the top of the head is interpreted as a conformable sequence of gelifucted sediment. This is the Fenit Formation. The raised beach is regarded as the Courtmacsherry Formation.

The Ballycroneen Formation

The Ballycroneen till is common along the east and south coasts as far west as Ballycroneen, Co. Cork (Fig. 1). This is the marly boulder clay of Wright and Muff (1904), who described it in detail. At the stratotype Wright and Muff described the following sequence:

4.	Stony loam and soil	0.6 m
3.	Red boulder clay	3.7 m
2.	Grey, marly boulder clay	0.9 m
1.	Lower head	1.1 m

The formation itself is a carbonate-rich (in its unweathered state), clay-rich till characterized by stone erratics of Irish Sea basin provenance, shell fragments, and, occasionally, Ailsa Craig microgranite.

On the south coast it is for the most part mutually exclusive with the Bannow Formation and the Ballyvoyle till, which is here interpreted to be incorporated in the Bannow Formation (see Mitchell et al., 1973). But it is overlain in its western extremity between Ballycroneen and Garryvoe by the Garryvoe Formation, and in east Waterford it is overlain by the Bannow Formation. At Clogga on the east coast in Co. Wicklow, the Ballycroneen Formation overlies the Clogga till unit, which is a till of inland provenance similar to the Bannow Formation and may be regarded as a member of that formation (Synge, 1964).

The stratigraphic relationships of these till units to the Ballycroneen Formation cannot be interpreted in terms of a normal chronostratigraphic sequence reflecting normal superposition of sediments, as the Ballycroneen till is not isochronous and, in part at least, probably youngs from west Cork back to the Irish Sea basin. It may also, in its lower part, young in the opposite direction. Thus one can only say that the eastern extremity of the Garryvoe Formation postdates the Ballycroneen Formation at the western extremity of the latter. The Bannow Formation seems to overlie it in south Counties Wexford and Waterford (Synge, 1964; Watts, 1959), but on the east coast near Arklow the Clogga till underlies it (Synge, 1977a).

North of Clogga as far as Loughshinny in north Co. Dublin, there is considerable disagreement as to the till stratigraphy in coastal areas, particularly in the areas of Shanganagh and Loughshinny (Synge, 1977a; Hoare, 1977; McCabe and Hoare, 1978). Synge (1977a) maintains that at both localities a till of inland or Midland provenance overlies a till of Irish Sea basin provenance, while Hoare and McCabe (Hoare, 1977; McCabe and Hoare, 1978) maintain that all of the tills at these sites are of Irish Sea basin

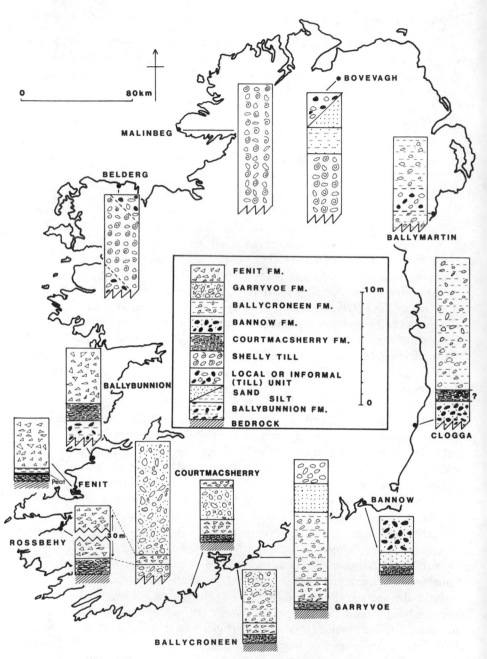

Fig. 1. Stratigraphic columns for the Quaternary lithostratigraphic stratotypes and other sites relevant to the text.

provenance. There is, however, general agreement that sites further inland demonstrate an inland till overlying an Irish Sea basin till (Ballycroneen Formation).

The sequence outlined by McCabe and Hoare (1978) further north in Counties Meath and Louth is more complex. They identified an alternating sequence which from the bottom upward is (1) Inland till, (2) Irish Sea basin till, (3) Inland till, (4) Irish Sea basin till, (5) Inland till as a northerly younging sequence from Dublin to Dunany Point, Co. Louth. From the second part of the sequence this was seen as reflecting two contemporaneous ice sheets, one spreading southeastward from the Midlands and another pressing in from the Irish Sea basin, with a zone of contact, or near-contact, which moved back and forth across the present coastline area in response to a nonsynchronous retreat pattern.

At Mell in Co. Meath, a glaciomarine sequence seems to lie between the lower inland till and the lower Irish Sea basin till (Drogheda and Gormanstown Formations, respectively). As there is doubt both as to its stratigraphic position and as to whether or not it is *in situ* (McCabe and Hoare, 1978), it is at present of very little use as an aid in unravelling what is in this area a complex stratigraphic sequence.

Further north in Counties Down and Antrim, calcareous shelly till with Ailsa Craig microgranite is again seen in complex relationship to till of inland provenance (Hill and Prior, 1968; Stephens *et al.*, 1975). In south Antrim and east Down the shelly till is overlain by a till derived from the Lough Neagh lowlands, but on the east coast of Antrim only one till can be distinguished with certainty. Whereas this till does not contain Ailsa Craig granite erratics, the lower till in south Antrim and east Down does contain small amounts (Hill and Prior, 1968). In south Down, two tills of Irish Sea basin provenance are separated by a till of inland provenance at Ballymartin on the coast of the Mourne Mountains (Stephens *et al.*, 1975). In north Antrim and north Co. Derry, a shelly till with Ailsa Craig granite erratics is the uppermost deposit. This deposit has been traced from Ballycastle to Coleraine and inland as far south as Armoy and Ballymoney.

But in northwest Derry (Bovevagh Till Formation) and Co. Donegal (Burnfoot Till Formation) (Colhoun, 1971) a shell-bearing till with associated flints underlies a till of southern inland provenance (Gelvin Formation). In southwest Co. Donegal at Malin Beg and on the north coast of Co. Mayo between Ballycastle and Belderg, shell-bearing tills are also seen. The Donegal deposit is called the Malin Beg Till Formation (Colhoun, 1973) and lies at the surface. The north Mayo till, the Belderg Till of Synge (1968), whether *in situ* or reworked, seems also to extend to the surface (Hinch, 1913; Synge, 1968). The shelly tills of north Mayo and southwest Donegal are of uncertain provenance (see Charlesworth, 1963), but those of

Inishowen and northwest Derry seem to be of northeastern provenance, probably of Scottish origin (Colhoun, 1971). It has been argued that the north Mayo tills relate to Scottish ice (Synge and Stephens, 1960), and the occurrence of shelly till at Malin Beg in southwest Donegal would seem to strengthen this interpretation (Colhoun, 1973). Thus the shelly tills of the Irish Sea and south coasts and those of the north and northwest coasts seem to relate to a general southerly movement or movements of Scottish ice.

It is clear, however, that the shelly till is diachronous, and, although it is seen in complex stratigraphic arrangement with other tills, there is no evidence that it represents more than one glaciation. Culleton (1978a) identified only one shelly till in south Co. Wexford and could not substantiate Mitchell's (1976) contradistinctive two till units. McCabe and Hoare (1978) interpreted the complex stratigraphic situation on the east coast in terms of a complex ice-front retreat in the Irish Sea basin. And, although Stephens *et al.* (1975) and Colhoun (1971) place some of the shelly tills of the northeast and northwest in an earlier glaciation, a radiocarbon date of 24,050 ± 650 bp from the shelly till at Glastry, Co. Down, and lack of any evidence of an interglacial in the sediments of Counties Antrim and Down suggest that the shelly tills are young and that the north Antrim sequence (Stephens *et al.,* 1975) might best be interpreted along the lines of McCabe's model for the east coast referred to above.

It is clear that the Ballycroneen Formation extends from Ballycroneen in Co. Cork to Coleraine in Co. Derry. The other shell-bearing tills of Derry, Donegal, and Mayo may form a separate formation (informally the Belderg formation). Owing to the diachronous nature of tills, it is not possible to define a top and bottom to the Ballycroneen Formation or any other till formation.

The Bannow Formation

This formation incorporates the Ballyvoyle till (Wright and Muff, 1904), the Bannow till (Synge, 1964), and the Clogga till (Synge, 1964). The formation is essentially that described by Culleton (1978b), and the Ballyvoyle till and Clogga till can be regarded as constituent members. Its extent is roughly coincident with the "till from Lower Palaeozoic shales, slates" outlined by Synge (1977b). It is dominantly composed of shales and slates but includes erratics of Leinster granite, Carboniferous limestone, and, occasionally, Devonian sandstone. At Clogga, Co. Wicklow, it is overlain by the Ballycroneen Formation, but on the coast of southeast Waterford the converse applies. Along much of the coast of Co. Waterford the Bannow Formation replaces the Ballycroneen Formation in the following sequence:

Till (Bannow Formation)
Head (Fenit Formation)
Raised beach (Courtmacsherry Formation)

The Garryvoe Formation

This is a till unit composed dominantly of Devonian sandstone and Lower Carboniferous sandstones and shales. This till was first described by Wright and Muff (1904). But the site at Garryvoe, where Farrington (1954) interpreted a sand unit separating it from the underlying Ballycroneen Formation as an outwash deposit associated with the latter, has gained precedence in the literature and is confirmed here as the stratotype. West of Ballycroneen the Garryvoe Formation takes the place of the Ballycroneen Formation in the raised beach--head--till sequence. At Ross Behy, Co. Kerry, the Drom Formation is the local equivalent of the Garryvoe Formation and occupies the same stratigraphic position *vis-à-vis* the raised beach and head deposits (Warren, 1978).

From Fenit to Ballybunnion in north Kerry the sequence is incomplete. Both the raised beach (Courtmacsherry equivalent) and the head are seen, but the overlying till unit is missing. There is no *in situ* till here overlying the Courtmacsherry Formation. This is an important hiatus as it implies that this area was not glaciated after the interglacial represented by the Courtmacsherry Formation.

The Stratigraphy of Inland Deposits

The tills of inland provenance exposed on the east and north coasts have no useful lithostratigraphic marker to which they can be related. Their position in relation to the Ballycroneen Formation can frequently be established, but as this is a diachronous unit, the tills cannot be correlated on this basis. Traditionally (see Mitchell *et al.*, 1973) the Ballycroneen Formation has been regarded as a deposit of a penultimate glaciation, and overlying tills regarded as deposits of the last glaciation. In this way the Ballycroneen Formation has been used as a stratigraphic datum.

In Counties Derry, Antrim, and Down, the uppermost tills have been normally regarded as relating to the last glaciation, but lower tills, particularly tills of Scottish and/or Irish Sea basin (North Channel) provenance, chiefly underlying tills of inland origin, are normally referred to an earlier glaciation (Colhoun, 1971; Stephens *et al.*, 1975; Mitchell *et al.*, 1973). Colhoun (1971) offered no evidence, stratigraphic or otherwise, as to why, for example, the Spincha (Till) Formation and/or the Bovevagh (Till) Formation of Co. Derry should relate to an earlier glaciation. The only direct reference to the interglacial which is assumed to have occurred between their deposition and the Gelvin (Till) Formation, which he placed in a later (Midlandian) glaciation, is as follows (Colhoun, 1971, p. 51):

> There is no direct evidence in this area for an interglacial period prior to the Late Sperrin Glaciation and after any of the consecutive Spincha, Bovevagh and Early Sperrin Glaciations which are considered to be Roeian (Munsterian/Saalian) age, but the possibility of a period of interglacial conditions must not be excluded.

This comment might well be extended to all of the deposits of northeast and northwest Ireland. And while one must agree that the *possibility* of an interglacial cannot be excluded, this possibility does not warrant its arbitrary inclusion.

Similarly, the association of the lower tills in Co. Down with the penultimate glaciation has no basis in stratigraphy, and, although Stephens *et al.* (1975) and Mitchell *et al.* (1973) indicate an interglacial separating upper and lower till suites, there is no evidence whatever for this. They recognize that this is so, and their argument is very similar to that quoted above for, having placed all of the till units exposed on the Mourne coast at Ballymartin in the penultimate glaciation on the basis of morphological expression and having cast some doubt on a date of 24,050 ± 650 ^{14}C years bp derived from shells from the Glastry lower shelly till, Stephens *et al.* (1975, p. 18) argue as follows:

> If, therefore, correlation is attempted between the lower shelly till at Glastry and the lower or both shelly tills on the Mourne coast, a Midlandian age may be indicated for them all, giving full weight to the C-14 date of 24,050 years BP on shells contained in the Glastry lower shelly till. If this C-14 date is ignored or regarded with some scepticism, then the matter of correlation is open to question.

The matter of correlation is indeed open to question, but there is no objective reason to discount either the date of the shells in the Glastry till or that the three till units at Ballymartin likely belong to the most recent glaciation.

In the northwest and west there was no stratigraphic datum, but the tills of inland provenance were separated on the basis of their morphological expression. However, as Charlesworth (1963) pointed out, there is no evidence that the tills west of the Ballycastle–Mulrany line in west Mayo (Synge, 1968) and in northern Inishowen (Stephens and Synge, 1965) are older than the last glaciation. The reasons for relegating the surface till in these areas to a penultimate glaciation relate to the facts first that the deposits are frost disturbed and second that there are no significant glacigenic morphological features in these areas. The first reason is based on a *non sequitur,* for frost disturbance indicates cold, not interglacial, conditions. The second reason is equally untenable, for there are, and one expects that there will be, large areas of till sheets and sandur debris within any glaciated area which will not display appreciable amounts of constructional features (see Flint, 1971).

This difficulty in establishing controlled stratigraphic sequences where a clear isochronous unit is lacking becomes even more acute inland owing to the paucity of exposures. Moraines have traditionally been used as marker horizons, and, although they can serve this function locally, attempts to identify particular moraines as isochronous units which can be correlated over large distances cannot succeed unless there are other precise stratigraphic controls.

The Dunany Moraine has been correlated by Synge (1970a) and others with the Moville Moraine in Inishowen and the Ballycastle–Mulrany Moraine in west Mayo. It was also correlated with the Fedamore Moraine (Fig. 2). There is no independent evidence to substantiate such correlations. They were regarded as confining moraines of the drumlin swarms, but there is no evidence that drumlinization was constrained by either time or event. Thus, the term Drumlin Advance (or Readvance) has no real meaning when applied on this grand scale.

It is significant that, in attempts to identify corroborating evidence that the moraines which make up the Southern Irish End-Moraine separate tills of different glaciations, both Finch (1971) and Synge (1966, 1970b) have, on analysis, merely demonstrated that the Southern Irish End-Moraine is not a reliable stratigraphic marker. Synge, in trying to demonstrate that the tills south of the moraine were more leached of limestone than those north of it, thus demonstrating a period of interglacial weathering between the deposition of one and the other, adjusted the "Younger Drift Limit" to coincide with the maximum extent of unleached till. He did not consider factors such as exponential distance decay in limestone carry-over onto more acidic rocks, change in acidity of adjacent and subjacent geology, or the groundwater conditions. Rather, he concluded that ice of the younger glaciation extended some distance south of the moraine before retreating and stablizing to form the moraine. Finch applied precisely the same principles to southeast Co. Tipperary, where he extended the limit of glaciation from Cahir to the foot of the Comeragh Mountains, because the till here, south of the morainic gravels, is not leached of limestone. The till is, however, resting directly on the limestone bedrock of the Suir valley at a point where that valley is open to the north, it does not come into contact with Devonian sandstone bedrock until it reaches the hills south of Clonmel, and limestone tends to have been weathered out of it from this point south. Warren (1979) has shown that limestone leaching is closely related to the lithology/petrology of the deposit and of adjacent and subjacent bedrock, probably reflecting groundwater acidity. It cannot, therefore, be used ex-

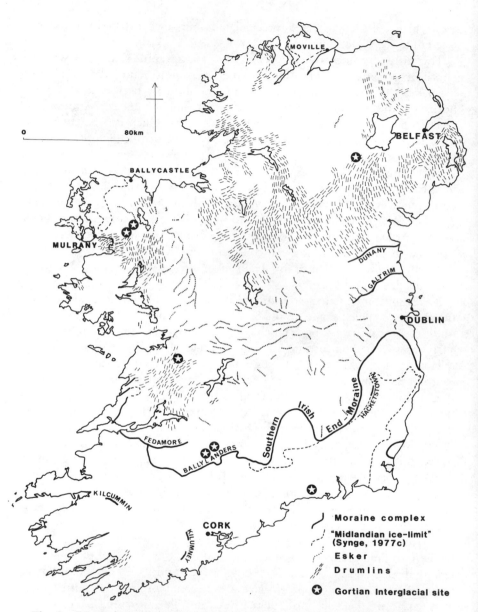

Fig. 2. Principal morphostratigraphic and related units used chiefly by Synge (1970a, 1977c), Mitchell *et al.* (1973), and Charlesworth (1928). Localities of Gortian Interglacial deposits are also indicated.

clusively as an indicator of age. Thus Finch (1971) has simply shown that limestone-dominated tills extend on either side of the morphological horizon (the Southern Irish End-Moraine). Warren (1980) has also shown that a similar till occurs on either side of this moraine on the southern side of the Shannon estuary, where there is no evident difference in degree of weathering of limestone clasts.

It is clear that morphostratigraphic units such as moraines are of very limited use and are impossible to correlate without other stratigraphic markers; those who appear to have used them did in fact use entirely different markers (degree of limestone leaching) when drawing important boundaries. Synge's (1977c) Midlandian limit does not follow the line of the Southern Irish End-Moraine (Fig. 2).

There is no attempt here to identify specific named units in the inland deposits as they are very little known and have not, by and large, been named in the literature. It is pertinent to note, however, that none of the glacigenic deposits or features inland have been demonstrated to be older than the most recent glaciation, and, on the evidence of the coastal sections, it seems reasonable to regard them as part of the general deglacial sequence of the most recent glaciation.

It seems probable, from the disposition of the moraine ridges and esker systems, that, in general, deglaciation progressed toward the Midlands from the south and east with the ice front taking up an almost north–south line and retreating generally westward. The retreat pattern of the western and northern margins of the ice sheet is not clear, but it seems as if there was a general southern retreat of the ice front from the northern coastal area to the Tyrone–Fermanagh area and the Lough Neagh basin (see Colhoun, 1970). Neither the number nor degree of ice advances and retreats is known, but interstadial deposits in Co. Fermanagh (see below) suggest at least two major glacial events.

BIOSTRATIGRAPHIC UNITS

Gort Biozone

At Boleyneendorrish near Gort in Co. Galway, a biostratigraphic sequence in organic mud reflects in its pollen spectrum a transition from herbs and grassland to open woodland advancing to deciduous forest with strong *Quercus* and regressing to coniferous forest with strong Ericales presence including *Rhododendron*. This is overlain by fluvial and glacial deposits. The stratigraphic sequence here reflects a transition through interglacial to glacial in climate stratigraphic terms.

The Courtmacsherry Formation is not seen north of Ballybunnion on the west coast nor is it seen on the east or north coasts (although it may occur at Clogga on the east coast). Thus, there is no clear lithostratigraphic datum to which the stratigraphy north of these points can be referred. The relative stratigraphic position of units so affected will depend to a large extent on the interpretation of the Courtmacsherry Formation as a raised beach deposited during an interglacial period. Both Bowen (1973) and Warren (1979) have argued that on a strict stratigraphic interpretation this must be regarded as the most recent interglacial. Mitchell (1976), *inter alios,* argued that both the Courtmacsherry Formation and the Gortian Interglacial deposits are products of a penultimate interglacial, while Warren (1979) has argued that the Gortian deposits (along with the Courtmacsherry Formation) relate to the most recent interglacial, probably Stage 5 of the deep-sea record (Emiliani, 1966).

The interpretation outlined by Warren (1979) is retained here for the reasons stated in that paper: there is neither biostratigraphic nor lithostratigraphic evidence of a later interglacial; there is no independent evidence that they relate to an earlier interglacial than the last; correlation on a biostratigraphic basis with the penultimate interglacial deposits established elsewhere is equivocal; Gortian Interglacial deposits on either side of the Southern Irish End-Moraine bear precisely the same relationship to the tills which overly them; and one cannot include in a stratigraphic sequence a unit for which there is absolutely no evidence while there are strong and acceptable alternatives which are based on sound stratigraphic principles. The primary stratigraphic model must be based on observed stratigraphic units. The organic Gortian Interglacial deposits and the Courtmacsherry Formation can be independently assigned to the last interglacial (see Bowen, 1973, and Warren, 1979), but Mitchell (1970) has also shown that the Courtmacsherry Formation at Fenit probably relates to the Gortian Interglacial.

Organic deposits determined palynologically to belong to this interglacial occur in Counties Galway, Mayo, Limerick, Waterford, and Tyrone. These all underlie till deposits. In only one case, at Fenit in Co. Kerry, is there an organic deposit of probable interglacial age that is not overlain by till; it is overlain by head (see Warren, 1981). Gortian deposits occur both north and south of the Southern Irish End-Moraine, but nowhere is there any stratigraphic marker which would substantiate the suggestion that the Gortian is of penultimate interglacial age and that the moraine separates deposits of two entirely different stages. This lends weight to the contention that all of the surface glacial and glaciofluvial deposits relate to a single glacial stage and that this is the most recent one.

Interstadial Deposits

At Derryvree in Co. Fermanagh (Colhoun *et al.*, 1972), fossil remains from a freshwater organic silt deposit have been examined. This deposit is sealed between two tills, an upper till (Maguiresbridge Formation) of northern provenance and a lower till (Derryvree Formation) of northwestern provenance. The organic deposits have given a ^{14}C date of $30,500 \pm \frac{1170}{1030}$ bp. Their contained floral and faunal remains indicate that they were deposited under interstadial rather than interglacial conditions. Interstadial is used in the loose sense of a period of relative warmth between two glacial events; it does not imply that an interstadial boreal climax has been recognized. In this instance there is no evidence of forest development, but a period of cool climate with severe winters has been interpreted (Colhoun *et al.*, 1972). The importance of the stratigraphic sequence is that there is no indication of interglacial deposits between the Derryvree Formation and Maguiresbridge Formation, thus both deposits must logically be attributed to the same cold period (see Mitchell, 1977).

A similar sequence is seen at Hollymount where an organic silt deposit occurs between two tills. The upper till belongs to the Maguiresbridge Formation, and it is likely that the lower till is a lateral equivalent of the Derryvree Formation (McCabe *et al.*, 1978). The floral and faunal remains in the organic silts indicate similar conditions as those at Derryvree but are dated >41,500 ^{14}C years bp. Once more the two tills at this site seem to relate to the most recent glaciation.

A date of $33,500 \pm 1,200$ bp was derived from a mammoth bone taken from a cave at Castlepook, Co. Cork (see Mitchell, 1976). Apart from the date which suggests interstadial conditions at this time, the bone is of no real stratigraphic significance as it was found in secondary position and its stratigraphic position relative to other Quaternary deposits in the area is unknown.

CHRONOSTRATIGRAPHIC
INTERPRETATION

The chronostratigraphic units proposed here are outlined in Fig. 3, as are the lithostratigraphic and biostratigraphic units on which they are based. The units between question marks are informal groups whose precise stratigraphic position is in doubt, and the units preceded by a question mark are formal units whose precise stratigraphic position is in doubt. These units may occur in more than one place, indicating their alternative stratigraphic positions.

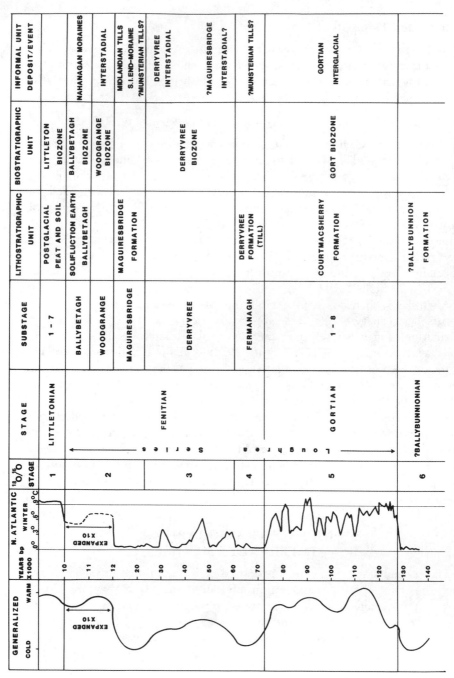

Fig. 3. The Irish Quaternary stratigraphic sequence as proposed here showing suggested correlations with the generalized deep-sea $^{18}O/^{16}O$ palaeotemperature curve (Emiliani, 1966) and the North Atlantic Ocean palaeotemperature curve for winter surface water, also based on the deep-sea $^{18}O/^{16}O$ record (Sancetta *et al.*, 1973)

The Ballybunnionian Stage

This stage is based on the Ballybunnion Till Formation. However, neither the top nor the bottom of this formation has properly been identified. A detailed examination of the stratotype area will probably reveal the bottom of the formation. This, however will not indicate the base of the stage as it is likely that cold conditions preceded the actual glacial event that produced the till. As it is in unconformable contact with the overlying raised beach deposits (Courtmacsherry Formation) its top cannot properly be seen. And, although it is probably reasonable to assume that the till here represents a glacial event immediately preceding the interglacial represented by the raised beach, we cannot with certainty say what period of time or number of events are represented by the line of unconformity.

The Gortian Stage

The Courtmacsherry (Raised Beach) Formation probably represents the sea level achieved during the climatic optimum of the most recent interglacial. As such it represents no more than a moment in geological time and should be treated as a chronostratigraphic horizon rather than a deposit representative of the interglacial. The Courtmacsherry sediments have been correlated with the Gortian Interglacial deposits, and there is general agreement on this matter (see Mitchell, 1976; Warren, 1979). The interglacial is represented by the palynology of the sediments at Gort, but Watts's (1967) zonation which is referred to here is based on two other interglacial sites, one at Kildromin and the other near Herbertstown, in Co. Limerick. Watts recognized six pollen zones (but see Watts, Chap. 8, this volume), G1 through G6, which reflect the transition from cool early interglacial conditions in its herb and *Betula* (?*nana*) pollen through a climax phase (*Pinus–Quercus*) to a telocratic phase (*Taxus–Abies–Rhododendron*). At all of the fully reported Gortian sites the sediments are "truncated" at the telocratic phase. Watts (1967) implies that this is probably due to erosion at this level; however, the coincidence of truncation at the same level must raise the possibility that the interglacial did not complete its course, that very rapid cooling at this point brought on the following glacial stage. Thus the upper Kilbeg *Pinus*-dominated sediments may represent, as Watts suggested, an interstadial, as might the peats at Fenit and Newtown. On the other hand, a sudden cooling bringing the period represented by zone G6 to a close might have induced soil instability and changes in sedimentation and have allowed late interglacial floras to persist, leaving only sporadic records. This is close to Mitchell's (1976) interpretation, and, although it begs the question of the definition of an interstadial, until evidence of a stadial emerges between what Mitchell refers to as deposits of phase IGWD (roughly Watts's zone G6) and those in IGWE and IGWF (ebbing and final phases), it is probably the best possible reconstruction.

Thus the Gortian Interglacial can be subdivided into eight zones, zones G1 through G6 of Watts (1967) and phases of IGWE and IGWF of Mitchell (1976), which should probably be referred to as zones G7 and G8. The Gortian Stage can be divided into eight substages each representing a zone (but see Watts, Chap. 8, this volume). They are here referred to as substages 1 through 8.

The Fenitian Stage

As the most recent cold stage of which we have evidence in Ireland, the Fenitian Stage has left a wide variety of sediments which are susceptible of considerable subdivision. Because sediments previously regarded as belonging to an earlier glaciation, as well as those previously regarded as the exclusive deposits of the most recent glaciation (the Munsterian and Midlandian, respectively, of Mitchell et al., 1973), are here regarded as the deposits of a single glaciation, neither of these names will be used to define the sediments of this stage. A new name is proposed, the Fenitian Stage. The stratotype is the coastal section between Fenit and Spa, Co. Kerry, where it is represented by the Fenit Formation. All tills overlying this formation on the south coast belong to this stage as do all deposits which stratigraphically overly the Courtmacsherry Formation or the Gort Biozone. Two interstadials and three stadials are recognized in this stage, and they are given the rank substage. Thus the earliest substage is the Fermanagh (cold) Substage based on the lower till at Derryvree. This is followed by the Derryvree Substage based on the Derryvree interstadial deposits, and this again is followed by the Maguiresbridge Substage which is based on the overlying Maguiresbridge (till) Formation. The extent of glaciation associated with either of these cold substages is not known with certainty. It is likely that ice associated with the Maguiresbridge Substage extended as far south at least as the Southern Irish End-Moraine. If it extended no further, then the tills to the south (formerly Munsterian) relate to the Fermanagh Substage. Alternatively, the "Munsterian" tills may relate to the Maguiresbridge Substage, and the Fermanagh glaciation may have been of much lesser extent.

Two further substages, both belonging to the lateglacial period, were first identified in Ireland from sediments at Ballybetagh, Co. Dublin, by Jessen and Farrington (1938). A warm phase and a final cold stage phase were labelled pollen zones II and III, respectively. Zone II has been named the Woodgrange Interstadial (Mitchell, 1976) from its stratotype at Woodgrange, Co. Down. It is here named the Woodgrange Substage. This warm interstadial period is now recognized to have been more complicated than had originally been thought, and both Watts (1977) and Mitchell (1976) characterize it as embracing three pollen assemblage zones. However, the stratigraphic approach of Jessen and Farrington (1938) and to some extent that of Morrison and Stephens (1965) obviate to some degree the problem of diachronous pollen assemblages. At Woodgrange, Co. Down, Singh (1970) introduced a transition zone (Zone I–II) between Zones I and II. Mitchell (1976) placed the base of his Woodgrange Interstadial slightly below the top of Singh's (1970) Zone I, and lithostratigraphically the unit extends from the base of Singh's (1970) light grey marl to the top of his dark grey clay–mud. This is the stratotype of the Woodgrange Substage.

The cold phase or stade associated with the Zone III sediments at Ballybetagh was proved to relate to a limited cirque glaciation in a recent exposure at Lough Nahanagan, Co. Wicklow (Colhoun and Synge, 1980). This was named the Nahanagan Stadial by Mitchell (1976). However, it is suggested here that, although the Nahanagan site clearly demonstrates cirque glaciation during this phase, it is a very poor stratotype. Accordingly, the stadial in question is here named the Ballybetagh Substage from the original site of Jessen and Farrington (1938) which is the stratotype. The lowest part of the solifluction earth which defines the *Salix herbacea* zone at Ballybetagh (Jessen and Farrington 1938) marks the base of the Ballybetagh Substage. This is probably slightly below the *Artemisia* pollen assemblage zone recognized by Watts (1977) at the same site, where it seems to be marked by his Cruciferae peak. Again, as pollen assemblage zones are likely to be diachronous and often site specific, the basis of this unit, following Jessen and Farrington (1938), is lithostratigraphic. Radiocarbon dates for the begining of this substage vary from about 11,500 to 10,600 bp. The date 11,480 ± 150 bp at the "transition of organic mud to solifluction clay at Zone II/III boundary" at Roddans Port (Godwin and Willis *in* Morrison and Stephens, 1965, p. 250) seems to mark it as defined here, although there is some question as to the reliability of this date (see Watts, 1977, p. 290). The upper part of the solifluction earth at Ballybetagh marks the top of the substage.

The top of the Zone III sediments at Ballybetagh also signifies the end of the Fenitian Stage, which is followed by the Littletonian Stage, the present postglacial period.

The Littletonian Stage

The name and stratotype of this stage are taken from Mitchell *et al.* (1973). The definition is altered slightly here as the base of the stage (maximum *Juniperus*) as defined by Mitchell is unsatisfactory and is likely to be diachronous. It is suggested here that, lithostratigraphically, the base of the Littletonian might be regarded as the base of the organic mud unit in which the *Juniperus* maximum was identified by Mitchell. Alternatively, the arbitrary date 10,000 radiocarbon years bp might be adopted. At the stratotype, Littleton, Co. Tipperary, the base of the stage is defined by Mitchell (1982) as the maximum of *Juniperus* pollen at a depth of 750 cm in the 1954 boring. The basal date selected by Mitchell from Roddans Port, Co. Down, was 10,130 ± 170 bp (but see Edwards, Chap. 12). This stage can be divided into substages based on pollen assemblage zones (as suggested by Godwin, 1975) recognized within the deposits of the Littletonian Stage (see Mitchell 1965, 1976). These would simply be named Littletonian Substage 1 through 7, corresponding to postglacial pollen assemblage Zones IV through X (Mitchell, 1965). However, chronostratigraphic correlations for this period can best be made using absolute dating techniques.

Glaciations

The chronostratigraphic framework presented here is designed as a chronological sequence suitable for the Quaternary strata of Ireland. The term glaciation is avoided in referring to the cold stages as in Ireland actual glacial events seem to have been relatively short lived, particularly during the last cold stage. For this reason it is particularly appropriate that the stratotype of the Fenitian Stage does not include any primary glacial sediments; they are almost all periglacial. In the Irish context the term glaciation should be used with reference to the glacial events that occurred during the cold stage in question. The term stade may be used informally with reference to the period during which such an event took place.

The Loughrea Series

Although the International Subcommission on Stratigraphic Classification (Hedberg, 1976) suggests that the Quaternary System is divided into two series, the Holocene Series and the Pleistocene Series, in practice the Holocene is classed as a stage in the Quaternary along with the various other warm and cold stages. It is suggested here that the most suitable basis for grouping Quaternary stages or for subdividing the Quaternary System to facilitate long-distance correlation and to lessen the risks attendant in correlating what might be homotaxial units is grouping the sediments of each warm stage (interglacial) with those of the succeeding cold stage (glacial). In Ireland we can recognize only one complete series (Gortian Stage–Fenitian Stage), one as yet incomplete series (Littletonian Stage), and the upper part of an earlier series (Ballybunnionian Stage). It is proposed to call the complete series the Loughrea Series after the barony of that name in which the stratotype of the Gortian Stage occurs. The base of the series is the base of the Gortian and the top is the top of the Fenitian (i.e., the top of the Ballybetagh Substage). It is suggested that international correlation at stage level should not be made unless correlation can be made at the series level from one stratotype to another.

CORRELATIONS

The stratigraphic model proposed here is very much an imperfect one. It is unlikely that it will ever be anything more, given that the discovery of a continuous depositional conformable sequence is obviated by the erosional nature of glacial ice and the unconformable nature of glacial sediments. It would cause more problems than it would solve to try to correlate this stratigraphic model with that commonly accepted for Britain or the Continent or with any other British or continental model, as the same problem (lack of continuity of sedimentation, particularly in the late Quaternary) applies there as well. What is needed is a common yardstick, a yardstick based on a record of continuous sedimentation—a marine sequence (see Hedberg, 1976). Reconstructed temperature curves based on $^{18}O/^{16}O$ ratio measurements have been available since 1955 (Emiliani, 1955). The method has been refined and the deep-sea record classified, correlated with onshore loose sediments, and in part dated, so that a reliable continuous record of such marine records now spans the last 900,000 years (see Kukla, 1977, for onshore correlations).

Figure 3 is set against Emiliani's (1966) generalized palaeotemperature curve based on the $^{18}O/^{16}O$ ratio. A second curve based on palaeoenvironmental analysis of planktonic foraminifera (Sancetta et al., 1973) is included. This is a more detailed curve with some ^{14}C dating control. This is particularly useful, for it provides detailed information on fluctuations of surface sea temperature in the North Atlantic quite close to Ireland (51°35′N, 21°56′W) for $^{18}O/^{16}O$ Stages 1–5 inclusive. The Loughrea Series is correlated with Stages 2–5 of the deep-sea record.

It would seem reasonable to relate the Derryvree Substage to Emiliani's Stage 3 as the date (30,500 years bp) for this substage is very close to the date (31,000 years bp) of one of the three short warm periods identified within Stage 3 by Sancetta et al. (1973). Thus the Maguiresbridge Substage probably relates to Stage 2 and the Fermanagh Substage to Stage 4 of the oxygen isotope record. At Fenit, the stratotype for the sediments of the last cold stage, an apparent lag layer or concentration of boulders a little more than half way up through the head unit coincident with reported ice-wedge casts (Mitchell, 1970; Warren, 1981), suggests stabilization of the earlier active periglacial mass movement process before the final part of the head was deposited. It is probable that this reflects somewhat warmer conditions during which some changes in vegetation, possible weather conditions, and runoff patterns inhibited active mass waste. The concentration of large boulders suggestive of a lag layer indicated that a slope-wash process operated but not at a sufficiently dynamic level to move larger erratics and that any freeze–thaw process that operated was not sufficiently extreme to fragment them. It is probable that the lower part of the head represents Stage 4 and the uppermost Stage 2 while the boulder concentration represents Stage 3.

The Courtmacsherry Formation and the Gortian Biozone deposits probably relate to Stage 5. Whether all of Stage 5 is represented or just the climatic optimum represented by Substage 5e (see Kukla, 1977) is unknown.

The lack of continuity in sedimentation makes it difficult to place the Ballybunnionian Stage in its stratigraphic context. It is likely that it relates to a recent cold stage in the oxygen isotopic classification, possibly Stage 6, but this correlation is based on little more than a "count from the top" stratigraphic principle. The lower till at Cuan Lathaí in Co. Kerry (Warren, 1977) and possibly the Clogga Till (Synge, 1977a) may also relate to cold stages earlier than Stage 5, but again lack of continuity in the sediments makes correlation impossible.

It seems clear from their stratigraphic position that all surface tills postdate Stage 5 (the Gortian), but there is no clear indication as to whether at their maximum extents, particularly on the south coast, they (the Ballycroneen Formation and the Bannow Formation, for example) relate to Stage 4 or 2, that is, whether they predate or postdate the Derryvree Substage. It is tempting to suggest that the deposits that were formerly attributed to the "Munsterian Glaciation" should now be attributed to the Fermanagh Substage because of its relative stratigraphic position. Although there is no unequivocal stratigraphic evidence to support this, and, to date, no surface sediments relating to Stage 4 have been identified in northern Europe (Kukla, 1977) the balance of local stratigraphic probability lends weight to this interpretation. The absence of surface till in the Fenit–Ballybunnion area indicates an area (yet undefined) which was not glaciated after Stage 5.

The Ballybetagh Substage and the Woodgrange Substage are clearly represented by the sudden extreme temperature changes near the top of Stage 2. The Littletonian Stage is represented by Stage 1.

CONCLUSIONS

Figure 3 represents the best available stratigraphic interpretation of the Quaternary sediments of Ireland. It is not in agreement with the normal morphological or morphostratigraphic models.

The author is acutely aware that all of the sediments now regarded as Gortian may turn out on further analysis to belong to separate interglacial stages; this is the problem of homotaxis. He is also aware that sea level may have been similar during different interglacials and that the stratigraphic correlation between the Courtmacsherry Formation and the Gortian sediments is somewhat tenuous. Nevertheless, it is felt that the model presented here is established on sound principles and capable of adjustment to new stratigraphic evidence.

The most profitable areas for new stratigraphic work must be in the palynology of the various Gortian sediments. These need to be examined to test further whether they belong to one or more separate periods. The Mell Formation needs to be dated. Proponents of the argument that the Southern Irish End-Moraine is the end moraine of the last glaciation must provide stratigraphic evidence. It is not sufficient to say that Gortian deposits at Kilbeg underlly penultimate glacial (Munsterian) sediments, therefore the Gortian is of penultimate interglacial age. This begs the question of the age of the Munsterian deposits. If evidence of an interglacial is not found overlying "Munsterian deposits" interstadial sediments should be sought. A recent trench (2–3 m deep) dug from Dublin to Cork during the laying of a gas pipeline failed to reveal either interglacial or interstadial sediments, although it was logged for its entire length by Geological Survey personnel (Creighton, 1984).

References

Bowen, D. Q. (1973). *Proc. Geol. Assoc.* **84,** 249–272.
Charlesworth, J. K. (1928). *Q. J. Geol. Soc. London* 84(2), 293–342.
Charlesworth, J. K. (1963). *Proc. R. Ir. Acad. Sect. B* **62,** 295–322.
Colhoun, E. A. (1970). *Ir. Geogr.* **6,** 162–185.
Colhoun, E. A. (1971). *Proc. R. Ir. Acad., Sect. B* **71,** 37–52.
Colhoun, E. A. (1973). *Ir. Geogr.* **6,** 594–609.
Colhoun, E. A., and Mitchell, G. F. (1971). *Proc. R. Ir. Acad., Sect. B* **71,** 211–245.
Colhoun, E. A., and Synge, F. M. (1980). *Proc. R. Ir. Acad., Sect. B* **80,** 25–45.
Colhoun, E. A., Dickson, J. H., McCabe, A. M., and Shotton, F. W. (1972). *Proc. R. Soc. London, Ser. B* **180,** 273–292.
Creighton, J. R. (1984). "Geological observation of the Dublin to Cork natural gas pipeline trench 1982," Open file report, 6 volumes. Geological Survey of Ireland, Dublin.
Culleton, E. B. (1978a). *J. Earth Sci. (Dublin)* **1,** 33–39.
Culleton, E. B. (1978b). *Proc. R. Ir. Acad., Sect. B* **78,** 293–308.
Emiliani, C. (1955). *J. Geol.* **63,** 538–578.
Emiliani, C. (1966). *J. Geol.* **74,** 109–124.
Farrington, A. (1934). *Proc. R. Ir. Acad., Sect. B* **42,** 173–209.
Farrington, A. (1949). *J. Glaciol.* **1,** 220–225.
Farrington, A. (1954). *Ir. Geogr.* **3,** 47–53.
Finch, T. F. (1971). *Sci. Proc. R. Dublin Soc.* **3B,** 35–41.
Flint, R. F. (1971). "Glacial and Quaternary Geology." Wiley, New York.

Geikie, J. (1914). "Antiquity of Man in Europe." Oliver & Boyd, Edinburgh.
Godwin, H. (1975). "History of the British Flora," 2nd Edn. Cambridge Univ. Press, Cambridge.
Harkness, R. (1869). *Geol. Mag.* **6,** 542–550.
Hedberg, H. D. (1976). "International Stratigraphic Guide." Wiley, New York.
Hill, A. R., and Prior, D. B. (1968). *Proc. R. Ir. Acad., Sect. B* **66,** 71–84.
Hinch, J. de W. (1913). *Ir. Nat.* **20,** 1–16.
Hoare, P. G. (1977). *Proc. R. Ir. Acad., Sect. B* **77,** 295–305.
Huddart, D. (1977). "South-east Ireland," Guidebook for excursion A14 INQUA X Congress. Geo Abstracts, Norwich.
Hull, E. (1871). *Geol. Mag.* **8,** 294–299.
Hull, E. (1878). "Physical Geology and Geography of Ireland." Edward Stanford, London.
Jessen, K., and Farrington, A. (1938). *Proc. R. Ir. Acad., Sect. B* **44,** 205–260.
Jessen, K., Andersen, S. T., and Farrington, A. (1959). *Proc. R. Ir. Acad., Sect. B* **60,** 1–77.
Kinahan, G. H. (1872). *Geol. Mag.* **9,** 265–268.
Kinahan, G. H. (1878). "Manual of the Geology of Ireland." C. Kegan Paul, London.
Kukla, G. J. (1977). *Earth Sci. Rev.* **13,** 307–374.
Lewis, H. C. (1894). "The Glacial Geology of Great Britain and Ireland." Longmans, Green and Company, London.
McCabe, A. M., and Hoare, P. G. (1978). *Geol. Mag.* **115,** 397–480.
McCabe, A. M., Mitchell, G. F., and Shotton, F. W. (1978). *Proc. R. Ir. Acad., Sect. B* **78,** 77–89.
Mitchell, G. F. (1948). *Proc. R. Ir. Acad., Sect. B* **52,** 1–14.
Mitchell, G. F. (1965). *Spec. Pap.—Geol. Soc. Am.* **84,** 1–16.
Mitchell, G. F. (1970). *Proc. R. Ir. Acad., Sect. B* **70,** 141–162.
Mitchell, G. F. (1971). *Sci. Proc. R. Dublin Soc.* **4A,** 181–199.
Mitchell, G. F. (1976). "The Irish Landscape." Collins, Glasgow.
Mitchell, G. F. (1977). *Philos. Trans. R. Soc. London, Ser. B* **280,** 199–209.
Mitchell, G. F. (1982). *J. Earth Sci. (Dublin)* **4,** 97–100.
Morrison, M. E. S., and Stephens, N. (1965). *Philos. Trans. R. Soc. London, Ser. B* **249,** 221–225.
Mitchell, G. F., Penny, L. F., Shotton, F. W., and West, R. G. (1973). *Spec. Publ.—Geol. Soc. London* **4,** 99 pp.
Sancetta, C., Imbrie, J., and Kipp, N. G. (1973). *Quat. Res. (N.Y.)* **3,** 110–116.
Singh, G. (1970). *Proc. R. Ir. Acad., Sect. B* **69,** 189–216.
Stephens, N., and Synge, F. M. (1965). *Proc. R. Ir. Acad., Sect. B* **64,** 131–153.
Stephens, N., Creighton, J. R., and Hannon, M. A. (1975). *Ir. Geogr.* **8,** 1–23.
Synge, F. M. (1964). *Ir. Geogr.* **5,** 73–82.
Synge, F. M. (1966). *In* "Soils of County Limerick" (T. F. Finch and P. Ryan, eds), Soil Survey Bulletin 16, pp. 12–20. An Foras Talúntais, Dublin.
Synge, F. M. (1968). *Ir. Geogr.* **5,** 372–386.
Synge, F. M. (1970a) *In* "Irish Geographical Studies in Honour of E. Estyn Evans" (N. Stephens and R. E. Glasscock, eds.), pp. 34–48. The Queen's Univ. of Belfast, Belfast.
Synge, F. M. (1970b). *Geol. Surv. Irel. Bull.* **1,** 65–71.
Synge, F. M. (1977a). *In* "The Quaternary History of the Irish Sea" (C. Kidson and M. J. Tooley, eds.), pp. 199–222. Seel House Press, Liverpool.
Synge, F. M. (1977b). *In* "Atlas of Ireland" (J. P. Haughton, ed.), pp. 18–19. Royal Irish Academy, Dublin.
Synge, F. M. (1977c). *In* "Atlas of Ireland" (J. P. Haughton, ed.), p. 21. Royal Irish Academy, Dublin.

Synge, F. M., and Stephens, N. (1960). *Ir. Geogr.* **4,** 121–130.

Watts, W. A. (1959). *Proc. R. Ir. Acad., Sect. B* **60,** 79–134.

Watts, W. A. (1967). *Proc. R. Ir. Acad., Sect. B* **65,** 339–347.

Watts, W. A. (1977). *Philos. Trans. R. Soc. London, Ser. B* **280,** 273–293.

Warren, W. P. (1977). *In* "South and South West Ireland," Guidebook for Excursion A15 INQUA X Congress (C. A. Lewis, ed.), pp. 37–45. Geo Abstracts, Norwich.

Warren, W. P. (1978). "The glacial history of the MacGillycuddy's Reeks and the adjoining area." Unpublished Ph.D. thesis, National Univ. of Ireland.

Warren, W. P. (1979). *Geol. Surv. Irel. Bull.* **2,** 315–332.

Warren, W. P. (1980). *Quat. Newsl.* **31,** 12–18.

Warren, W. P. (1981). *Biul. Peryglac.* **28.**

Wright, W. B. (1914). "The Quaternary Ice Age," 1st Edn. MacMillan, London.

Wright, W. B. (1937). "The Quaternary Ice Age," 2nd Edn. MacMillan, London.

Wright, W. B., and Muff, H. B. (1904). *Sci. Proc. R. Dublin Soc.* **10,** 250–324.

CHAPTER 4

GLACIAL GEOMORPHOLOGY

A. Marshall McCabe

School of Environmental Sciences
University of Ulster at Jordanstown
Newtownabbey, Northern Ireland

INTRODUCTION

Although glacial landforms in Ireland have been studied for well over 100 years, our knowledge of them remains incomplete. Most of the early workers of the Geological Survey were careful observers, especially of striae pattern and esker orientation, but it is only in the last 50 years or so that the complexities of the Quaternary have been recognized by workers such as J. K. Charlesworth (1929, 1939), A. Farrington, (1934, 1942, 1947, 1953, 1957), F. M. Synge (1970, 1977), and G. F. Mitchell (1972). From this work it was recognized that the glacial landforms of lowland Ireland (Fig. 1) could be grouped broadly into two major provenances associated with two Quaternary cold phases now termed the Munsterian (older) and Midlandian (younger) (Mitchell, 1972). In southern Ireland, glacigenic Midlandian forms are delimited by the Southern Irish End-Moraine, to the south of which Munsterian drift relics occur at the surface in a belt from west Co. Wexford to Co. Kerry (Fig. 1). Independent centres of ice dispersion from the mountain masses, largely located on the margins of the island, have also been identified (Farrington, 1934, 1966; McCabe and Hoare, 1978; Synge, 1968; Warren, 1979a).

Although the aerial expression of major landform units is known, recent studies indicate that landform–sediment associations are more complex than hitherto realized (Cohen, 1979; Dardis, 1981; McCabe and Hoare, 1978; Synge, 1979). The broad distribution of glacial landforms in Ireland has been described by Synge (1969) and Davies and Stephens (1978) and will not be reviewed here. This chapter presents a restricted, spatial view of landform distribution, followed by case studies of major landform assemblages and morphosedimentological units which tend to recur in particular environments. Where possible, particular attention will be devoted to the morphology, internal structure, and formative environmental factors of selected landforms. A strict morphological approach has been avoided since it is hoped that future investigations will place an increasing emphasis on the sedimentology of glacial landforms.

GENERAL PATTERNS OF ICE-SHEET GROWTH AND DECAY

The spatial distribution of glacial landforms and deposits has been translated from a wide variety of sources into general maps depicting major directions of ice flow (Figs. 1 and 2). These diagrams are based on the concepts of fairly stable lowland ice axes with radiating flow lines ending at specific moraines or drift limits. However, ice sheets are dynamic, involving shifts in the main areas of ice dispersion, which involve profound changes in the form of the ice mass and location of the ice margin (McCabe, 1969; Synge, 1969). An alternative glacial growth model may involve a series of large, temporally unstable ice domes separated by lower areas of faster moving ice. Interpretations of the nature and pattern of drift deposits, especially in northern and western areas, require a dynamic view of ice-sheet growth and decay rather than a static one. Eventually it may be possible to assign identifiable changes in the ice-sheet form to climatic changes and glaciological variations in the ice itself.

It is generally accepted that the well-defined drift association known as the Southern Irish End-Moraine represents a major southern limit of late Midlandian ice (see Bowen, 1973; Warren, 1979b) (Figs. 1 and 2). Similar ice limits are known in the uplands of Counties Mayo and Donegal and pass laterally into deposits associated with ice lobes that moved off the present coast along major tectonic embayments into glaciomarine environments. In eastern Ireland the initial pattern of ice retreat to the north and west was accomplished by downwasting and backwasting. Equilibrium standstills of the ice front are marked by frontal moraine systems which course west to east or southwest to northeast for long distances (Fig. 1). In east-central and western areas, the well-defined morainic complex limiting the drumlin fields has been interpreted as a readvance feature related to a deterioration in climate at about 17,000 years bp (Synge, 1969). In eastern Ireland the rate of ice dissipation increased considerably after the formation of the Dunany–Carlingford–Killard Point moraines as few frontal moraine lines occur in

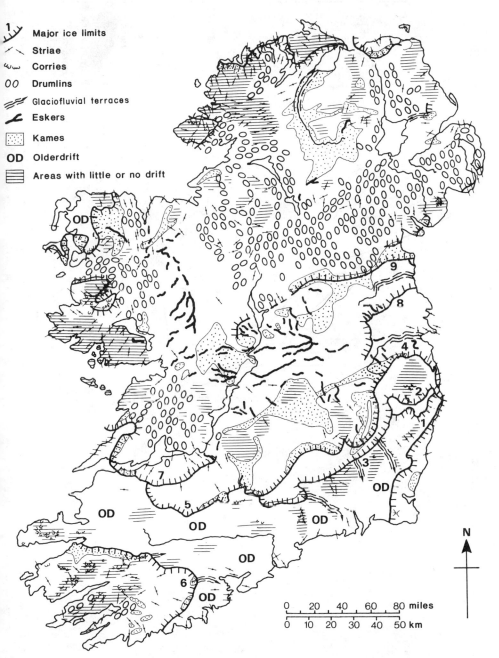

Major ice limits
Striae
Corries
Drumlins
Glaciofluvial terraces
Eskers
Kames
OD   Olderdrift
Areas with little or no drift

Fig. 1.   Quaternary geology of Ireland (redrawn from Synge, 1979).

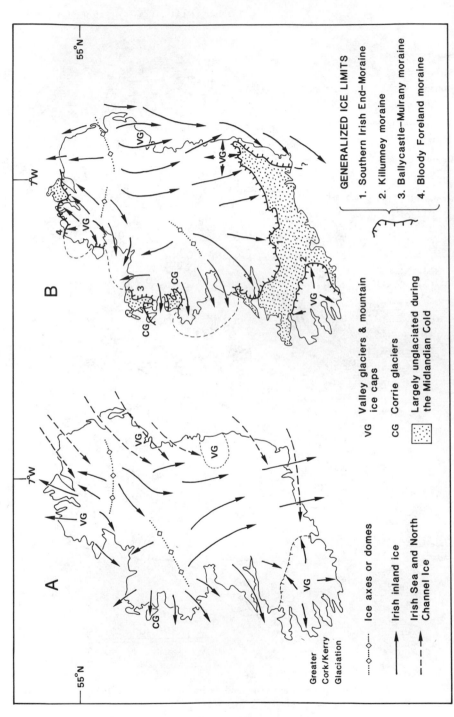

Fig. 2. General directions of ice-sheet movement in Ireland during the Munsterian (A) and late Midlandian (B) cold stages, based on the published works of Charlesworth, Creighton, Colhoun, Farrington, McCabe, Mitchell, Stephens, and Synge.

GENERALIZED ICE LIMITS

1. Southern Irish End-Moraine
2. Killumney moraine
3. Ballycastle–Mulrany moraine
4. Bloody Foreland moraine

VG   Valley glaciers & mountain
         ice caps

CG   Corrie glaciers

⬚   Largely unglaciated during
         the Midlandian Cold

Greater
Cork/Kerry
Glaciation

········◇········   Ice axes or domes

───▶   Irish inland ice

──▶   Irish Sea and North
           Channel Ice

Counties Down, Armagh, and Monaghan and north Co. Cavan. Instead, deglacial phenomena consist largely of sediment gravity flow deposits draped over drumlins and other associated glacigenic landforms. Clearly the drumlin moraine events may not be exactly synchronous in all areas, especially if they relate to specific ice-sheet and terrain conditions rather than widespread climatic deterioration.

In the west, deglacial patterns are complex (many problems which still exist are highlighted later). In the north of Ireland, deglaciation started with the break up of Scottish and Irish ice in the vicinity of the Armoy moraine (Stephens et al., 1975). Ice wasted south toward centres of dispersion located immediately south of Lough Neagh, in the Omagh Basin and in Co. Donegal (Colhoun, 1970; Stephens et al., 1975; Dardis, 1980). Ice-directed drainage features conducted meltwater through the Sperrins, and deposition along the mountain fringes was largely subglacial, ice-marginal, and deltaic in character.

Future reconstructions of ice-marginal environments and the nature of deglaciation must be based on systematic observation and analysis of the sedimentology and morphology of glaciofluvial and associated landform assemblages. For example, the sedimentary parameters of the glaciofluvial and glacigenic landscape of central Ireland are poorly documented. Also, in the north of Ireland and in the central plain, deglacial landforms are best assessed in terms of subglacial and ice-marginal deltaic sedimentation related to ice-sheet characteristics. Clearly, models of terrestrial, glaciolacustrine, and glaciomarine landform assemblages must take account of the complexity of ice-marginal processes and sediments found in present-day glacierized environments (see Boulton, 1972; Boulton and Paul, 1976).

GLACIAL FORMS IN LOWLAND AREAS OF MUNSTERIAN DRIFT

Although Warren (1979b) has questioned the precise age of the Munsterian drifts, there seems little doubt that they comprise unique landform–sediment associations both in northwest Mayo and south of the Southern Irish End-Moraine which are morphologically distinct from Midlandian associations. In general, the drift pattern is characterized by discontinuities, deep weathering profiles, and a subdued or low surface drift relief which probably reflects weathering over a long period of time (Finch and Synge, 1966). The Munsterian drifts have probably been subjected to periglacial modification much longer than the Midlandian drifts to the north (Synge, 1979).

Little morphological research has been undertaken in areas of older drift apart from some notable exceptions, such as Synge's (1968) work in Co.

Mayo. This shortcoming and the possible periglacial modification of the original forms make it difficult to assess the glacial morphology in detail. At present one can only contrast the fresh Midlandian topography dominated by morphologically distinct moraines, kettleholes, eskers, and kames with a Munsterian zone characterized by a general absence of such elements and dominated by deeply weathered, discontinuous sheets of genetically un-differentiated tills.

The nature of Munsterian gravels may be seen at Burncourt, northeast of Mitchelstown, where a long, linear ridge of reworked glacial gravels rest directly on the Carboniferous limestone along one of the Suir tributaries. There is a complete absence of unweathered limestones in the deposit, indicating that it is old and has undergone intensive chemical weathering. At other localities along the Shannon estuary, sections in the Midlandian moraines clearly demonstrate that the younger Midlandian ice ploughed into the weathered Munsterian drift, forming a series of push moraines.

GLACIAL FORMS IN LOWLAND AREAS OF MIDLANDIAN DRIFT

Drumlins and Related Forms

Traditionally, drumlins have been described as circular mounds to well-developed streamlined till forms with blunt stoss ends and gentle, elongated lee slopes (Close, 1867). However, this classic view must be reexamined for the following reasons. First, drumlins are simply an element within a con-tinuum of streamlined forms composed either of drift or rock in any com-bination. Second, drumlins occur in close juxtaposition with transverse till ridges (rogen moraine) which are often drumlinized at their summits. In this respect large-scale models of landscape patterns have been identified back from limiting moraines (see Muller, 1974). Third, drumlin form is highly variable. Megadrumlins exist with superimposed small-scale drum-lins on their flanks. In addition, drumlin long profiles often depart from classic profiles. Fourth, recent studies indicate that many drumlins are not composed solely of lodgement till. Their internal geometry is highly orga-nized with sediment transformation from a stoss-end glacigenic till nucleus to stratified lee-end deposits (Dardis *et al.*, 1984).

As a group, drumlins are unique forms in the Irish glacial landscape in that they tend to occur on the lowlands below 150 m OD and are concen-trated within major ice limits collectively termed the Drumlin Readvance Moraine (Synge, 1969) (Fig. 1). In most cases drumlin long axes occur at right angles to the limiting moraines and have been used to reconstruct the final directions of ice movement. However, the precise relationship between drumlin–till agglomeration and the relative location of the end moraines is not known. There is no doubt that optimum drumlin development in the Irish lowlands has occurred in areas of thick drift derived from relatively soft sedimentary strata. In this respect the origin of most drumlin debris is local, so that the associated erosional, transportational, and depositional mechanisms were accomplished over a distance of about 2–4 km. In areas such as east Co. Down, well-formed drumlins have a substantial component of fine-grained Irish Sea drift in contrast to the mainly rock-cored forms developed on hard Silurian strata of central and western Co. Down and east Co. Armagh. Similar contrasts in gross drumlin composition are known from all the major swarms across Ireland. Various studies (Vernon, 1966; Hill, 1971, 1973) have attempted to isolate such contrasts within drumlin fields by identifying density zones. According to Hill, the Co. Down drumlins show nonrandomness at several scales with a general density decline from the centre near Lough Neagh toward the margin of the field. Alternating zones of high and low density occur normal to ice movement and with a tendency for clustering at a linear scale of 1 km. However, it is difficult to account for such variations because of the complexity of the subglacial ice drift system which involves drift availability, basal velocity variations, particle-size variations, pore water pressures, and time-dependent thermal contrasts. In addition, external drumlin forms are the end products of a time-dependent system and do not fully reflect the conditions responsible for till agglomeration, drumlin genesis, or geotechnical change.

The diversity of drumlin structure is not in doubt but genesis is. The sedimentological and structural variability indicates that drumlin origin must be viewed in terms of erosion, transport, and deposition over a short distance within a dynamic subglacial system, which included lodgement, melt out, and meltwater activity. The origin of Irish drumlins may be examined under six main headings based on internal structure, sedimentology, and associated formative environmental factors.

Drumlins Containing a Core of Older Till

Certain drumlins in Co. Fermanagh contain an *in situ* organic horizon separating two regional till sheets (McCabe and Hoare, 1978). At Derryvree this organic layer has been dated to $30,500 \pm {}^{1170}_{1030}$ years bp (Colhoun *et al.*, 1973) (Fig. 3). At this site the lower till and organic silts have been eroded, and it is only the upper or Maguiresbridge till which is drumlin-forming, enveloping the core of older, dated drift. It is possible that elements of the older drift landscape acted as obstacles which enhanced till agglomeration prior to drumlin streamlining.

A

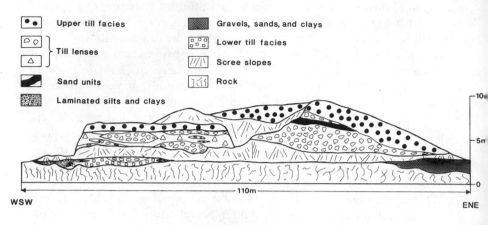

B

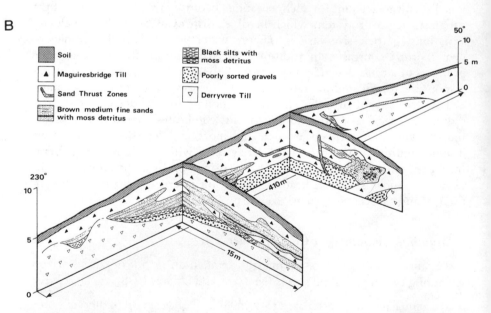

Fig. 3.   The internal geometry of two Irish drumlins. (A) Glover's Quarry drumlin, Moneymore, Co. Antrim. (B) Derryvree drumlin, Co. Fermanagh (with acknowledgment to G. F. Dardis).

4. GLACIAL GEOMORPHOLOGY 75

Drumlins Consisting of Multiple Tills

In contrast to the Fermanagh examples which involved erosion of preexisting drift, Hill (1971, 1973) envisaged that the drumlins of south Counties Antrim and Down originated mainly by deposition, with obstacles not playing a major role in their initial development. He, like Vernon (1966), considered that drumlins are concentrated in high density belts normal and parallel to ice flow. It was suggested that these patterns relate to systematic variations in flow at the ice-sheet base. However, such ideas of periodicity do not explain drumlin stratigraphy. For example, in some areas the lower till is assumed not to be drumlin forming and in others it is, even though an upper till carapace exists. In addition, the presence of sand and gravel interbeds in both tills is not considered as sedimentologically important. In such cases an examination of units in terms of lodgement and/or melt-out relationships may provide a more realistic model of drumlin-forming conditions at the ice-rock or ice–sediment interface.

Drumlins Composed of Banded or Foliated Till

Field evidence suggests that parts of some drumlins form by till accretion, layer upon layer, on top of an earlier core (Davies and Stephens, 1978). Drumlin tills on the north shore of Galway Bay are distinctly banded and reflect subglacial processes of alternate till melt out and erosion or simply halts in deposition of successive waves of till. Banding, therefore, represents successive waves of buried till microsurfaces rather than additions of large concentric sheets. However, such banding should not be confused with stress relief, syneresis joint patterns, or postdepositional permafrost effects.

Drumlins with Sand and Gravel Cores

Drumlins with sand and gravel cores are known to occur along the floor and flanks of the Poyntz Pass tunnel valley of Co. Armagh (Dardis and McCabe, 1983). They generally consist of three stratigraphic associations (Fig. 4). The lowest consists of rip-block and lodgement till. The core comprises complex sequences of glaciolacustrine turbidite sands, deltaic sands and gravels, and sediment gravity-flow deposits. The sand core is overlain by the drumlin-forming till carapace formed by a combination of melt-out and sediment gravity-flow processes. The surface units were remobilized during the final stages of drumlinization. This facies arrangement indicates that the bulk of the sediments formed in water-filled cavities associated with a major meltwater escape route beneath a debris-laden ice sheet (Fig. 4). The absence of glaciotectonics in the sections suggests that the drumlins may not result from glacial overriding of proglacial outwash.

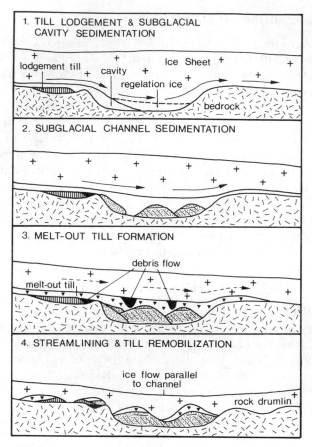

Fig. 4. Simplified reconstruction of major ice–rock and ice–water interface environments prior to drumlinization. The bulk of subglacial cavity sedimentation occurred before drumlinization, but some remobilization occurred during final drumlin formation (with acknowledgment to G. F. Dardis).

Drumlins with Lee-Side Stratified Deposits

Widely spaced drumlins are characterized by tongues or mounds of stratified sediments at their lee ends, especially in drumlins of parabolic shape. Such drumlins lack the classic long profile and are highest toward their lee-end leading point (Dardis *et al.*, 1984). The lee-side deposits have not been deformed by glaciotectonics and form an integral part of the drumlin sediment sequence. A well-defined sediment transformation sequence from proximal till to distal sands and silts occurs. This widespread pattern indicates that hydraulic processes operating along the drumlin at the base of an ice sheet play an important role in the formation of certain varieties of drumlins.

Drumlin Distribution and Permeable Bedrock

Boulton (1975) has shown that newly deposited till, when saturated with water, could not provide sufficient internal shear stress to survive high basal shear pressures at the ice–sediment interface. Clearly, till must undergo a geotechnical change (dewatering) to increase its internal strength to facilitate streamlining and to allow the form to survive as a morphological entity (see Menzies, 1982). Such (dewatering) escape routes probably occur within the subglacial ice debris system but could also be related to permeable bedrock. Colhoun and McCabe (1973) have described solutional systems of the type which may have facilitated meltwater escape at depth in the Irish lowlands.

There is no unifying theory of drumlin genesis at present. Clearly, any theory which emerges must not only explain drumlin distribution and internal structure but also relate to glaciological variables responsible for till agglomeration. Dardis's (1980, 1981) work in Co. Tyrone is useful in that he attempts to relate drumlin initiation and evolution to changes in the thermal regime of the ice and associated patterns of sedimentation.

Deglacial Landforms

Introduction

The major phases of ice-sheet decay already outlined comprise a wide range of largely depositional landforms formed in close association with wasting ice masses. The major features are often continuous across country for great distances (Figs. 1 and 2), and have been interpreted as marking important regional ice limits such as the Southern Irish End-Moraine or the Drumlin Readvance Moraine (Charlesworth, 1928; Farrington, 1953; Synge, 1968; McCabe, 1979). Detailed reconstructions of ice-marginal depositional environments emphasize that genetic interpretations of deglacial landform assemblages should be based largely on assemblages of sedimentary characteristics rather than on morphology alone (Fig. 5). This approach is adopted because many ice-marginal situations are dynamic and the original morphology is likely to be altered by various forms of penecontemporaneous and later erosion. The origins of deglacial landforms described in this section are based on terrestrial and subaqueous depositional models with reference to parameters such as the nature of meltwater/debris inputs, ice-sheet condition and configuration, and local topographic constraints.

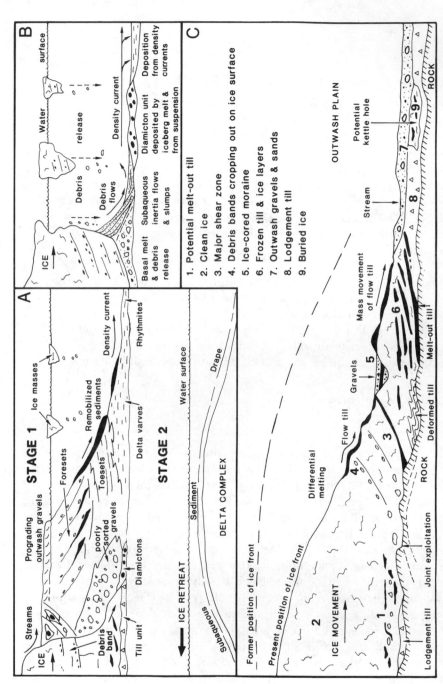

Fig. 5. Generalized models of ice-marginal deposition. (A) Classic Gilbert-type delta deposition followed by ice retreat. (B) Grounded ice-shelf zone and subaqueous deposition. (C) Composite section through a hypothetical ice-sheet margin.

B

Water surface

Water

Ice masses

release

Debris release

Debris flows

Density current

Deposition from density currents

Diamicton unit deposited by iceberg melt & from suspension

Subaqueous inertia flows & slumps

ICE

Basal melt & debris release

C

1. Potential melt-out till
2. Clean ice
3. Major shear zone
4. Debris bands cropping out on ice surface
5. Ice-cored moraine
6. Frozen till & ice layers
7. Outwash gravels & sands
8. Lodgement till
9. Buried ice

OUTWASH PLAIN

Potential kettle hole

Stream

ROCK

A

STAGE 1

Streams

Ice masses

Prograding outwash gravels

Remobilized sediments

Foresets

Density current

Toesets

poorly sorted gravels

Rhythmites

Diamictons

Water surface

Drape

Delta varves

STAGE 2

Sediment

DELTA COMPLEX

ICE

Debris band

Till unit

Subaqueous

ICE RETREAT

Former position of ice front

Present position of ice front

Differential melting

Flow till

Gravels

Mass movement of flow till

5

6

Melt-out till

4

3

Deformed till

ROCK

ICE MOVEMENT

2

1

Lodgement till

Joint exploitation

ROCK

Eskers and Associated Forms

A typical esker may be regarded as a linear accumulation of gravels and sands deposited largely by meltwater within confining ice walls. Esker ridges are widely distributed in Ireland in diverse topographic situations and may be found in close association with almost any type of frontal moraine except those of submarine origin along the eastern seaboard (Fig. 1). The most common morphological form of an esker is that of a long, narrow sinuous ridge which usually meanders. Others may be mesalike or occur as beaded, separated conical shaped hills. Complex esker morphologies include ridges with flanking outwash, ridges bordered by elongate marginal kettles, dendritic patterns, and anastomosing patterns. The large esker systems that have been investigated in some detail are the Tullamore–Daingean system (Farrington and Synge, 1970; A. M. McCabe, unpublished data), the Dunmore–Ballyhaunis system (A. M. McCabe, unpublished data), the Trim system (Synge, 1950), and the Evishanoran system (Charlesworth, 1926; Gregory, 1925; Dardis, 1980). The broad questions of esker sedimentation for each esker system will be considered in terms of the concepts of Banerjee and McDonald (1975), and possible modes of formation outlined.

The Tullamore–Daingean System.   During deglaciation the large ice dome which had covered Ireland along a northeast to southwest axis shrank into a series of smaller domes and lobes, creating local ice and hydrological divides. One such ice divide at Lough Rea has determined the overall location of the two major esker systems in the Irish lowlands, with drainage escaping eastward along the Tullamore–Daingean system and northward along the Dunmore–Ballyhaunis route (Figs. 1 and 6). The Tullamore–Daingean system consists of steep-sided ice-contact ridges occurring in a dendritic pattern, which narrows from the river Shannon eastward to Daingean (Fig 6A). At the Derries the complex ends as a single ridge which merges into glaciofluvial and deltaic gravels covered largely by raised bog. Westward toward Tullamore the esker ridges are punctuated by north-to-south transverse gravel ridges and hummocks of ice-marginal origin. The form is, therefore, time transgressive, and, in all probability, the esker traces acted as subglacial feeder channels to ice-marginal gravels associated with westward backwasting of the ice front. The facies associations are complex in the ridges, but tend to occur as alternating, stacked sequences of disorganized matrix to clast, boulder gravels, and plane-bedded and rippled sands. Matrix-supported diamictons often occur along the ridge flanks and suggest melt out or collapse of an adjacent ice cover. Collapse of enclosing ice walls is also indicated by the occurrence of large, glacially faceted boulders set in a finer gravel matrix. Fines are generally absent, although lenses

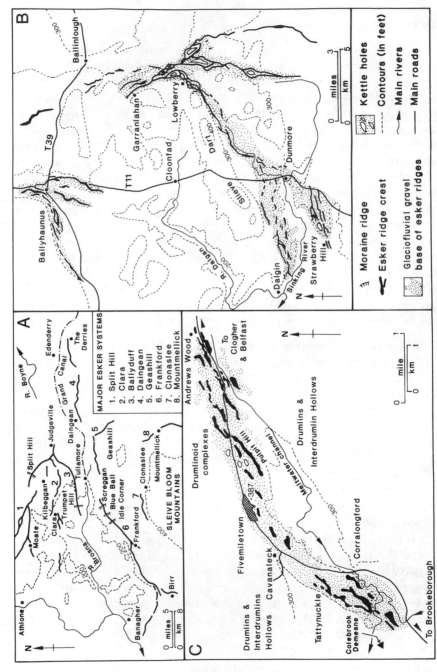

Fig. 6. Examples of esker systems. (A) Central Plain. (B) Dunmore–Ballyhaunis, Counties Galway and Roscommon. (C) The Clogher valley, Co. Tyrone.

of pebbly mudstone often form a distinctive facies interbedded with pebbly gravels. Contortions, minor faulting, and arched bedding may be attributed to removal of ice support. These facies associations suggest that the ridge forms accumulated largely within closed conduit conditions and reflect repeated surges of high density heterogeneous suspensions separated by phases of normal in-channel sedimentation (see Banerjee and McDonald, 1975; Ringrose, 1982).

The Dunmore–Ballyhaunis System.   This system may be traced from Dalgan on the low ground south of the Slieve Dart ridge as west-to-east braided ridges which generally bend north around the eastern margin of the Slieve Dart inlier toward Ballinlough and Ballyhaunus, a distance of about 29 km (Fig. 6B). The northward erratic carriage of Galway microgranites and Slieve Dart felsites indicates that meltwater flow patterns were controlled by the disposition of ice domes to the southwest and east and by the east–west topography of the Slieve Dart ridge. It is probable that meltwater exploited ice weaknesses induced by topography and ice thinning related to a decreasing ice supply from opposing domes. The facies associations in the esker system indicate that it is probably polygenetic in origin. The anastomosing ridges are often separated by elongated depressions (kettles) and are composed chiefly of ripple-bedded sand and gravel. Parts of the complex are deltaic and are composed of large-scale foresets overstepping bottomsets, especially at Dunmore and Ballyhaunus. Certain small, steep-sided esker ridges are suggestive of subglacial feeder channels to the delta portions. However, it is likely that most of the sediments accumulated within a fairly thin ice mass in open segments which ended in ice-contact deltaic environments. This conclusion is reinforced by the absence of continuous downstream sequences of vertically stacked sequences of sheet and cross-bedded or parallel-bedded sand and gravel units (see Banerjee and McDonald, 1975).

The Trim System.   This consists of a series of individual, steep-sided ridges joining the Galtrim delta moraine at right angles (Synge, 1950; Fig. 7A). It is clear from their form and internal characteristics that their traces acted as subglacial feeder channels to the stationary ice margin which impounded glacial Lake Summerhill. Most examples feed directly into the ice-contact face of the moraine, but others rise over the moraine crest, indicating deposition beneath an overhanging ice shelf or cliff. Morphologically, the system is time transgressive to the northwest with clearly defined beads related to minor standstills of the ice margin (Synge, 1950).

The Evishanoran System.   This complex in central Co. Tyrone illustrates the difficulties encountered when using esker as a genetic term to

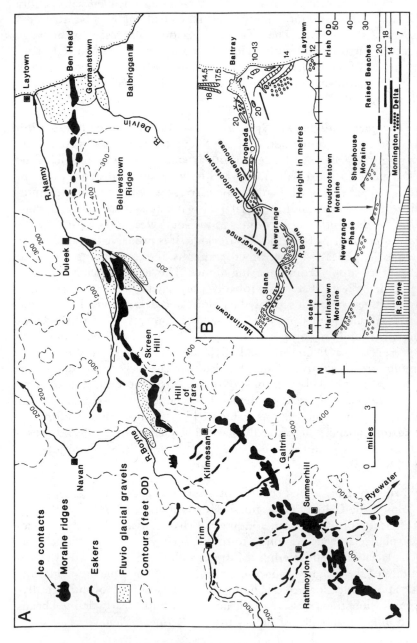

Fig. 7. (A) Gravel moraine ridges, eskers, and outwash features in eastern and central Co. Meath. (B) Ice limits, moraines, and glaciofluvial terraces in the lower Boyne valley, Counties Meath and Louth. (Based on unpublished work by A. M. McCabe and F. M. Synge).

describe winding linear ridges. Charlesworth (1926) and Gregory (1925) have discussed this example at length in terms of subglacial esker or moraine. Both failed to highlight sedimentological variations along the feature in relation to the position of supporting ice walls or to differentiate between ice-contact slopes and primary slopes controlled by delta foreset deposition. Dardis (1980) has shown on the basis of morphology and internal structure that the continuous ridge, usually known as the Evishanoran esker, accumulated in three major depositional environments. The northern segment acted as a subglacial feeder channel to the Davagh delta–moraine complex. The central portion where it crosses the east–west Ballinderry depression is deltaic in structure and is related to ice-marginal deposition at an ice cliff as the ice mass withdrew eastward toward Lough Neagh. Genetically, this section is similar to the cross-valley moraine sequences which occur across the depression to the east. The southern portion of the complex is probably subglacial in origin and relates to ice wastage across the local watershed.

The examples considered indicate that the Irish eskers are polygenetic in site of formation, morphology, internal structure, facies relationships, and palaeocurrent variability. Clearly, future work should concentrate on sedimentological approaches in order to develop models of environments of formation and to aid reconstructions of regional deglacial events.

Kame and Kettle Moraines

This term has been used to describe the disorganized, hummocky moraine topography which results when a broad zone of an ice mass disintegrates *in situ* and deposition occurs over an ablating ice surface. Typically the surface landscape includes isolated mounds, discontinuous linear ridges, plateau mounds, terrace fragments, and conspicuous kettle holes. The non-oriented forms are sometimes collectively termed disintegration moraine and may occur in broad zones up to 2 km in breadth bordering former ice-marginal positions. The detachment of ice masses implies downwastage of ice rather than backwastage. The resulting changes in surface morphology, supraglacial deposition, and opening of cavities in the ice lead to a varied suite of sediments largely of glaciofluvial origin. Stratified sediments range from poorly sorted boulder gravels associated with high-energy flow to cross-bedded sands and laminated silts formed in lower energy conditions. Diamicton flow units often interfinger with washed sediments. Fault and slump structures are common, reflecting compaction or the melting out of buried ice masses. Detailed study of these sedimentary sequences reflects the history of ice melting and varied types of sedimentation patterns (see Lawson, 1979).

Kame and kettle sequences occur as major units along many sections of the Southern Irish End-Moraine and sometimes alternate with other moraine types along major ice limits further north. Most of the Irish kame and kettle moraines are composed of glaciofluvial sediments and consist of sub-parallel undulating ridges and mounds with intervening kettle holes. It is possible that many of these so-called moraines are inverted fluviolacustrine moulds of former ice-cored moraines. Certainly, the absence of distal outwash spreads associated with many kame and kettle moraines supports this concept.

Delta Moraines

The deposits and landforms associated with the southern limit of the Late Midlandian ice on the northwestern flanks of the Wicklow Mountains have provided researchers (Farrington, 1957; Farrington and Mitchell, 1973; Synge 1979; Cohen, 1979) with excellent field sites to study the evolution of landforms associated with glacial Lake Blessington. At a maximum, the glacial lake was 24 km long and 5 km wide. To the east the lake was bounded by the rising ground of the Wicklow Mountains with local ice masses, while to the north and west the elongated slate ridge of Brittas/Toor and lowland-based ice acted as the barrier. Breaches along this ridge allowed limited penetration of glacier ice and the concentration of meltwater inputs into the basin. The main overflow channels were located around Hollywood and Toor at the southern end of the basin. The main evolutionary stages of the complex are based on height relationships between delta moraines, meltwater channels, and morphological–sedimentological associations. The resulting deposits occur as an extremely thick sedimentary sequence, especially along the eastern flanks of the ridge between Brittas in the north and Pollaphuca in the south.

Cohen (1979) has described sedimentary succession around the main exposures near Blessington village. Cohen's model envisages ice-contact slopes to the west and lakeward slopes to the east which correspond to foreset dips. The delta at 272 m OD corresponds to one major lake level. The delta surface is uneven due to subaerial and meltwater erosion and is draped by a till-like unit. The glaciolacustrine sediments are divided broadly into foreset and bottomset sequences. Delta progradation is shown by coarsening-up sequences and interdigitation of foreset and bottomset deposits. The foresets dip between 15 and 35° lakeward and have a textural range from boulder gravel to coarse sand, although occasional large boulders occur. The succession shows rapid spatial variations in sediment type which reflect changes in the direction and magnitude of meltwater input. It was concluded that the main formation processes were avalanche and grain flow. However, the presence of fine-grained rhythmites within the foreset sequence indicates that discrete types of mass flow under low energy conditions occurred. Remobilization of delta front sediments and consequent slurry flow is probably responsible for the presence of discontinuous diamicton units. The bottomset sediments contain a wide range of sedimentary features including A, B, and S type climbing ripples, parallel laminated units, structureless beds, dune cross-bedding, graded beds, flame structures, and dropstones. Cohen concluded that the only mechanism to account for the wide distribution of these sequences is bottom-flowing density currents; they may be due to high density inflowing meltwater or may be slump generated. The entire sequence shows normal faulting toward the unsupported, lakeward side of the delta.

Clearly the sediments of the so-called Blessington moraine accumulated in a high energy, ice-marginal delta environment. Different sediment types reflect the operation of distinct processes over the range from mass movement to grain flow in a dynamic environment complicated by changes in the sources of meltwater input. Morphologically, the delta surface is not flat as expected from classic studies, but this is due to the diversity of ice-marginal deposition and subsequent erosion. The model outlined is simplified and does not take into account regional variations in lake levels and changes in ice-marginal positions and outlet channels which have been demonstrated by Synge (1979).

Linear Gravel Moraines

The pattern of lowland ice retreat has often been reconstructed from the pattern of frontal, linear gravel-ridge complexes which characteristically occur as continuous features for many kilometers. One of the best known examples occurs at Ben Head in eastern Co. Meath and marks the first major stabilization in the northward retreat of the ice margin from the Wicklow Mountains (McCabe, 1973). The moraine consists of two distinct sections. From the coast at Ben Head inland to the Hill of Tara, the feature occurs as a east–west ridge bounded by steep ice-contact slopes on the north (Fig. 7). The southern or ice-distal slopes are kettled and grade into an extensive spread of outwash gravels around Gormanstown. The dominant ridge sediments are disorganized boulder gravels with limited amounts of fines. The ice-contact nature of the ridge is emphasized by the presence of large southward-rising shear planes developed in sand on the north side of the ridge. West of the outwash spread, small proglacial channels have steepened the distal ridge slopes. From Tara Hill west to Enfield the moraine changes in orientation to northeast to southwest (Synge, 1950). Unlike the eastern section, this section, the Galtrim section, is characterized by a well-defined esker system and delta moraine associated with proglacial Lake

Summerhill. This example illustrates that major depositional environments change along an ice margin and that esker formation occurs only west of Tara Hill where the ice front stood in ponded water.

Subaqueous Moraines

The southern limit of drumlins in east-central Ireland is well marked by a complex of ridge moraines which cross the present coast at various points between Clogher Head to the south and Killard Point to the north (McCabe and Hoare, 1978) (Figs. 1 and 8). After deglaciation from the Clogher Head limit, the topographic constraints of the Carlingford and Mourne Mountains created three distinct ice lobes located in Dundalk Bay, Carlingford Lough, and south-east Co. Down (Fig. 8). Correlations of these ice limits are based on the location of moraine lines and sedimentary sequences and on the pattern of penecontemporaneous lateglacial raised beaches (McCabe, 1979; Stephens and McCabe, 1977; Synge, 1977).

Although the Dunany Point, Rathcor, Kilkeel, and Killard Point moraines are morphologically similar to terrestrial-based gravel moraines, recent work indicates that they were formed in close accociation with grounded and partially floating ice-sheet margins (McCabe et al., 1984) (Fig. 5B). The large ridge at Dunany consists of major diamicton units separated by composite units of pebbly gravel, current-bedded sands, and subaqueous flow phenomena. Since each unit prograde southward in a stacked sequence, a single-crested ridge results. On the opposite side of the Bay at Rathcor, similar units occur in an off-lap relationship toward the west, resulting in a series of smaller moraine ridges. It is, therefore, probable that the similarities on either side of the bay reflect deposition from the same ice lobe but from different glaciological situations. The till facies vary from massive (glacigenic) to stratified (debris flows) and are interbedded with gravel, silt, and clay units deposited by subaqueous high- and low-density sediment gravity flows. The Kilkeel coast moraines occur as a series of nodes composed of massive and laminated tills with interbedded lenses and sheets of gravels, sands, silts, and pebbly mudstones. The till nodes represent subaqueous deposition at successive grounded ice-margins, and the intervening depressions are infilled with stratified sequences, showing transition from subaqueous debris flow to low-density turbidity current deposits. Iceberg facies occur as pebbly mudstones and dropstones. Inland, the ice shelf facies is replaced by terrestrial-based gravel facies.

The Killard Point moraine consists of three major facies associations (McCabe et al., 1984). Initial ice-proximal sedimentation occurred beneath an ice-shelf grounding zone, resulting in deposition of till over red, massive

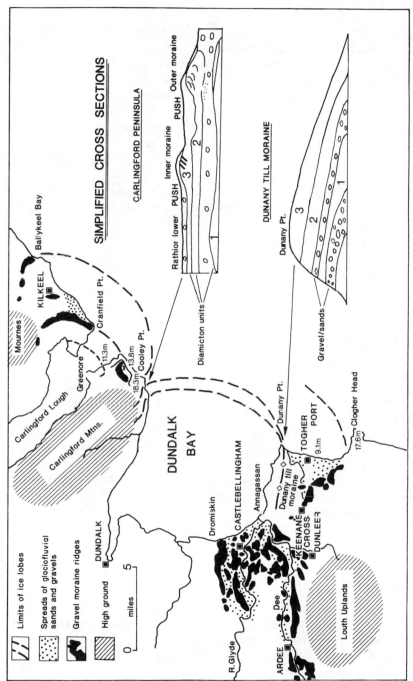

Fig. 8. Deglacial features associated with the decay of the Dundalk Bay ice lobe.

clay units. Remobilization of the glacigenic till resulted in stratified tills which interfinger with pebbly mudstones formed chiefly by iceberg sedimentation. Subsequent sedimentation resulted in the deposition of a complex of prograding, low-angle, submarine fans. The fan deposits show a lateral facies transition from high-density sediment gravity flows in the form of channelled boulder and cobble gravel to low-density turbidity current sands and silts. Massive clay units are interbedded at all levels in the complex. The internal structure of the moraines, while exhibiting many sedimentary characteristics typical of terrestrial fluviatile environments (scour-and-fill structures, cross-bedding, and ripple-drift cross-laminations), contain a number of distinct facies associations with sedimentary features which are not truly characteristic of subaerial outwash sedimentation (sediment gravity flows, dropstone facies, pebbly mudstones, laminated tills, massive clays, water-escape and load structures, grain flows, and turbidites). This suggests that they are subaqueous in origin and are, in all probability, glaciomarine, on the basis of shoreline evidence (Stephens and McCabe, 1977) and sparse fauna. Depositional models of this type (Fig. 5B) can account for many of the late Pleistocene "glacial" successions exposed along the eastern coast of Ireland (see Huddart, 1977, 1981; Thomas and Summers, 1981).

Glaciofluvial Terraces

The pattern and sequence of glaciofluvial terraces and associated ice limits have been used in the Foyle and Boyne basins as an aid in the reconstruction of patterns of deglaciation (Fig. 8). In both areas extensive terraces formed in rock-bounded depressions as the ice margin wasted up valley. Colhoun (1970, 1972) recorded at least 10 major stages in ice recession along the Foyle depression between Londonderry and Newtownstewart. Each cross-valley ridge was thought to represent a stabilization of the ice margin and is associated with terrace fragments sloping north. However, before such reconstructions are considered, each terrace must be mapped and levelled accurately in association with an examination of the facies associations along the feature. In the Lower Boyne valley, F. M. Synge and A. M. McCabe (unpublished data) have identified 7 major ice limits, each associated with a distinct moraine–terrace couplet, showing the westerly recession of the ice toward Navan and Kells (Fig. 8). Typically, terrace profiles flatten eastward where their levels were controlled by various lateglacial sea levels in the Boyne estuary. Some terraces end in delta tongues composed of large-scale, multiple foreset gravel sequences. Other terraces are clearly related to lateglacial raised beaches (McCabe, 1979).

GLACIAL LANDFORMS IN MOUNTAIN
AREAS

The main mountain masses of Donegal, Connemara, Cork–Kerry, the Mournes, and the Wicklows, which are located mainly on the periphery of the island, undoubtedly owe much of their scenic attraction to intense erosion and deposition by independent centres of ice dispersion during repeated cold stages of the Quaternary. Most of these mountain masses exhibit all the well-known features of alpine glaciation (Farrington, 1934, 1938, 1944, 1949, 1953, 1966; Lewis, 1974; Stephens et al., 1975; Synge, 1963, 1968, 1973; Warren, 1979a). Not all areas have been investigated in the same detail: there is a general lack of data from Co. Donegal, whereas research over the last 50 years in the Wicklows has revealed an ever increasing complexity of events and landforms. Most researchers attempt to identify the number of phases of ice growth and the limits of autochthonous cirque and valley ice masses on stratigraphic, erratic, and morphological criteria. Farrington (1934, 1942, 1944) and Synge (1973) have identified at least three major phases of alpine glaciation in addition to localized corrie (cirque) development in the Wicklows. Smaller expansions and subsequent retreat patterns of ice associated with restricted valley and corrie systems have been identified on the basis of aerial distribution of drift, drift limits, topographic controls, and moraine morphology. Similar types of evidence have been used by Farrington (1953) and Synge (1968) in Co. Mayo, by Warren (1979a) in southwest Ireland, and by Stephens et al. (1975) in the Mournes to reconstruct phases of corrie and valley glacier development.

Researchers have long been interested in the relationships between lowland and mountain-based ice masses. This focusses attention on the limiting drift features of the mountain ice and the erosive effect of extraneous ice. For example, Warren (1979a) has shown that extraneous Midlandian ice from Kenmare penetrated through the MacGillycuddy's Reeks in the southwest. In contrast, Stephens et al. (1975) consider that lowland ice did not penetrate the high Mournes during the Midlandian. The complex relations between mountain and lowland ice have been clearly demonstrated by Synge (1979) and Cohen (1979) in the Wicklows, where, in certain parts, lobes of local ice can be correlated with advances of general ice, although lobes from different centres reached their greatest extent at different times.

It has proved difficult to date accurately the sequence of mountain glaciations, especially those associated with ice expansion from corries. Many authors have recognized the outer moraines and suites of recessional moraines associated with local mountain glaciations. In some cases it was noted that the outer moraines tended to be more weathered and degraded than the inner moraines. On this basis, tentative correlations have been proposed among different suites of moraines assigned to different phases of the Quaternary. As Warren (1979a) pointed out, however, such interpretations must be viewed with caution since not all corrie and valley ice systems will necessarily respond in the same way to similar inputs either in their formative or in later weathering phases. Furthermore, morphological differences will inevitably result from differences in the size and shape of accumulation zones and outlets, rock type, modes of ice transport and deposition, and the degree of entrainment of older detritus; and it is also likely that genetic types of ridge features were highly variable, ranging from ice-cored moraines to protalus ramparts (see Warren, 1979a). On the basis of threshold heights of corries and on regional slopes of distinct lowland ice sheets, Synge (1968) suggested that the corries of west Mayo belong to distinct phases of the Quaternary. Another approach, using periglacial features in conjunction with the stratigraphic record to isolate nonglacial phases between lowland and mountain phases of glacial activity, was adopted by Hoare and McCabe (1981). Finally, the Turlough Hill pumped storage electricity scheme provided an excellent opportunity to examine the morphology and deposits within the limiting block moraines of Lough Nahanagan, Co. Wicklow, when the impounded lake level was lowered artificially (Colhoun and Synge, 1980). The large outer cirque moraine at this site has been correlated with Farrington's (1944) Athdown phase. Within this limit, the inner cirque moraines consist of ice-pushed lake clays and protalus granite blocks and record the final ice recession back to the corrie head. When initially exposed, detailed micromorphological features such as grooved drift were preserved and resembled those found in recently deglacierized areas. Organic silts from the ice-pushed lake clays suggest that renewed corrie activity occurred in this site during the period 11,000–10,500 bp. However, such evidence does not imply renewed corrie glacierization over a wide area. In fact the final "ice push" structures may relate to pressure from a prolatus component rather than to development of substantial corrie ice. Nevertheless, this site demonstrates that further datings and correlations of features of alpine glaciation must be based on evaluations of biostratigraphic evidence in conjunction with morphology.

CONCLUSIONS AND FUTURE WORK

Research on the glacial geomorphology of Ireland has generally been conducted on the basis of individual rather than cooperative effort. As a result, there is no coordination of research which would aim to obtain an overview of the problems that undoubtedly exist. Although the landform distribution is known in outline, very few researchers have concentrated their efforts on the processes responsible for landform genesis. Many workers have tended to concentrate solely on the morphology of glacial landforms, but it is increasingly obvious that this approach is unsatisfactory and in the future more attention should be given to the internal structure of landforms if understanding of landform genesis is to progress. Similarly, individual landforms should not be treated as distinct entities with a particular uniqueness, but one should look for landform associations and attempt to relate the whole to ice-sheet characteristics and dynamics. In this context it is paramount that field geomorphologists experience techniques and pursue knowledge from many other branches of earth science, especially sedimentology (Eyles et al., 1983; Maill, 1978, 1983). Glacial geomorphologists must also develop theoretical and practical knowledge of existing ice masses, especially their climate, mechanisms of erosion and deposition, structure, and form changes in order to appreciate fundamental processes. Finally, if a coordinated programme could be established, it may be possible to identify meaningful landform patterns and associated processes over the island and then hope to develop dynamic models of landform genesis and ice-sheet growth and decay.

References

Banerjee, J., and McDonald, B. C. (1975). *Spec. Publ. Soc. Econ. Paleontol. Mineral.* **23,** 123–153.

Boulton, G. S. (1971). *In* "Till: a Symposium" (R. P. Goldthwait, ed.), pp. 41–72. Ohio State Univ. Press, Columbus, Ohio.

Boulton, G. S. (1972). *J. Geol. Soc. London* **128,** 361–393.

Boulton, G. S. (1975). *In* "Ice Ages Ancient and Modern" (A. E. Wright and F. Moseley, eds.), pp. 7–42. Seel House Press, Liverpool.

Boulton, G. S., and Paul, M. (1976). *Q. J. Eng. Geol.* **9,** 159–194.

Bowen, D. Q. (1973). *Proc. Geol. Assoc.* **84,** 249–272.

Charlesworth, J. K. (1926). *Geol. Mag.* **63,** 223–225.

Charlesworth, J. K. (1928). *Q. J. Geol. Soc. London* **84,** 293–344.

Charlesworth, J. K. (1939). *Proc. R. Ir. Acad., Sect. B* **45,** 225–295.

Close, M. H. (1867). *J. R. Geol. Soc. Irel.* **1,** 207–244.

Cohen, J. M. (1979). *In* "Moraines and Varves" (C. Schluchter, ed.), pp. 357–367. A. A. Balkema, Rotterdam.

Colhoun, E. A. (1970). *Ir. Geogr.* **6,** 162–185.

Colhoun, E. A. (1972). *Proc. R. Ir. Acad., Sect. B* **72,** 91–137.

Colhoun, E. A., and McCabe, A. M. (1973). *Proc. R. Ir. Acad., Sect. B* **73,** 165–206.

Colhoun, E. A., and Synge, F. M. (1980). *Proc. R. Ir. Acad., Sect. B* **80,** 25–45.

Colhoun, E. A., Dickson, J. H., McCabe, A. M., and Shotton, F. W. (1973). *Proc. R. Soc. London, Ser. B* **180,** 273–292.

Dardis, G. F. (1980). *In* "Field Guide No. 3, Co. Tyrone, Northern Ireland" (K. J. Edwards, ed.). I.Q.U.A.

Dardis, G. F. (1981). *Ann. Glaciol.* **2,** 183.

Dardis, G. F., and McCabe, A. M. (1983). *Boreas* **12,** 263–278.
Dardis, G. F., McCabe, A. M., and Mitchell, I. (1984). *Earth Surf. Processes Landforms* **9,** 409–424.
Davies, G. L. H., and Stephens, N. (1978). "Ireland." Methuen, London.
Eyles, N., Eyles, C. H., and Maill, A. D. (1983). *Sedimentology* **30,** 393–410.
Farrington, A. (1934). *Proc. R. Ir. Acad., Sect. B* **42,** 173–209.
Farrington, A. (1938). *Proc. R. Ir. Acad., Sect. B* **45,** 65–71.
Farrington, A. (1942). *Proc. R. Ir. Acad., Sect. B* **47,** 279–291.
Farrington, A. (1944). *Proc. R. Ir. Acad., Sect. B* **50,** 133–157.
Farrington, A. (1947). *Ir. Geogr.* **1,** 89–97.
Farrington, A. (1949). *J. Glaciol.* **1,** 220–225.
Farrington, A. (1953). *J. Glaciol.* **2,** 262–267.
Farrington, A. (1957). *Ir. Geogr.* **3,** 216–222.
Farrington, A. (1966). *Ir. Nat. J.* **15,** 226–228.
Farrington, A., and Mitchell, G. F. (1973). *Ir. Geogr.* **6,** 543–560.
Farrington, A., and Synge, F. M. (1970). *In* "Irish Geographical Studies" (N. Stephens and R. E. Glasscock, eds.), pp. 49–52. The Queens Univ. of Belfast, Belfast.
Finch, T., and Synge, F. M. (1966). *Ir. Geogr.* **5,** 161–172.
Gregory, J. W. (1925). *Geol. Mag.* **62,** 451–458.
Hill, A. R. (1971). *Geogr. Annaler* **53A,** 14–31.
Hill, A. R. (1973). *Ann. Assoc. Am. Geogr.* **63,** 226–240.
Hoare, P. G., and McCabe, A. M. (1981). *Biul. Peryglac.* **28,** 57–78.
Huddart, D. (1977). "South East Ireland," Guide Book for excursion A14 INQUA X Congress. Geo. Abstracts, Norwich.
Huddart, D. (1981). *Quat. Newsl.* **38,** 28–41.
Lawson, D. E. (1979). C.R.R.E.L. Report 79–9.
Lewis, C. A. (1974). *Sci. Proc. R. Dublin Soc., Ser. A* **5,** 207–235.
Maill, A. D. (1978). *Con. Soc. Petrol. Geol. Mem.* **5,** 597–604.
Maill, A. D. (1983). *J. Sediment. Petrol.* **53,** 477–491.
McCabe, A. M. (1969). *Ir. Geogr.* **6,** 63–77.
McCabe, A. M. (1973). *Proc. R. Ir. Acad., Sect. B* **73,** 355–382.
McCabe, A. M. (1979). *In* "Field Guide to East Central Ireland" (A. M. McCabe, ed.), pp. 1–10. Quaternary Research Association, Dublin.
McCabe, A. M., and Hoare, P. G. (1978). *Geol. Mag.* **115,** 397–413.
McCabe, A. M., Dardis, G. F., and Harvey, P. (1984). *J. Sediment. Petrol.* **54,** 716–730.
Menzies, J. (1979). *Earth Sci. Rev.* **14,** 315–359.
Mitchell, G. F. (1972). *Sci. Proc. R. Dublin Soc., Ser. A.* **4,** 397–413.
Muller, E. H. (1974). *In* "Glacial Geomorphology" (D. R. Coates, ed.), pp. 187–204. Binghampton, New York.
Ringrose, S. (1982). *In* "Research in Glacial, Glacio-fluvial and Glacio-lacustrine Systems" (R. Davidson-Arnott, W. Nickling, and B. D. Fahey, eds.), pp. 117–137. Geo Books, Norwich.
Stephens, N., and McCabe, A. M. (1977). *In* "Quaternary History of the Irish Sea" (C. Kidson and M. J. Tooley, eds.), pp. 179–198. Seel House Press, Liverpool.
Stephens, N., Creighton, J. R., and Hannon, M. A. (1975). *Ir. Geogr.* **8,** 1–23.
Synge, F. M. (1950). *Proc. R. Ir. Acad., Sect. B* **53,** 99–110.
Synge, F. M. (1963). *Ir. Geogr.* **4,** 397–403.
Synge, F. M. (1968). *Ir. Geogr.* **5,** 372–386.
Synge, F. M. (1969). *In* "Quaternary Geology and Climate," pp. 89–92. National Acad. Sciences, Washington, D.C.

Synge, F. M. (1970). *In* "Irish Geographical Studies" (N. Stephens and R. E. Glasscock, eds.), pp. 34–48. The Queens Univ. of Belfast, Belfast.

Synge, F. M. (1973). *Ir. Geogr.* **6,** 561–569.

Synge, F. M. (1977). *In* "Quaternary History of the Irish Sea" (C. Kidson and M. J. Tooley, eds.), pp. 199–222. Seel House Press, Liverpool.

Synge, F. M. (1979). *In* "Field Guide to East Central Ireland" (A. M. McCabe, ed.), pp. 40–48. Quaternary Research Association, Dublin.

Thomas, G. S. P., and Summers, A. J. (1981). *Quat. Newsl.* **34,** 15–18.

Vernon, P. (1966). *J. Glaciol.* **6,** 401–409.

Warren, W. P. (1979a). *In* "Moraines and Varves" (C. Schlüchter, ed.), pp. 223–236, A. A. Balkema, Rotterdam.

Warren, W. P. (1979b). *Geol. Surv. Irel. Bull.* **2,** 315–332.

CHAPTER 5

PERIGLACIAL FEATURES

Colin A. Lewis

Department of Geography
University College
Dublin, Ireland

INTRODUCTION

Ireland has a rich and varied periglacial heritage, and periglacial processes, on a very restricted scale, still occur in some upland areas. In 1894 Kinahan described block moraines and boulder–littered slopes that are still occasionally enlarged by the movement of debris downslope over snow patches. Saul (personal communication) has also seen small active polygons on the summit plateau of the Comeragh Mountains in Co. Waterford. Periglacial features are widespread and include examples of most periglacial phenomena. Unfortunately, there have been few detailed studies of periglacial features in Ireland, and only pingo remnants and the casts of former ice wedges have attracted appreciable interest in the published literature. Other features have been mentioned cursorily, although Colhoun (1971) has presented an account of the periglacial phenomena of the Sperrin Mountains, as have Hoare and McCabe (1981) of east-central Ireland. Quinn (1975) emphasized the role of periglaciation in western Iveragh, Co. Kerry. More recently, Mitchell (1977), Lewis (1978), and Warren (1981) have attempted to assess the efficacy of past periglacial processes in Ireland and to describe their effects upon the landscape.

One of the first references to a periglacial feature in Ireland was probably Kilroe's description, in 1908, of what was apparently an ice-wedge cast in glacial gravels near Londonderry. Kilroe appreciated that the feature had

95

formed in association with frozen ground, but failed to give the correct explanation for its genesis. Four years earlier, in 1904, Wright and Muff had described extensive head deposits that overlie raised beach deposits on the south coast. They knew that head resulted from frost shattering and subsequent downslope movement, but their main interest was with the beach rather than with the periglacial deposits. Neither the incorrect explanation of the origin of the ice-wedge cast proffered by Kilroe nor the limited interest in the head deposits by Wright and Muff was surprising. Anderson's classic paper on solifluction (the main process by which the head was moved into position) did not appear until 1906, while Leffingwell's seminal work on ice wedges was not published until 1915.

PERIGLACIAL FEATURES

Periglacial areas are characterized by features that form under the influence of frost action. Permafrost may or may not occur in such areas. Some of these features form remarkable landforms, such as altiplanation surfaces, nivation cirques, pingos, or solifluction terraces. Others, although reflected on the surface, are well developed as subsurface structures. They include ice wedges and such cryoturbation forms as festoons and involutions, which may be evidenced on the surface by various forms of patterned ground.

Within Ireland at least 18 types of features that may have formed under periglacial conditions have been identified. These are block fields, cryoturbation, erected and fractured pebbles, festoons, garlands, head, ice-wedge casts, involutions, nivation ridges (protalus ramparts) and snow-patch hollows (nivation cirques), pingo remnants, rotational slumps, polygons, solifluction terraces, ordinary and stratified screes, stripes, and tors. Some debris fans and cones have also been ascribed to periglacial origins, and there is limited evidence for loess deposition under periglacial conditions. While some writers have differentiated between different forms of cryoturbation (such as festoons and involutions), others have been less meticulous. The term *cryoturbation* therefore includes a number of different features, and its usefulness is limited to showing that frost action has, at some time or other, occurred beneath the former ground surface. Similarly, the terms *head* and *solifluction deposits* have sometimes been taken as synonymous.

The distribution of periglacial phenomena that have been identified in Ireland is shown on Fig. 1, but the map should be treated with considerable caution. Although it appears that there is a paucity of periglacial features in the Midlands, as Mitchell (1977) has argued, it could be maintained that these are areas of ignorance where few have yet searched for such features. This is clearly epitomized by studies of ice-wedge casts.

In 1966 Farrington reported an ice-wedge cast in beach deposits at Courtaparteen, Co. Cork. Two years later Bryant reported ice-wedge casts

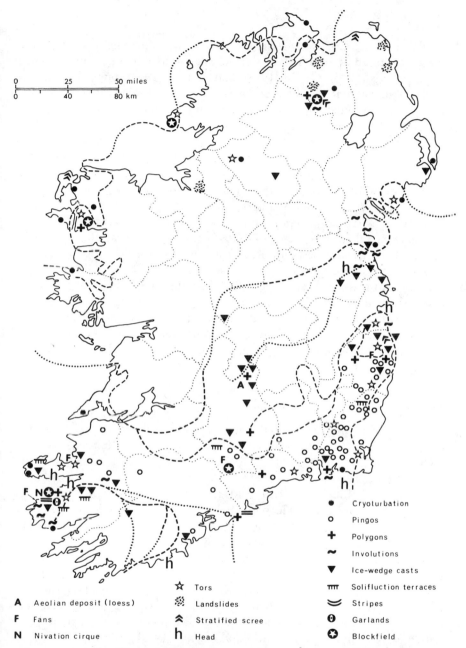

	●	Cryoturbation	
	○	Pingos	
	+	Polygons	
	~	Involutions	
	▼	Ice-wedge casts	
	⊤⊤⊤	Solifluction terraces	
☆	Tors	≋	Stripes
⁂	Landslides	θ	Garlands
⋀	Stratified scree	✪	Blockfield
h	Head		

A   Aeolian deposit (loess)

F   Fans

N   Nivation cirque

Fig. 1. Periglacial features in Ireland. Based on Lewis (1978), with additional information on pingos (from the unpublished work of P. O'Callaghan) and on features in east-central Ireland (based on Hoare and McCabe, 1981). The heavy dotted and dashed lines indicate Midlandian ice limits according to Synge (1977).

near Waterville, Co. Kerry (Bryant, 1968). In succeeding years Colhoun (1971) identified them in the area of the Sperrin Mountains, while Mitchell (1970), Lewis (1974), Quinn (1975), and Warren (1977) found them in Co. Kerry. During that period there was only one systematic study of ice-wedge casts elsewhere in Ireland, namely, by Lewis (1977) in northeast Co. Wicklow. The resultant map, showing ice-wedge casts in the Sperrins, in Co. Kerry, and in northeast Wicklow, did no more than reflect the areas studied. In 1979 Lewis examined a north–south transect in the south Midlands. He found that ice-wedge casts were widespread in outwash deposits, thereby proving that ice wedges had been far more common than had previously been envisaged. Similarly, until Mitchell drew attention to them in 1971, it was not generally appreciated that pingo remnants exist in Ireland. Since then, O'Callaghan's (1981) analysis of aerial photographs of the south of Ireland has indicated that pingo remnants exist at over 50 localities in that area.

Periglacial features may be classified according to their mode of formation and fossil morphology. Within Ireland they broadly divide into mass movement, patterned ground, thermokarst, niveomorphic, fluvial, and aeolian phenomena. Some features, understandably, may be included in more than one category, so that the classification adopted here should not be regarded as rigid.

Mass Movement Phenomena

Mass movement phenomena include a variety of features, such as cambering, landslides, solifluction terraces and lobes, stratified and ordinary screes, block streams, ploughing blocks, and rock glaciers, not all of which have yet been recorded in Ireland.

In 1861 J. B. Jukes described head deposits in occasional sections along the south coast of Ireland. Head is usually angular debris of local origin that results from a variety of different mass movement processes, but in Ireland particularly from solifluction. The size of particles within the deposit may vary considerably, from angular blocks of a few metres in length to fine material of clay size.

In 1904 Wright and Muff showed that head is ubiquitous along much of the south coast of Counties Cork, Waterford, and parts of Wexford. The head usually overlies a raised beach deposit (known as the Courtmacsherry Raised Beach) and is commonly covered by glacial debris. Much of the fascination of Wright and Muff's research was that they studied an area that was affected, at least in part, by two different ice sheets. To the west of Cork Harbour an advance of local ice from the interior of the county had overrun the coast, depositing a local till on top of head. In some cases, as at

Courtmacsherry, Co. Cork, a further period of periglacial activity had succeeded the retreat of the ice, leading to the deposition of a second layer of head. The two heads, with the local till sandwiched between them, were designated "Upper Head" and "Lower Head," respectively.

East of Cork Harbour the coastlands were affected by a mass of ice that apparently originated to the north of the Irish Sea basin. After moving south through that basin and crossing the lowlands of east Co. Wexford, the ice extended at least as far west as Ballycroneen, Co. Cork. There Wright and Muff (1904) recorded its deposits as overlying local head that, in its turn, covered a raised beach deposit. Superimposed on the Irish Sea till they found a local till derived from an advance of ice from west Co. Cork.

Recent studies have shown that there are extensive head deposits along the inner shores of Dingle Bay, particularly between Ross Behy and Cuan Lathaí. They also occur along the western portion of Corca Dhuibhne and at Fenit on the north shore of Tralee Bay. At Fenit, organic muds interbedded with the basal silts of the head have a radiocarbon age in excess of 42,000 bp (Mitchell, 1970). The Ross Behy deposits are particularly impressive and exceed 30 m in thickness (Lewis, 1967; Warren, 1977, 1981). At Cuan Lathaí, some 2 km west of Ross Behy, they apparently overlie a glacial till derived from Glen Behy, while nearer Ross Behy the head is itself overlain by a more recent till, also derived from glacial advance from the glen (Warren, 1977). It is likely that, while ice occupied Glen Behy, periglacial slope processes were occurring beyond the glacial limits on the north side of the Knockatinna ridge, causing head accumulation along what is now the south shore of Dingle Bay.

Although head deposits occur elsewhere in Ireland, apart from along the south coast and around Dingle Bay and adjacent areas to the north, there are no known deposits that are comparable in thickness to those of Ross Behy. McCabe (1969) has reported a depth of 3 m of head underlying (and therefore pre-dating) till near Lisnaskea in Co. Fermanagh. Since he believed the till to date to the "Late Weichsel," it is reasonable to assume that the head predates the late Midlandian ice advance in that area. Further north, in the Sperrins, Colhoun (1971) has recorded head deposits resting on till of the late Sperrin glaciation, which he considered to date to the late Midlandian glaciation. Further south, in Co. Wicklow, Lewis has seen up to 3 m of coarse angular blocky head at altitudes around 300 m on the eastern (west-sloping) rim of the Calary plateau. This deposit was overlain by what appeared to be slopewash deposits. The site lies outside the supposed limit of the Midlandian age local ice caps of the Wicklow Mountains but within the limits of the earlier mountain ice advance (Farrington, 1934). Since the early advance has been equated with the Munsterian, the head deposits may date either to the latter part of that stage or to any part of the Midlandian.

Many other exposures within Ireland show head deposits, and it is obvious that the periglacial processes responsible for head formation have been widespread throughout the country on a variety of different occasions. Indeed, Farrington (1966) has even reported head underlying the Courtmacsherry Raised Beach.

Head deposits, although widespread, appear to be best developed in southern areas. The earliest predate the Courtmacsherry Raised Beach. At Fenit, head deposition began before 42,000 years bp, presumably predating the Castlepook Interstadial (Mitchell, 1977), although it may be of much greater antiquity. By comparison, at least some head deposits in the Sperrin Mountains apparently postdate the late Midlandian glaciation of that area and may equate with the more recent Nahanagan Stadial (Colhoun, 1971).

Solifluction was defined by Anderson (1906) as the slow flowage from higher to lower ground of masses of waste saturated with water. The term was later refined to necessitate frost action that causes water within the waste to alternately freeze and thaw, thereby displacing the waste downslope. Sometimes this process is called gelifluction. Solifluction may take a variety of forms. Sometimes lobes of debris move downslope either individually or as lobate and wavy terraces. In other cases, lines of terraces that, in plan, appear straight fronted descend downslope. In section the leading edge of such terraces may be curved, as if the terraces are rolling downslope.

Ever since Wright and Muff (1904) studied the head deposits of the south coast it has been apparent that solifluction has made a significant contribution to the landscape. Morrison and Stephens (1965) at Roddans Port in Co. Down, showed that there were two phases of lateglacial solifluction activity, dating to the Nahanagan Stadial and to the end of the cold phase that preceded the Woodgrange Interstadial. At the Old Head on the southern shore of Clew Bay, Co. Mayo, the process last took place around 10,000 bp as is indicated by radiocarbon dating of organic lenses incorporated within a solifluction deposit (Mitchell, 1977). Similarly, at Drumurcher in Co. Monaghan, a grey muddy, sandy silt was dated to 10,515 ± 195 bp (Coope et al., 1979). It is known that solifluction also occurred on previous occasions, one of which is suggested by the deposits reported by McCabe (1969) to underlie till near Lisnaskea, Co. Fermanagh. Dardis et al. (in press) report a date of 32460 ± 270 bp for organic detritus within laminated sediments of possible lacustrine origin that are interbedded with solifluction deposits at Greenagho, near Belcoo in Co. Fermanagh.

Lee et al. (1964) drew attention to the intricate microrelief that occurs at certain localities in the Galtee Mountains, on the borders of Counties Limerick, Tipperary, and Cork. They observed that at altitudes between 685 and 460 m the surface deposits probably resulted from solifluction and hill-wash effects and that solifluction had resulted in a stepped terrain with a riser of 0.3–0.5 m and a tread of approximately 0.6 m.

Other studies have shown that solifluction terraces are common else-where. Many of them are considerably larger than those described from the Galtees. In 1974, Lewis recorded a microrelief of solifluction terraces on the western side of the Owenmore Valley east of Lough Cruttia, in the penin-sula of Corca Dhuibhne, Co. Kerry. Similar features exist on the eastern slopes of Croghan Mountain, between Croghan and the Wicklow Gap, in Co. Wexford. Quinn (1975, 1977) mapped well-developed solifluction ter-races on the southern slopes of Knocknadobar in Co. Kerry, where they form the lowest of a spectacular altitudinal sequence of periglacial features. She has also shown how, on the north-facing slopes of that mountain, stone polygons and stripes merge into solifluction lobes. Of equal interest are the deposits of Glen Gaddagh, on the north side of MacGillycuddy's Reeks, also in Co. Kerry. Warren has recorded terraces up to 4 km long, composed of both glacial deposits and solifluctional debris, that curve along the eastern side of that valley. He interpreted them as lateral morainic terraces. "The fact that they are partly composed of soliflual material need not alter this conclusion, for it is reasonable to assume an accumulation of hill slope debris along an ice margin." The uppermost terraces, altitudinally, and particularly those furthest to the northeast, he believed to be purely solifluc-tional: "...some of the features ... are typical both in plan and cross-section of stone-banked solifluction lobes ... being fronted by large blocks" (War-ren, 1979, p. 232).

One of the apparent results of solifluction and other processes has been the formation, or exhumation, of tors. The tors of the Wicklow Mountains are both spectacular and well known and have received occasional mention in the literature for almost a century. Nevertheless, no systematic study has been made of them, and details of their formation and morphology are still unknown. Cursory examination shows that Wicklow Mountain tors are most common away from the central granite plateau that forms the core of the uplands. This is divided from an encircling rim of hills to the east by a lower level plateau. The encircling upland rim is composed partially of granites and partially of aureole and country rocks. While tors exist on the granites of this margin area, as on the rounded uplands that overlook the boggy sources of the Dargle River, they also exist on the adjacent schists of the rim uplands, such as on Djouce Mountain. The tors on Three Rock and Two Rock Mountains vary in height from 3 to 4 m and are particularly well known. Hoare and McCabe (1981) suggested that the plan shape of the Three Rock Mountain tors was controlled by prominent joint directions in the rock. Tors also exist in intrusive rocks along the summits of the ridge that culminates in Croghan Mountain, overlooking the town of Arklow on the border of Counties Wicklow and Wexford. The mountain itself is crowned with a jagged tor, with smooth-surfaced slopes of solifluction deposits extending downhill to the east, where they later assume a terraced form. Similarly the tors of the Blackstairs Mountains along the Wexford–Carlow border, which Synge (1979) believed to be preglacial Tertiary landforms but which may be periglacial, lend interest to that upland. In Co. Down, Colhoun has suggested that the remarkable summit tors of the granitic Mourne Mountains originated under the periglacial conditions that prevailed when Scottish ice covered all but the highest peaks. He has shown that well-developed tors occur in the quartzose (Annalong) granites and those in the western granites are less well developed. He observed that the Ballagh and Kilkeel granites did not produce tors because their felspars undergo rapid chemical decomposition (see Evans, 1967). Tors occur on many other upland areas in Ireland, but their systematic investigation is awaited.

The magnificent landslides that characterize both the northeastern edge of the Antrim basaltic plateau and its northwestern corner, at Binevenagh overlooking Lough Foyle, have diverted attention from less spectacular processes, such as cambering. Rotational slumps are also widespread in the Sperrins (Colhoun, 1971), and in Glen Car (Whittow, 1974) on the southeastern side of the Benbulbin limestone escarpment, near Sligo. In all three cases, glacial erosion undercut the hill slopes, while less competent rocks lie beneath upper and stronger strata. Whether these landslides should be considered as entirely of periglacial origin is debatable, but they certainly postdate the retreat of glacial ice from the slopes from which they were derived. Davies and Stephens (1978) have pointed out that, at least in Antrim, the slumped blocks do not deform postglacial raised beaches and the blocks must have been stable for most of the postglacial. It is probable that the landslides date to the recession of the late Midlandian ice and predate the Woodgrange Interstadial. No such large landslides have been described elsewhere in Ireland, possibly because of the absence of suitable geological conditions for their formation.

Screes are widespread and occur in virtually all the upland areas of the country. They are particularly prominent on quartzite, as on the flanks of Great Sugar Loaf and Little Sugar Loaf in Co. Wicklow, and on Errigal in Co. Donegal. They are also pronounced on the flanks of some of the Carboniferous uplands, as on Ben Bulbin in Co. Sligo, where the scree is largely derived from the Ben Bulbin shales. Screes occur at all altitudes, even down to sea level, as at Slieve League in Co. Donegal and around the

Brandy Hole on Bray Head in Co. Wicklow. In some cases garlands can be identified within the screes, as, faintly, on the western flank of Little Sugar Loaf. Quinn (1977) has reported that on Knocknadobar in Co. Kerry scree occurs on slopes which exceed 34°, but insufficient work has been completed to identify maximum and minimum angles throughout Ireland for scree accumulation. Most screes are, at present, largely vegetated and subject to little development. It is likely that many originated under periglacial conditions. They are commonly seamed by vertical gulleys, which carry water off the rock face above the scree. Debris is still transported via some gulleys under avalanche or flood conditions, and in the mid 1970s arcuate levée formation occurred at the base of gulleys in the Knockmealdowns in Co. Waterford.

Farrington (1947) used the existence of scree cover to suggest that areas in the Iveragh peninsula of Co. Kerry had not been glaciated, but the argument was poor, since many areas that have obviously been glaciated possess extensive screes. It is likely that most screes postdate the retreat of glacial ice from their immediate vicinity, although the scree material incorporated within the Lower Head along parts of the south coast shows that scree formation is not merely a postglacial phenomenon. In all probability, screes formed on a number of occasions in the past, but the unconsolidated scree material was easily removed by glacial advances, so that few patches of scree remain to testify to any other than the postglacial periods of formation.

Stratified scree, in which rudely sorted layers of coarser and finer debris are deposited, also exists, as in the Rocky Valley on the north slope of Great Sugar Loaf in Co. Wicklow; near sea level on the Mullet of Mayo, north of Belmullet (S. Watson, personal communication); and at the Giant's Causeway in Co. Antrim. Colhoun (1971) has reported what seem to be stratified scree deposits on the Sperrins. Apart from the probability that all these stratified screes postdate the retreat of glacial ice from their vicinity, nothing further is known as to either their age or mode of formation. Dylik (1960) believes that stratified screes result from periglacial slope wash.

Block fields occur on many summit areas, as on the plateau to the west of the Knockmealdowns in Co. Waterford (Lewis, 1976), on the uplands of the Corraun Peninsula in Co. Mayo (Farrington, 1953), and on Knocknadobar in Co. Kerry. In both the latter cases they are associated with stone polygons (see Quinn, 1977), but blockstreams as such have not been described. Neither have clearly identifiable ploughing blocks been reported, although some stones in upland areas do have a roll of turf around their downslope side.

Rock glaciers have not been identified, although aerial photographs sug-

gest that blocky arcuate morainelike deposits on the eastern face of Lugnaquillia, in the Wicklow Mountains, may prove to be the remnants of such a
feature (McGrory, 1980).

Patterned Ground Phenomena

Polygonal patterns, of a variety of different origins, have been reported
from a number of sites in Ireland, as have the subsurface features responsible
for most of them. The same applies for stripes and garlands. Additionally,
subsurface festoons and involutions, which may not necessarily have had
any surface expression, also occur in section.

Small, sorted polygons that are still active under present climatic conditions exist on the summit plateau of the Comeraghs in Co. Waterford.
North of the summit of Knocknadobar, Co. Kerry, sorted polygons also
occur (Quinn, 1977), and since they are largely unvegetated it may be that
they are not entirely dormant. Much larger stone polygons, caused by frost
action on the block field areas on the summit plateau of that mountain,
appear to be entirely remnant. Like stone polygons elsewhere in Ireland,
these features are largely unvegetated, although this is partly due to their
exposed and windswept position and to the lack of fine material suitable for
plant growth.

Colhoun (1971) discovered well-developed sorted stone polygons on the
summit plateau of the Sperrin Mountains. The polygons are located at
altitudes above 420 m on slopes of less than 2°. They vary from 1 to 3 m in
diameter and occur within block field detritus. Since they are covered by 1
to 2 m of blanket bog, from which they are currently being exhumed, they
must predate bog formation. They postdate the disappearance of the local
ice cover from the area, which occurred during the late Midlandian. Quinn
(1975) found that polygons on Knocknadobar occur on north-facing slopes
with angles that do not exceed 5°. Sections showed that the polygons there
average 24 cm in depth and 3.5 m in diameter. When slopes exceed 5°,
polygons are replaced by stone stripes and garlands. Other fossil polygons
have been reported from such upland areas as the Corraun peninsula in Co.
Mayo (Farrington, 1953) and on the western flanks of the Wicklow Mountains (Farrington and Mitchell, 1973). In all probability, in upland areas
patterned ground is widespread although largely hidden by the extensive
deposits of Littletonian blanket bog.

On lowland areas, Lewis (1979) has discovered relic sorted polygons on
gravel deposits near Cahir and Roscrea, in Co. Tipperary. They occur
within larger networks of ice-wedge polygon casts and are apparently reflected below the surface by patterns of cryoturbation. The polygons are
indicated in section by ice-wedge casts, the remnants of the former ice

wedges responsible for the polygonal networks. Ice-wedge casts are well developed in glaciofluvial gravels near Knockgraffon, north of Cahir and adjacent to the River Suir (Fig. 2). The casts of ice-wedge polygons also exist in Co. Wexford, where Mitchell (1976) has identified them at Broom-hill Point. Aerial photographs suggest that casts of ice-wedge polygon networks also occur southeast of Roundwood in Co. Wicklow, where there are excellent exposures of ice-wedge casts (Lewis, 1977). Apart for those at Broomhill Point, the sorted polygon remnants so far discovered on the lowlands were found because deposits that otherwise cap them had been removed to facilitate extraction of the gravels in which they exist. It is likely that many other areas of patterned ground have yet to be revealed.

Most of the ice-wedge casts found in Ireland evidence epigenetic ice-wedge formation, but the infill of wedges near Roscrea and Cahir, Co. Tipperary, indicates flooding of the surface of the deposits in which the ice wedges developed. Since these wedges formed in glaciofluvial gravels on outwash plains, it is reasonable to suggest that ice-wedge development at both locations took place while the nearby ice sheets were still melting and while a limited amount of deposition was still occurring on the outwash deposits. Pollen grains, all of which are nonarboreal, enclosed within a wedge cast at Roscrea suggest that the wedges formed in a harsh environment, with little vegetation present, as might be expected near the margins of a major ice sheet. Since these ice-wedge casts are located within the maximum limits of the late Midlandian ice advance they must postdate that advance, but since they may relate to the melting of the late Midlandian ice

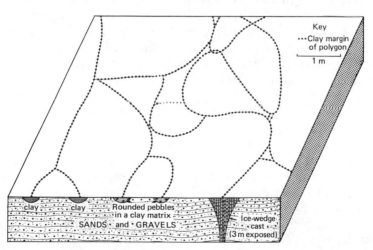

Fig. 2. Detail of polygon patterns exposed in 1978 at Knockgraffon, Co. Tipperary. From Lewis (1978).

they probably predate the succeeding Woodgrange Interstadial (Lewis, 1979).

Ice-wedge casts have been found in many areas of Ireland, apart from those already mentioned. Warren (1977, 1981) has recorded them from the lowlands north of the MacGillycuddy's Reeks in Co. Kerry, McCabe (1977) from the coast of Co. Louth and the Maguiresbridge area of Co. Fermanagh (McCabe et al., 1978), Colhoun (1971) from the Sperrins, Lewis (1978) from the vicinity of Gouganbarra in Co. Cork and from numerous localities in northeast Wicklow (Lewis, 1977). At all these places the wedges postdate the retreat of the last glacial ice from their vicinity, and Colhoun, Lewis, and Warren suggest that the formative ice-wedges might have existed during the Nahanagan Stadial. At Moneystown in Co. Wicklow, ice-wedge casts of varying width and depth are visible. They probably formed during a period of cold, vigorous, but variable climate, with thin winter snow cover (the concomitant of ice-wedge formation), characterized by abrupt falls of temperature well below freezing in winter (Lewis, 1977). Hoare and McCabe (1981) have suggested that some of the ice-wedge casts of east-central Ireland formed prior to the Woodgrange Interstadial and that their infills were subsequently modified in the Nahanagan Stadial.

At Courtaparteen in Co. Cork, Farrington (1966) recorded an ice-wedge cast in the Courtmacsherry Raised Beach, covered by the Lower Head deposits. This suggests that permafrost, necessary for wedge formation, predated initiation of the head deposits. At Ballymakegoge, near Fenit in Co. Kerry, Mitchell (1970) found organic material stratigraphically above raised beach deposits that are probably the local representative of the Courtmacsherry Raised Beach. Since these organic deposits were radiocarbon dated and found to be older than 42,000 bp, the ice-wedge cast in the beach deposits at Courtaparteen must be even older. Mitchell interpreted the organic deposits as of Gortian age and ascribed the overlying Lower Head deposits to the Munsterian. Within this head, and capped by deposits that include Upper Head, ice-wedge casts exist. These are obviously older than the Nahanagan Stadial, the most recent time when conditions could conceivably have been suitable for head formation.

Recent marine erosion has exposed the remains of polygons and stripes at Garryvoe in Co. Cork (Farrington and Stephens, 1964). These developed in the frost-shattered limestone bedrock and are overlain by till deposited in an advance of ice from the Irish Sea Basin. Although their age is not known, they are probably the oldest features of patterned ground to have been found in Ireland. Stripes have been recorded only from this site and from surface deposits at Knocknadobar in Co. Kerry.

Cryoturbation features, including involutions, festoons, and erected and fractured pebbles, have been reported at many sites in Corca Dhuibhne and

Iveragh in Co. Kerry, at the south coast of Wexford, and at a few localities
on the coasts of Co. Louth, Co. Down, Inishowen in Co. Donegal, northwest Co. Mayo, at the mouth of Killary Harbour, and at Carrickaholt in
Co. Clare (see Lewis, 1978). They have also been noted at a few inland sites,
as in the Sperrins (Colhoun, 1971). They are well developed in association
with ice-wedge casts at Knockgraffon, near Cahir in Co. Tipperary. At that
site, their formation appears to have been contemporary with that of the ice
wedges. The uppermost sands and gravels on either side of the wedge casts
show marked evidence of cryoturbation, but there is no evidence that any of
the wedge-cast deposits have been cryoturbated. In relation to other evidence this suggests that cryoturbation at this site predates the Woodgrange
Interstadial (Lewis, 1979). In the Finglas Valley, near Waterville in Co.
Kerry, Bryant (1974) obtained radiocarbon dates of 11,170 ± 170 and
11,950 ± 200 bp for organic material overlying involuted gravels, which
provides a minimum age for the features at that site. Elsewhere, cryoturbation features may be more recent. It is likely that further studies will show
that cryoturbation occurred widely in Ireland on a number of different
occasions.

Thermokarst Phenomena

The remnants of pingos, which are dome-shaped hills of ice within local
sediments, were first recorded in Ireland by Mitchell in 1971. He described
essentially circular and semicircular ramparts of debris, presumably formed
by the overlying material slumping down the sides of the melting pingos
and thereby reflecting their outlines. The continuous ramparts enclose a
central depression. In the Camaross area of Co. Wexford (Mitchell, 1973)
the ramparts rise 2–3 m above the water level in the enclosed basins.
Organic sediments and solifluctional debris floor the basins. Where pingo
remnants exist on sloping ground, the ramparts tend to elongate downslope
and in extreme cases form an open U shape with no enclosing rampart on
the upslope side.

Two types of pingos are known to exist in areas of modern periglacial
activity. Closed system pingos "occur almost exclusively in areas of continuous permafrost, often in alluvial lowlands with little vertical relief. They
are brought about by the local downward aggradation of permafrost into a
previously unfrozen zone" and tend to form "when permafrost advances
into a shallow lake basin which is either being drained or is infilling"
(French, 1976, p. 98). By comparison, open system pingos form

in areas of thin or discontinuous permafrost where surface water can penetrate into
the ground and [circulate] in unfrozen sediments. Where this water rises towards the
surface it freezes to form localized bodies of ice which force the overlying sediments

upwards. Thus, open system pingos usually develop in distinctive topographic situations, such as in valley bottoms or in lower valley side slopes, where hydrostatic head or pressure can develop sufficiently to provide a constant and regular water supply to the freezing plane" (French, 1976, pp. 95–96).

Mitchell (1973) concluded that the features in the Camaross area are the remnants of open system pingos. They are located on a poorly drained slope composed of soliflucted till with a surface inclination of about 2°. On the valley floor itself the pingo remnants become sparse. In other words, the pingo remnants exist at precisely that zone on the valley side at which one might expect groundwater to seep to the surface. Mitchell showed that, at Camaross, there was evidence for more than one generation of pingo development. The remains of more recent pingos were superimposed on the outlines of older pingos.

Pingo remnants are now known to exist at over 50 sites in Ireland (O'Callaghan, 1981). Initially it was thought that they formed only beyond the limits of the ice sheets of the last glaciation (Synge, 1973), but Mitchell (1973) found remnants near Ballaghkeen in Co. Wexford that lie just inside the limit of the ice sheet of the last glaciation. Pollen analyses suggest that at Camaross, and at Meenskeha near Millstreet in Co. Cork (Coxon, 1983), pingo basins did not accumulate sediments until the end of the Nahanagan Stadial. This led Mitchell to infer that conditions were cold enough for the formation of pingos during that stadial but that some pingos may be older.

Spectacular pingo ramparts exist at Snugborough, some 5 km north of Arklow in Co. Wicklow beside the road that leads to Redcross. The largest feature there has a rim that rises some 3 m above the level of the present basin floor (Huddart, 1977). Even more impressive examples exist on the slopes of the valley floor at Clonroche, in Co. Wexford. A drainage ditch cut through a number of converging pingo remnants clearly shows the ramparts' sand and gravel composition. On the inner side the deposits dip down into the pingo basin but lie horizontally where the rampart reaches its crest. The structure of the outer side of the ramparts is not clearly evident. In contrast to the ramparts, the basins are floored with finer grained sediments. although the ditch did not reveal the base of the deposits and they may overlie coarser material.

P. O'Callaghan's (personal communication) analyses of aerial photographs suggest that, except in Co. Wexford, pingo remnants are confined to areas that lie outside the maximum limits of Midlandian advance as defined by Synge (1977). No pingo remnants, for example, have yet been found in the Midlands. Nevertheless, O'Callaghan's analysis was restricted to south and southeast Ireland, and further research may well identify pingo remnants elsewhere.

Niveomorphic Phenomena

Niveomorphic phenomena comprise features that owe their existence mainly to frost-induced processes that occurred in association with snow masses. They include aggradational features, such as protalus ramparts (or nivation ridges), that result from debris sliding downslope over the surface of a snow mass to accumulate at its foot to form a ridgelike landform.

At Coomnahincha, on the western slopes of Knocknadobar in Co. Kerry and overlooking Coonanna Harbour, there is a spectacular nivation cirque. This is similar to those described from the Urals by Botch (1946) or from mid-Wales by Watson (1966). It probably formed by regelation around the head of a snow bed and is associated with solifluction downslope, as well as with supranival sliding of talus material down over the surface of the snow bed. Quinn (1975, 1977) has shown that a protalus rampart is located beneath, and parallel to, the backwall, while further downslope there are spectacular terraces of soliflucted material. A section on the east side of Coonanna Harbour shows 2.5 m of periglacial fan debris and soliflual deposits, derived from the Coomnahincha nivation cirque, on top of till. Quinn ascribed the till to a glacial advance from the south, with ice spilling north into Dingle Bay via the harbour. Quinn suggested, on the basis of Bryant's findings in the Waterville area (1968), that the till predates the late Midlandian ice advance in Iveragh and could have an earlier Midlandian or even Munsterian age. On this basis the deposits associated with the nivation cirque could date to any age more recent than that of the last Munsterian ice advance. Alternatively, they may be no older than the late Midlandian.

Warren (1979) has mapped accumulations of angular blocks in the cirques of Cummeenmore and Cummeengrin on the north side of MacGillycuddy's Reeks in Co. Kerry and noted their resemblance to protalus ramparts. He stated (p. 234) that "... they owe their genesis to the occurrence of a solid canopy over which frost riven blocks could slide to accumulate in a pile to the bottom," but he was unwilling to define the crystalline state of the medium responsible for their formation. As far as he was concerned they might be either glacial or periglacial in origin. Similarly ambiguous accumulations of debris exist elsewhere in the mountains of Co. Kerry, as on the south side of Moll's Gap beside the road from Killarney to Kenmare. In Co. Wicklow, linear accumulations of boulders occur in the upper reaches of the Glenmacnass valley head, in the Wicklow Mountains. Both Warren (1970) and Pritchard (1981) have ascribed them to the sliding of debris down over a mass of snow. Accurate dating of these features is not yet possible, but, as with the protalus rampart described by Colhoun (1981) in the Mourne Mountains, they must date to the latter part of the Midlandian, probably to the Nahanagan Stadial. Kinahan (1894) reported that, in exceptionally severe winters, frost-riven stones still accumulate to form ramparts at the base of snow patches in the cirques of Lugnaquillia in the Wicklow Mountains and in upland areas of Co. Galway.

Altiplanation surfaces which have developed in association with snow beds have been reported from Knocknadobar by Quinn (1975, 1977).

Aeolian Phenomena

Studies outside Ireland have shown the significance of wind action in periglacial environments, and wind-blown deposits and ventifacts testify to its former importance in areas of relic periglacial phenomena. In spite of this there has not yet been any systematic study of former aeolian periglacial activity in Ireland. In 1979 Lewis suggested on the basis of colour and particle size, that the infill of an ice-wedge cast existing in glaciofluvial outwash sediments at Roscrea in Co. Tipperary was at least partly composed of loess. He concluded that, in conjunction with palaeobotanical evidence, it was obvious that aeolian processes operated during the time of wedge formation, suggesting a cold, dry, and largely unvegetated environment. Since numerous outwash plains dating to the retreat of the Midlandian ice have been identified, it is likely that further aeolian deposits remain to be discovered.

Contemporary Periglacial Processes

Contemporary periglacial processes appear to be limited mainly to the development of small sorted polygons on certain upland areas (as in the Comeragh Mountains of Co. Waterford), to the occasional accumulation of talus material on screes, and to protalus rampart accumulation under occasional severe winter conditions. Needle-ice formation and consequent erosion remains a common winter occurrence, as do other forms of frost heaving of a minor nature. During the winter of 1978–1979, roads in upland areas, as on the Calary plateau on the eastern edge of the Wicklow Mountains, experienced considerable damage due to frost heaving, especially where snow cover had been removed from the road surface by wind action. Such road damage is fortunately relatively rare and was last previously recorded on an appreciable scale in the winter of 1962–1963.

PERIGLACIAL PHASES

Although there is great debate about the age of many Quaternary deposits in Ireland, it is known that periglacial activity occurred on a number of different occasions. This is shown by the stratigraphic position of periglacial phenomena, such as the head deposits that exist above and below the Courtmacsherry Raised Beach and the ice-wedge cast that Farrington (1966) recorded within that beach at Courtaparteen. Solifluction deposits have also been widely recorded in lateglacial sequences, relating to two distinct cold stadia. At sites such as the Old Head in Co. Mayo and Drumurcher in Co. Monaghan, absolute dating of these deposits has been undertaken. Nevertheless, the age of many periglacial features, including such landforms as protalus ramparts, pingo remnants, and the casts of ice-wedge polygon networks, is not known with any great accuracy. In spite of this the broad sequence of phases of periglacial activity is known (Fig. 3).

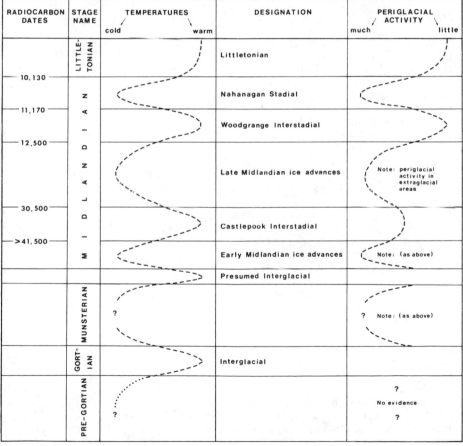

Fig. 3. Phases of periglacial activity in Ireland. (See Lewis, 1978, for further discussion of the stratigraphic position of periglacial phenomena.)

CONCLUSIONS

Periglacial features are widespread in Ireland and have been reported mainly from southern areas, beyond the supposed limits of the last major ice advance. Relatively little attention has been paid to them, and there have been few systematic studies of periglacial phenomena. In particular, little attention has been paid to the interior of Ireland and especially to the Midlands. Further investigation may well show that periglacial features are even more commonly distributed than is at present realized. Future research might concentrate on the study of patterned ground, aeolian deposits, tors, solifluction features, stratified scree, and the work and effects of fluvial processes in a periglacial environment.

References

Anderson, J. G. (1906). *J. Geol.* **14,** 91–112.
Botch, C. G. (1946). *Bull. Geol. Soc. U.S.S.R.* **78,** 207–234.
Bryant, R. H. (1968). "A Study of the Glaciation of South Iveragh, Co. Kerry." Unpublished Ph.D. thesis, Univ. of Reading, Reading.
Bryant, R. H. (1974). *Proc. R. Ir. Acad., Sect. B* **74,** 161–178.
Colhoun, E. A. (1971). *Proc. R. Ir. Acad., Sect. B* **71,** 53–71.
Colhoun, E. A. (1981). *Ir. Geogr.* **14,** 85–90.
Coope, G. R., Dickson, J. H., McCutcheon, J. A., and Mitchell, G. F. (1979). *Proc. R. Ir. Acad., Sect. B* **79,** 63–85.
Coxon, P. (1983). *Ir. Assoc. Quat. Stud. Newsl.* **6,** 9.
Dardis, G. F., Mitchell, W. I., and Hirons, K. R. (in press). *J. Earth Sci. (Dublin)*
Davies, G. L. H., and Stephens, N. (1978). "Ireland." Methuen, London.
Dylik, J. (1960). *Biul. Peryglac.* **8,** 31–41.
Evans, E. E. (1967). "Mourne Country." Tempest, Dundalk.
Farrington, A. (1934). *Proc. R. Ir. Acad., Sect. B* **42,** 173–209.
Farrington, A. (1947). *Ir. Geogr.* **1,** 89–97.
Farrington, A. (1953). *J. Glaciol.* **2,** 262–267.
Farrington, A. (1966). *Sci. Proc. R. Dublin Soc.* **A2,** 197–210.
Farrington, A., and Mitchell, G. F. (1973). *Ir. Geogr.* **6,** 543–560.
Farrington, A., and Stephens, N. (1964). *In* "Field Studies in the British Isles" (J. A. Steers, ed.), pp. 445–461.
French, H. M. (1976). "The Periglacial Environment." Longman, London.
Hoare, P. G., and McCabe, A. M. (1981). *Biul. Peryglac.* **28,** 57–78.
Huddart, D. (1977). "South East Ireland." Geo Abstracts, Norwich.
Jukes, J. B. (1861). *Mem. Geol. Surv. Irel.,* Sheets 185 and 186.
Kilroe, J. R. (1908). *In* "The Geology of the Country around Londonderry," pp. 41–69. Mem. Geol. Surv. Irel., Dublin.
Kinahan, G. H. (1894). *Ir. Nat.* **3,** 235–240.
Lee, J., Finch, T. F., and Ryan, P. (1964). *Ir. J. Agric. Res.* **3,** 175–187.
Leffingwell, E. de K. (1915). *J. Geol.* **23,** 635–654.
Lewis, C. A. (1967). *Ir. Geogr.* **5,** 293–301.
Lewis, C. A. (1974). *Sci. Proc. R. Dublin Soc.* **A5,** 207–235.

Lewis, C. A. (1976). *Ir. Geogr.* **9,** 18–29.
Lewis, C. A. (1977). *Sci. Proc. R. Dublin Soc.* **A6,** 17–35.
Lewis, C. A. (1978). *J. Earth Sci. R. Dublin Soc.* **1,** 135–142.
Lewis, C. A. (1979). *Ir. Geogr.* **12,** 10–24.
McCabe, A. M. (1969). *Ir. Nat. J.* **16,** 232–233.
McCabe, A. M. (1977). *In* "South East Ireland," Guidebook for excursion A14 INQUA X Congress (D. Huddart, ed.). Geo Abstracts, Norwich.
McCabe, A. M., Mitchell, G. F., and Shotton, F. W. (1978). *Proc. R. Ir. Acad., Sect. B* **78,** 77–89.
McGrory, D. (1980). "Scree Slope Study in the South-East Wicklow Glens." Unpublished B.A. dissertation, Univ. Coll. Dublin (N.U.I.).
Mitchell, G. F. (1970). *Proc. R Ir. Acad., Sect. B* **70,** 141–162.
Mitchell, G. F. (1971). *Nature (London)* **230,** 43–44.
Mitchell, G. F. (1973). *Proc. R. Ir. Acad., Sect. B* **73,** 269–282.
Mitchell, G. F. (1976). "The Irish Landscape." Collins, London.
Mitchell, G. F. (1977). *Philos. Trans. R. Soc. London, Ser. B* **280,** 199–209.
Morrison, M. E. S., and Stephens, N. (1965). *Philos. Trans. R. Soc. London, Ser. B* **241,** 221–255.
O'Callaghan, P. (1981). *Ir. Assoc. Quat. Stud. Newsl.* 2.
Pritchard, P. (1981). "The Deglaciation of the Glenmacnass Valley, Co. Wicklow." Unpublished B.A. dissertation, Univ. Coll. Dublin (N.U.I.).
Quinn, I. (1975). "Glacial and Periglacial Features in North West Iveragh, Co. Kerry." Unpublished M.A. thesis, Univ. Coll. Dublin (N.U.I.).
Quinn, I. (1977). *In* "South and South West Ireland" (C. A. Lewis, ed.), pp. 29–35. Geo Abstracts, Norwich.
Synge, F. M. (1973). *Ir. Geogr.* **6,** 561–569.
Synge, F. M. (1977). *In* "South and South West Ireland" (C. A. Lewis, ed.), pp. 4–8. Geo Abstracts, Norwich.
Synge, F. M. (1979). *In* "Prospecting in Areas of Glaciated Terrain 1979," pp. 1–7. Institution of Mining and Metallurgy, London.
Warren, W. P. (1970). "Cirque Glaciation in the Wicklow Mountains." Unpublished B.A. dissertation, Univ. Coll. Dublin (N.U.I.).
Warren, W. P. (1977). *In* "South and South West Ireland" (C. A. Lewis, ed.), pp. 37–45. Geo Abstracts, Norwich.
Warren, W. P. (1979). *In* "Moraines and Varves" (Ch. Schluchter, ed.), pp. 223–236. Balkema, Rotterdam.
Warren, W. P. (1981). *Biul. Peryglac.* **2B,** 239–246.
Watson, E. (1966). *Biul. Peryglac.* **15,** 79–101.
Whittow, J. B. (1974). "Geology and Scenery in Ireland." Penguin, London.
Wright, W. B., and Muff, H. B. (1904). *Sci. Proc. R. Dublin Soc.* **10** (N.S.), 250–324.

CHAPTER 6

COASTAL EVOLUTION*

Francis M. Synge†

Quaternary Section
Geological Survey of Ireland
Dublin, Ireland

INTRODUCTION

The idea of permanent sea level throughout time has long been rejected. Evidence of earlier marine coastlines higher than that of the present is not uncommon in the form of cliffs, erosion platforms, and, more rarely, beach and offshore deposits. On the other hand, phases of lower sea level are represented by drowned coastal deposits and terrestrial peats associated with forest beds. The presence of marine shells a considerable distance inland does not always indicate the former presence of marine waters, for there are many instances where former ice sheets moved inland, transporting material many kilometres from the sea floor.

The oldest recognizable shoreline, that of the Carboniferous Sea, declines eastward from the Burren of Co. Clare and Benbulbin in Co. Sligo to the flanks of the Wicklow Hills (Charlesworth, 1963). At a lower level, the edge of the Chalk (Cretaceous) Sea in all probability also declines eastward. Later, warping seems to have occurred in the opposite direction, for the peneplanes on the surface of the limestone at varying heights up to 120 m OD near Dublin pass westward across the island to decline beneath the waters of Galway Bay without any discontinuity (Corbel, 1957).

Much clearer evidence of former coastlines occurs below 50 m sea level, particularly within the ranges 11–20 and 4–6 m mean sea level (MSL) along

*This paper was drafted by Francis Synge while fighting his final illness. Its appearance is a tribute to him as a committed scientist.
†Deceased.

the southeast and south coasts of Ireland. These are so fresh in appearance that they are considered to be Pleistocene rather than Tertiary in age. As they both appear to lie beneath the youngest major till sheet, they are not likely to postdate the last interglacial warm stage. Evidence that the level of the sea dropped even lower is furnished by the presence of drowned river valleys extending seaward to depths below -100 m MSL. Such a low level was last attained during the maximum extent of most of the world's larger ice sheets between about 18,000 and 12,000 years ago (Milliman and Emery, 1968). This drop of sea level extended down to its lowest level, at about -130 m MSL, approximately 15,000 years ago, before returning almost to the present level about 5,000 years ago.

Such a eustatic fall of sea level would have exposed a considerable extension of the land surface out on the shallow shelves that border the continents. As a result the shallower seas and straits would have been converted into "bridges" or peninsulas joining island groups to the mainland. From this one would expect that a bathymetric map should give a clear representation of the position of the coastline during phases of lowered sea level. But this is not always the case, because those areas once covered by thick, extensive ice sheets subsided under their weight. Such isostatic depression meant that the rising sea level which occurred during deglaciation could flood areas of sea floor at an earlier date than would be expected from the depths indicated on the charts.

During deglaciation crustal uplift in those areas formerly deeply buried by the ice sheet proceeded more rapidly than the rise of sea level. Therefore, along the coast a descending succession of raised shorelines is now visible, each marking a phase when the rate of sea level rise coincided with that of land uplift. But outside the limits of the ice sheet, where land uplift was minimal or absent, only a rise of sea level has been recorded. Here the glacial shorelines have been formed at lower levels and were therefore subsequently drowned. From this we can readily visualize a heavily glaciated zone where all the former shorelines are raised; an outer zone where all the shorelines have been drowned; and an intermediate zone where emergent shorelines were later drowned after uplift had slowed. The resulting pattern of relative sea level change based on dated index points of known altitude form the basis of maps compiled to show changes of the position of the coastline throughout the past 30,000 years or so.

COASTAL FORMS

Coastlines are readily recognized by the type of landforms associated with them, either as erosional features such as cliffs, caves, stacks, and marine platforms, or as depositional features such as beach ridges, deltas, estuarine flats, and beach terraces. All such forms relate directly to one or other specific marine level such as high water mark at spring tide (HWMSpr), low water mark at neap tide (LWMNps), and mean sea level (MSL). In order that all coastal forms may be related to a uniform level approximating to mean sea level, all measurements for altitude must be related to the same datum. Three datums have been used in Ireland based on Dublin (OD), Belfast (BOD), and Malin Head. The Malin Head datum, based on mean sea level for a number of stations around the coast, is very close to the Belfast datum, which is 2.9 m above the Dublin datum. The Malin Head datum is intended to replace the other two, which are locally based, and to render them obsolete. Furthermore, it should be noted that mean tide level has often been taken for mean sea level; this is not strictly correct as mean tide level is not at the same height everywhere and can vary by as much as 6.1 m.

Shoreline features develop speedily in unconsolidated materials, as can be observed on the margins of any newly created reservoir. But they are best developed along coasts exposed to prolonged consistent wave attack. In the case of a falling sea level, shoreline terraces of beach material are abandoned in a descending stairway, whereas in the case of a rising sea level, the bulk of the beach deposit migrates from lower to higher level, growing all the time. In some cases, raised beaches may be stranded on older, rock-cut shore platforms. For this reason, shore cliffs cut in bedrock cannot always be relied upon to mark the youngest marine episode in an area; they could have survived one or more transgressions of the sea.

Shorelines are made up of a number of different landforms that may sometimes be misinterpreted. For instance, the most level part of a shoreline terrace is often taken as that part a little below the high water mark. But below the low water mark a second flat terrace commonly occurs, and in estuarine situations a very level sheet of mud, silt, and/or fine sand infill can extend for kilometres at a very low gradient. In certain valley systems extending across the trend of the isobases, raised terraces of such estuarine material might be mistaken for raised beaches. But a closer examination would show that the grain size of the deposit, unlike that of a true shore deposit, diminishes seaward.

DATING OF SHORELINES

The precise dating of old shorelines has always been problematical, either because the altitude of the datable material cannot be related to sea level or Ordnance Datum or because the range of error of the dates obtained is too high. Organic materials such as peat, wood, or humus have been used to obtain ^{14}C dates. When such material is interbedded with marine deposits it is generally interpreted to indicate a temporary regression of the sea. Therefore, a date from the lowermost layer of the organic bed would date the beginning of the regression of the sea, while one from the top of that bed would date the onset of the subsequent transgression. Such an interpretation assumes no erosion of the surface layers since deposition. But changes in base level need not have occurred; for example, incursion of the upper marine layer might have resulted from the breaching of a protective beach ridge by storm waves. The true interpretation depends on the evidence from other sites in the neighbourhood.

Marine shells can also be used for dating purposes. Radiocarbon and amino acid techniques are more commonly used for dating than others. Littoral species are most useful in assessing the position of the sea level. Offshore shell banks cannot be used for this purpose, although they can be used for dating minimum sea level heights in certain circumstances.

In the following sections the shoreline sequence from the end of the Tertiary Period up to the present day is described. Three distinct phases of shoreline development have been recognized: during interglacial stages; during final deglaciation; and during postglacial times, embracing the whole of the Holocene (Littletonian) up to the present day.

INTERGLACIAL SHORELINES

Throughout south and southeast England on land surfaces that have never been glaciated, there are remnants of a marine erosion surface at 183–198 m (600–650 ft) British OD, carrying pockets of sand and shingle with a shelly marine fauna of Red Crag age. The degraded cliff line belonging to this formation has been observed along the flank of the North Downs (Mottershead, 1977). It also occurs in southwest England (Kidson, 1977). The westward continuation of this Lower Pleistocene surface should be expected. Perhaps the steep slopes that margin a surface at about 120 m (400 ft) OD—the coastal lowland of east Co. Wicklow—date from this time; the difference in height could be due to warping. On this coastal platform a rich assemblage of Red Crag shells in glacial outwash gravel has been described from Killincarrig (McMillan, 1938). The relatively undamaged condition of the shells suggests that they cannot have been transported very far from their original position by the ice. Because the fauna represented has a marked northern aspect, it has been suggested that the Irish Sea at that time had no connection with southern waters. Similar truncation of a highland plateau by a coastal platform occurs in Co. Sligo. Here the Benbulbin escarpment (527 m OD) seems to have been cliffed at about the 200 m OD level.

Indications of lower levels of marine erosion are obscure above about 20 m MSL. In the unglaciated south of England this level has been generally associated with the last interglacial. In Ireland, a cliff notch at about 10 m MSL has been traced from Dublin to Waterford Harbour. Between Dublin and Wicklow Head this level has been upwarped by as much as 10 m (Fig. 1). Invariably the associated marine rock platform is overlain by glacial tills and is commonly ice moulded and striated. The extension of freshwater deposits of the last interglacial right down to present sea level in the south of England (West, 1968) suggests that the water surface had already started to fall toward the end of this warm stage. The distinctive low-level shoreline bordering both sides of the Celtic Sea at 4–6 m MSL is placed in the latter part of this interglacial. Known as the Courtmacsherry Raised Beach from its stratotype (type site) on the south coast of Co. Cork, it occurs at several localities beneath sand rock buried by up to 12 m or more of head deposit (Wright and Muff, 1904). On the top of the latter a marine till has been noted at a number of localities as far west as Ballycroneen, near the entrance to Cork Harbour. At other places, a rubbly local drift caps both beach and sand rock. This has been regarded as a facies of the inland (Munsterian) till, although in many of the sites examined by the author the upper deposit would appear to have reached its present position by solifluction. However, detailed local studies are needed to confirm this. Both east of Kilmore Quay, Co. Wexford, and west of Ballycroneen, a capping of till is associated with the subsequent erosion of the beach and striation of the rock platform by land ice. If the capping of till between Kilmore Quay and Ballycroneen has been soliflucted into its present position, then the Courtmacsherry Raised Beach (levelled at 3–4 m MSL in Co. Wexford) was formed after powerful streams of inland ice crossed the coast (Munsterian phase), and before weak marine ice (Ballycroneen phase) moved inshore from the Celtic Sea. A later stream of ice removed the beach in Co. Wexford (Irish Sea phase), and local inland ice also removed parts of the beach in the vicinity of Cork Harbour, when sea level had fallen below that of the present day.

Dating these different ice movements and associating them with particular sea levels is a matter of some difficulty. Only the position of the Ballycroneen till offers some clues in this direction. This peculiar ice movement on the Irish coast is paralleled by a similar one along the north side of the Cornubian peninsula in southwest England. At Fremington in north Devon, marine till has been emplaced around Bideford Bay up to 60 m OD (British) on top of shingle at 18 m OD (British) and is believed to be either

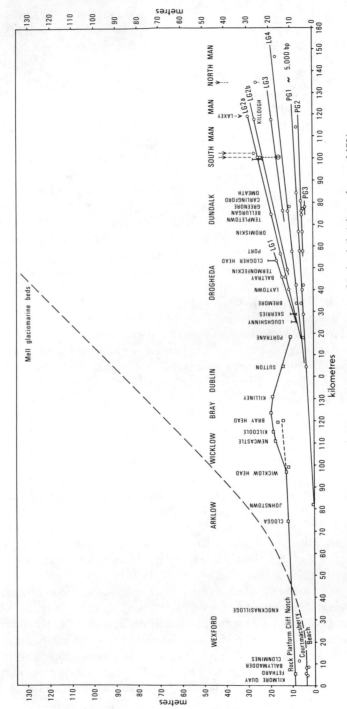

Fig. 1. Shoreline displacement diagram for the east coast of Ireland (heights relate to MSL).

marine (Stephens, 1966) or glaciofluvial (Kidson and Wood, 1974). As the latter interpretation is difficult to accept, owing to the complete absence of erratics, Stephens's view that this is a raised beach seems more likely. I would suggest that the evidence implies that till of the last cold period rests upon a beach of last interglacial age at this locality. A later beach at much the same level truncates the glacial sequence, and could be only slightly younger. The assumption that ice of the last cold stage could not possibly have reached north Devon is quite unwarranted. Why could not an early glacial surge have extended ice down the Irish Sea to expand into the Celtic Sea before sea level had started to fall very much? This would explain the signs of weak glaciation observed along the north side of the Cornubian peninsula (Stephens, 1966) including the emplacement of the rather anomalous St. Erth shelly (boulder?) clay and shelly gravels (Mitchell, 1965) and the glacial limit around the north edge of the Scilly Isles (Mitchell and Orme, 1967). The largest erratic boulders, of possible Scottish origin, extend along the coast a considerable distance up the English Channel and also to the coast of Brittany. At Saunton, at the entrance to Bideford Bay, shelly marine material has been dated $33,200 \pm \frac{2,800}{1,800}$ bp (Kidson, 1974).

Extension of glacial ice during the boreal to subarctic conditions in the middle part of the last cold period is not inconsistent with high sea level, provided large continental ice sheets had not developed. Thus we can see that the Courtmacsherry Raised Beach shows indications of cold conditions from the presence of ice-rafted erratics. Interdigitation of head deposits with beach gravel has also been interpreted as implying cold conditions (Farrington, 1966). Perhaps, too, the extension of the Courtmacsherry Raised Beach platform into the sheltered channels of Cork Harbour need not be explained by erosive action of small waves over a long period of time, but by the action of intertidal freeze–thaw activity over a relatively short time.

GLACIAL SHORELINES

During the middle part of the last cold (Midlandian) stage sea level dropped significantly. But before this fall of sea level had reached its full extent, the ice advanced, or, more probably, readvanced, from the North Channel shortly after $24,050 \pm 650$ bp according to the ^{14}C date obtained from marine shells in a lower till at Glastry in north Co. Down (Davies and Stephens, 1978). As the ice front subsequently receded northward in the Irish Sea, the rising sea level flooded coastal areas as far north as Clogher Head and the Isle of Man. The highest shoreline (LG1) has been traced north from present sea level at Dublin to an ice-marginal delta at 17–18 m MSL in the Boyne Valley, 4 km west of Drogheda (Synge, 1977). The southern, drowned continuation of this shoreline would appear to correspond to the line of the Kish Bank and other sand and gravel banks that extend parallel to the present Irish coast at a distance of about 10 km between Howth Head and the Tuskar Rock (Synge, 1977).

On the northern outskirts of Drogheda, at Mell, glaciomarine beds occur at 25–35 m with *Similipecten greenlandicus* as the dominant species. But are these beds *in situ*? They might have been deposited as a frozen erratic block fortuitously deposited with the bedding in a horizontal position. The original investigators of this site considered that the deposit is *in situ* (Colhoun and McCabe, 1973); if so, it would represent 120–150 m of isostatic down-warping, as this species of shellfish lives today at water depths of 90–110 m in cold saline waters under high-arctic conditions such as east Greenland and the Barents Sea.

The marine limit at 20–21 m MSL associated with the final disappearance of glacier ice from the Boyne Valley can now be dated by means of the organic material found near the bottom of a kettle hole in the Orrisdale moraine on the Isle of Man, because the same shoreline (LG1) lies close by, outside this glacial limit. The organic material yielded five dates ranging from 18,900 to 18,400 bp (Shotton and Williams, 1971). According to the time–depth sea level curve for the northeast coast (Carter, 1982), there was an average fall of 1 m every 300 years. Since the Bride Hills moraine, cutting across the northern tip of the Isle of Man, corresponds with a drop of the marine limit by about 10 m from shoreline LG1 to LG2, a halt of the ice front of about 3,000 years is implied. The Bride Hills moraine corresponds with the innermost belt of moraine marking the limit of the Irish drumlin field between Kells and the entrance to Strangford Lough. From this evidence, therefore, the ice front started to recede from this moraine at about 15,000 bp, the same time as the eustatic fall of sea level reached its lowest point to expose a proposed broad land bridge on the bed of the Celtic Sea connecting Brittany to the Cornubian peninsula and the south coast of Ireland (Fig. 2). At this time an ice-dammed lake occupied the Irish Sea basin, with a fluvial outlet extending southwest to the coast at −130 m MSL near the edge of the continental shelf.

Comparison between index points in northeast Ireland (Carter, 1982) and those in southwest Scotland (Jardine, 1982) suggests that the ice in the North Channel finally melted about 14,000 years ago so that the Irish Sea once again became a gulf of the Atlantic, while the Celtic Sea land bridge still closed the southern approach. The general eustatic rise of sea level narrowed the land bridge until it was confined to the Bardsey Ridge, a feature consisting largely of morainic material that links the Wicklow coast to the Lleyn peninsula in Wales. Shortly after this time (about 12,000 years

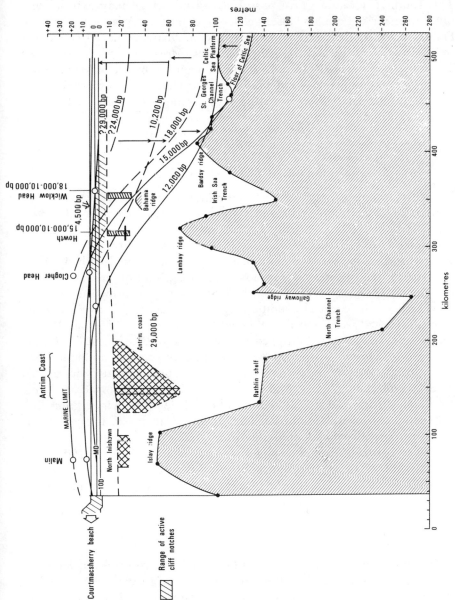

Fig. 2. Eustatic levels in relation to the floor of the Irish Sea.

ago), the "bridge" was breached during the Allerød interstadial (Fig. 2). This southern location of a "postglacial" land bridge may explain why much of the Irish flora has closer links with that of the Atlantic seaboard of France and Spain than with northwest Britain. Man may have first entered Ireland by this route, and the giant Irish deer (*Megaloceros giganteus*) almost certainly did. Presence of giant Irish deer remains in the Isle of Man are also significant, although this island would appear to have been severed from the Cumbrian mainland as late as ~10,200 bp, some 2,000 years after Ireland became isolated.

POSTGLACIAL SHORELINES

Uplift continued long after the disappearance of the ice sheet and even continues today, although at a much slower rate than previously. By about 13,000 bp, the ice had virtually gone from Ireland, although a substantial mass appears to have survived in the west of Scotland. At this time the sea waters had become warmer, and their levels are associated with shoreline "c," which stands some 7–8 m below the marine limit on the east coast of Co. Antrim (Synge and Stephens, 1966). Shoreline "b" formed about 1,000 years earlier. These milder conditions lasted until approximately 12,000 bp when the ice had receded well into the sea lochs of western Scotland. By this time relative sea level had dropped almost to that of the present day on the coast of Co. Down; at Roddans Port marine clay reaching no higher than 1.8 m MSL is overlain by a grey clay containing organic matter dominated by *Salix* pollen and ^{14}C dated to 12,100 ± 150 bp (Morrison and Stephens, 1965). This marine level, "d," lies about 6 m below the marine limit "b."

Relative sea level continued falling in the northeast as isostatic uplift continued apace, while at the same time it was rising eustatically in the southwest corner of Ireland. Here in Bantry Bay, freshwater Allerød deposits occur at −57 m MSL (Stillman, 1968). At the close of the last cool, or Younger *Dryas,* phase and in the early part of the succeeding pre-Boreal phase (about 10,000–9,000 bp), sea level stopped falling in the northeast owing to the cessation of rapid uplift. Before relative sea level started to rise again, a distinct shoreline appears to have formed at about −13 m British OD in the English Channel off Plymouth (Morey, 1976).

The culmination of the subsequent Holocene (Littletonian) transgression occurred at approximately 6,000 bp in the northeast after the coast had already been occupied by Mesolithic peoples (Movius, 1940; Woodman, 1978), reaching about +3 m (average maximum sea level), a figure estimated from 34 index points (Carter, 1982). Pumice, possibly borne by currents from an Icelandic source (Binns, 1967), has been recorded in this raised beach in Lough Swilly (Praeger, 1896). Along the north coast a subsequent regression of 5 m was followed by a second transgression up to +3 m during the last 1,000 years (Carter, 1982). Along the Mourne coast of Co. Down two Littletonian shorelines, at 7–8 m MSL (PG/1) and 3–4 m MSL (PG/2), respectively, have been identified (Stephens and McCabe, 1977). These two different levels, however, may represent the same shoreline formed under different tidal and storm conditions (Carter, 1982). Together with an intermediate shoreline at 5–6 m on the Carlingford peninsula, they may be part of the very recent sequence because they lie below the possible level of present high tide surges (see Carter, 1982).

The intricacy of the dated transgression/regression sequences is furnished by four important occupation sites on the northeast coast. The most northerly of these occurs at Cushendun, where Movius (1940) carried out excavations to establish the relationship between sea level and Mesolithic occupation sites. These show that the glacial deposits are overlain by woody fen peat at 0.2–0.1 m MSL, ^{14}C dated to 8,410 ± 140 bp, beneath a lower lagoon silt (0.5–0.2 m MSL) and underlying shallow water gravels (4.2–0.5 m MSL). Above this unit an upper tidal lagoon silt (5.0–4.2 m MSL) is dated 7,670 ± 140 bp beneath a gravel bay-mouth bar reaching 7.9 m MSL topped by humus, sand and stones, 0.5 m thick; at the base of the topmost humus layer a Neolithic hearth was found. Behind this bar, a river terrace at 3.5 m MSL and 1.0 m thick overlies silts containing peat and wood; a sample of the latter gave a date of 4,740 ± 100 bp. It appears likely that this terrace is graded to a sea level postdating, and at a lower level than, the highest Littletonian beach (Mitchell and Stephens, 1974).

At the next site, Carnlough, Co. Antrim, about 15 km further south, a marine transgression laid down beach deposits up to −1.7 m MSL. This was later covered by sands and thin lagoonal clays capped by a possible solifuction layer. Above this, 15–25 cm of compact peat yielded a date of 11,630 ± 170 bp. Signs of freeze–thaw structures in overlying blue–grey sandy silt point to severe climatic conditions before an upper peat layer, 20–30 m thick, was formed. This latter bed contains abundant Salix fragments and yielded a ^{14}C date of 8,110 ± 125 bp. The top of the upper peat at 5.5 m MSL stands below the level of the highest Littletonian beach ridge at 10 m MSL, but no stratigraphical continuity has yet been established between these two horizons (Prior et al., 1981). This sequence shows that one main regression intervened between about 12,000 and 8,000 bp, during which severe climatic conditions were present around 11,000–10,000 bp.

At Ringneill on Strangford Lough, sandy organic silt at 2 m (HWMSpr) dated 7,500–7,345 bp is covered by occupational debris with charcoal dated 5,380 ± 120 bp beneath beach sand and shingle extending up to 7 m OD. On top of the raised beach terrace a hearth dated 3,680 ± 120 bp is buried by shell midden dated 2,660 ± 110 bp. This sequence is truncated by a shore notch at 3 m OD. Therefore, in this case, the main transgression amounting to 5 m is bracketed between about 5,500 and 3,560 bp (Stephens and Collins, 1960). At Woodgrange, also on Strangford Lough, raised beach sand reaching 4 m OD is bracketed between organic deposits dated 6,565 ± 105 bp and 3,245 ± 85 bp (Dresser *et al.*, 1973), in fairly close agreement with the previous site, taking into account a possible hiatus or erosional break at the transgression contact.

No extensive work has yet been carried out on the dating of the magnificant suite of 33 raised beach ridges that fill the embayment of the Fane River south of Dundalk. Here two main systems can be recognized, both recurved by breaking waves from the southeast, with an inner suite trending northwest and an outer one trending due north (Fig. 3). On the shores of Dublin Bay cemented shelly raised beach has been stranded on the interglacial beach platform at 3–4 m MSL beneath hillwash derived from till (Stephens and Synge, 1958). Further west the crest of the raised tombolo at 4 m MSL has been ^{14}C dated, using marine shells, to 4,830 ± 140 bp and 4,460 ± 140 bp (J. J. Donner, personal communication). Some 34 km further south, at Leamore, a shingle bar was breached at the time of the pollen zone VII/VIII transition, reckoned to be about 5,000 bp (Mitchell and Stephens, 1974). These dates emphasize the fact that the Littletonian marine limit declines in altitude from north to south down the east coast of Ireland from 10 m MSL at Carnlough to an estimated 3.75 m MSL at Leamore and also becomes younger by about 1,000 years. This diachronous effect results from the fact that the younger transgression was greater in amplitude in the south where rapid uplift was absent.

On the west coast, the Littletonian raised beaches have not been investigated in any detail. Around Sligo Bay three beach levels (6–7 m, 5–6 m, and 3 m MSL) have been recorded by the author. Each of these is associated with a rich fauna dominated by *Littorina littorea, L. obtusata, Cardium edule, Ostrea edulis,* and *Mytilus edulis.* Extensive beds of *Ostrea edulis*-containing charcoal layers reach 5.5 m MSL at Culleenamore, where they overlie a bed of *Littorina littorea* resting on a limestone till. A stratigraphic record of the transgression was visible at Cummeen, showing marine clay bands interdigitated with a freshwater shelly mud and overlain by shelly marine clays. The sequence indicates that a freshwater lake in Sligo Bay was inundated by a marine transgression. At the culmination of this transgression marine waters may have extended up the Garavogue River to enter Lough Gill, which at present stands at 3.8 m MSL. Both the lake shore and the islets are bordered by a low raised beach 0.6 m above the present high water level, and fragments of marine shells have been found at Lahanagh on the southern shore (Fig. 4).

Fig. 3.   Raised beach ridges (between points A, B, and C) at the Fane estuary, Co. Louth (photograph by permission of the Director of the Geological Survey of Ireland).

At the head of Clew Bay shelly beach material is banked against a cliff notch cut in drumlins up to 0.7 m above HWMSpr. Similar features occur around the head of Galway Bay and in the inner parts of the Kenmare River and Bantry Bay. Apart from these areas, postglacial raised beaches are absent on the west coast. Drowning rather than uplift dominates the coast of the south of Ireland. In fact, tidal gauges show that the sea floor of the

Fig. 4. Raised shoreline features around Sligo Bay.

Raised Beach deposits

Raised Oyster Beds

Rock Scarps

Drumlins

Eskers

Gravel Spreads

Glacialmeltwater channels

Area formerly covered by sea

Glacial Striae

Lough Gill

Aghamore

△ 900

Garavoge River

SLIGO

Carns Hill
△ 392

Carrowroe

Railway

Carrowmore

Sligo Harbour

Oyster Island

Knocknarea
△ 1078

The Glen

BALLISODARE

Coney Island

Culleenamore

Ballisodare Bay

Ballinlig

N

km
0 1 2 3

Celtic Sea is at present subsiding (Valentin, 1954) as this area lies across the zone of marginal downwarp which surrounds the sites of the former ice sheets. Subsidence has now extended as far north as Dublin, where a rate of 0.04 m per century has been recorded. Further north, however, the present movement is one of uplift—0.1 m per century at Belfast and 0.25 m per century at Malin Head. These few tidal records are important in that they give the only reliable basis for the construction of isobases.

CONCLUSIONS

Throughout the Quaternary Period we have seen that fairly extreme fluctuations of relative sea level took place. The early part of the record is obscure owing to the destruction and burial of the earlier shorelines by the activity of the ice sheets. Yet scattered clues have survived here and there which enable us to reconstruct the main pattern of sea level change. It is evident that at the beginning of the period, the Red Crag times, the sea level approached 200 m OD but dropped in each succeeding interglacial until eventually, during the last one, the sea fell gradually from 12–14 to 3–4 m MSL.

For the last cold (Midlandian) stage the record is much more complete, as shore marks outside the retreating ice front are clearly visible. Datable material in the form of organic terrestrial deposits in, and related to, marine sedimentary horizons is available for establishing index points. Even though this record is somewhat fragmentary, we can establish a pattern which reveals a high eustatic level (at 4–6 m MSL), followed by a drop to a minimum (at about −130 m MSL) about 15,000 years ago and then a eustatic rise back to the former level by about 5,000 bp. The low level was occasioned by the late expansion of the North American ice sheet and the persistence of the large ice mass in the Baltic, although by then that part of the ice sheet occupying the Irish Sea had already receded as far north as the Isle of Man. As a result, the proposed wide and extensive "land bridge" between Ireland, Cornwall, and Brittany was established while an ice-dammed lake was impounded in the Irish Sea basin above a seabed that had been isostatically downwarped. As the following eustatic rise of the sea finally breached the land bridge shortly after 12,000 bp, there was only a limited opportunity for the early immigration of a temperate flora and fauna after the start of the Allerød Interstadial.

During further deglaciation, isostatic rebound was rapid in the northern part of Ireland, with relative sea level dropping at the rate of 0.3 m per century, an indication that isostatic rebound considerably outstripped the eustatic rise. The rapid uplift ceased about 10,000 bp when relative sea level in the northeast stood at −13 m MSL, which is the equivalent of the eustatic level at −30 m for the stable areas. Such a limited drop in the level of the sea was quite insufficient to establish any land connection with Britain at that time, contrary to opinions expressed in the literature (Mitchell, 1976).

Fluctuations in the Littletonian rise of sea level have been documented in Ireland by the analysis of some 40 index points. These show that the transgression maximum occurred between 7,000 and 6,000 bp in the northeast. But southward this shoreline declines in altitude below that dated 5,000–4,500 bp, which figures as an important transgression between Strangford Lough and Dublin Bay. The view that isostatic recovery was already complete by that date is now considered doubtful as tidal gauge records indicated that uplift still continues at present at Malin Head and Belfast, along with a compensatory sinking movement at Dublin (R. W. G. Carter, personal communication).

The suggestion that Littletonian sea levels on the east coast may have reached 3.5 m MSL (Mitchell and Stephens, 1974; Synge, 1980) could be erroneous as these are within recent and present storm levels. Furthermore Carter (1982, p. 22) is of the opinion that "sea-level indicators within 1.5× to 2× the storm wave height, on an isostatically stable coast, are unsatisfactory for constructing time/depth sea-level graphs." However, it should be pointed out that detailed analysis of the landforms and their stratigraphy in this zone can sometimes enable the investigator to distinguish older fossil features from those that are recent.

The story of sea level change during the Littletonian is still far from complete. Doubtless, future study of offshore sequences containing drowned peats and forest beds will furnish much additional information.

References

Binns, R. E. (1967). *Acta Borealia A: Scientia* **24**, 1–64.
Carter, R. W. G. (1982). *Proc. Geol. Assoc.* **93**, 7–23.
Charlesworth, J. K. (1963). "Historical Geology of Ireland." Oliver & Boyd, Edinburgh and London.
Colhoun, E. A., and McCabe, A. M. (1973). *Proc. R. Ir. Acad., Sect. B* **71**, 211–245.
Corbel, J. (1957). *Institut des Études Rhodaniennes de l'Université de Lyon: Mémoires et Documents* **12.**
Davies, G. L. H., and Stephens, N. (1978). "Ireland." Methuen, London.
Dresser, P. Q., Smith, A. G., and Pearson, G. W. (1973). *Proc. R. Ir. Acad., Sect. B* **73**, 53–56.
Farrington, A. (1966). *Sci. Proc. R. Dublin Soc.* **2A**, 197–219.
Jardine, W. G. (1982). *Proc. Geol. Assoc.* **93**, 25–42.
Kidson, C. (1974). *In* "Field Handbook, Easter Meeting 1974—Exeter" (A. Straw, ed.), pp. 43–44. Quaternary Research Association, Exeter.
Kidson, C. (1977). *In* "The Quaternary History of the Irish Sea" (C. Kidson and M. J. Tooley, eds.), pp. 257–298. Seel House Press, Liverpool.

Kidson, C., and Wood, R. (1974). *Proc. Geol. Assoc.* **85,** 223–237.
McMillan, N. F. (1938). *Proc. Liverpool Geol. Soc.* **19,** 255–266.
Milliman, J. D., and Emery, K. O. (1968). *Science* **162,** 1121–1123.
Mitchell, G. F. (1965). *Proc. Geol. Assoc.* **76,** 345–366.
Mitchell, G. F. (1976). "The Irish Landscape." Collins, London.
Mitchell, G. F., and Orme, A. R. (1967). *Q. Jl. Geol. Soc. London* **123,** 59–92.
Mitchell, G. F., and Stephens, N. (1974). *Colloq. Int. C.N.R.S.* **219,** 115–125.
Morey, C. R. (1976). *Field Studies* **4,** 345–368.
Morrison, M. E. S., and Stephens, N. (1965). *Philos. Trans. R. Soc. London, Ser. B* **249,** 221–255.
Mottershead, D. N. (1977). *In* "The Quaternary History of the Irish Sea" (C. Kidson and M. J. Tooley, eds.), pp. 299–320. Seel House Press, Liverpool.
Movius, H. L. (1940). *Proc. R. Ir. Acad., Sect. C* **46,** 1–84.
Praeger, R. L. (1896). *Proc. R. Ir. Acad.* **20,** 30–54.
Prior, D. B., Holland, S. M., and Cruickshank, M. M. (1981). *Ir. Geogr.* **14,** 75–84.
Shotton, F. W., and Williams, R. E. G. (1971). *Radiocarbon* **13,** 131–156.
Stephens, N. (1966). *Biul. Peryglac.* **13,** 141–156.
Stephens, N., and Collins, A. E. P. (1960). *Proc. R. Ir. Acad., Sect. C* **61,** 41–77.
Stephens, N., and McCabe, A. M. (1977). *In* "The Quaternary History of the Irish Sea" (C. Kidson and M. J. Tooley, eds.), pp. 179–198. Seel House Press, Liverpool.
Stephens, N., and Synge, F. M. (1958). *Proc. R. Ir. Acad., Sect. B* **59,** 19–28.
Stillman, C. J. (1968). *Sci. Proc. R. Dublin Soc.* **3A,** 125–127.
Synge, F. M. (1977). *In* "The Quaternary History of the Irish Sea" (C. Kidson and M. J. Tooley, eds.), pp. 199–222. Seel House Press, Liverpool.
Synge, F. M. (1980). *Bull. Assoc. Étude Quat.* **17,** 77–79.
Synge, F. M., and Stephens, N. (1966). *Trans. Inst. Br. Geogr.* **39,** 101–125.
Valentin, H. (1954). *Petermanns Geogr. Mitt.* **98,** 103–108.
West, R. G. (1968). "Pleistocene Geology and Biology." Longmans, London.
Woodman, P. C. (1978). "The Mesolithic in Ireland: Hunter–Gatherers in an Insular Environment." BAR Brit. Ser. No. 58, Oxford.
Wright, W. B., and Muff, H. B. (1904). *Sci. Proc. R. Dublin Soc.* **10,** 250–324.

CHAPTER 7

SOIL FORMATION

Edward B. Culleton
Michael J. Gardiner

An Foras Talúntais
Dublin, Ireland

INTRODUCTION

If land classification is evidence of an interest in soils then the eighth century AD could be the starting point for such investigations in Ireland. In that century the Brehon Laws listed three categories of cultivable and three of uncultivable land for the purpose of determining land value (Mac-Niocaill, 1971). Over a millenium later (1830–1853) the Griffith valuation was carried out for similar purposes, with instructions to the valuators to bear in mind the nature and depth of the soil and the quality of the subsoil (Anonymous, 1853). In 1848, Sir Robert Kane's project for a soil survey of Ireland involved a comparison of the value of the soil as determined by Kane's chemical analysis with the practical value as attributed by Griffith. Kane's coloured soil maps were deposited in the Museum of Irish Industry about 1862 but were subsequently mislaid (Simington and Wheeler, 1945). In 1907, Kilroe's "Soil-Geology of Ireland" drew attention to the relationship between the glacial geology, soils, and agriculture in the country.

A link was forged between Sir Robert Kane and the modern soil survey of Ireland at the Kane Centenary Symposium in 1944, when P. H. Gallagher, of University College, Dublin, made an impassioned plea for the establishment of such a survey. Gallagher himself had already carried out some pedogenetic studies on the role of mobile colloidal humus in the podzolization process (1942). Subsequently, Gallagher and Walsh (1942) divided Irish soils into two broad genetic groupings on the basis of the silica:sesquioxide ratios of their clay fractions.

THE QUATERNARY HISTORY OF IRELAND 133

In 1968, Gallagher's dream of earlier years was realized with the setting up of the National Soil Survey under the aegis of An Foras Talúntais (The Agricultural Institute). In a relatively short time this survey provided information on the nature and distribution of the major soils in the country, thereby giving a stimulus to and a framework for further pedogenetic studies.

Understandably perhaps, podzolization processes and clay movement have always exerted a peculiar fascination for Irish pedologists. Following the earlier studies of Gallagher and Walsh, Diamond et al. (1965) noted podzolization in the upper horizons of grey-brown podzolic soils in east Co. Galway; Ryan and Walsh (1966) sought to elucidate the genesis of podzolized soils developed in limestone gravels in the Midlands; clay movement in soils beneath the Midland peats was studied by Hammond (1968); and Conry et al. (1972) published the results of a study on the genesis and characteristics of brown podzolic soils in the southwest. The genesis of three soils formed mainly from limestone in Co. Kildare was investigated by Collins et al. (1975): the occurrence of brown podzolic soils within predominantly brown earth areas developed on Lower Palaeozoic shale in Co. Wexford was studied by Gardiner and Culleton (1978); a detailed analysis of several podzols developed in granite, schist/shale, and dolerite was carried out by O'Dubhain and Collins (1981); and in 1981, Cruickshank and Cruickshank described a possible development sequence for humus–iron podzols in Co. Antrim, based on radiocarbon dating and pollen analysis. Detailed investigations were carried out on the high cation exchange properties of the basalt-derived soils in northeastern Ireland by McAleese and McConaghy (1957, 1958), and man's influence on soils has been studied by Conry (1971, 1972a,b).

At a broader level, the National Soil Survey has published soil bulletins and maps at a scale of 1:126,720 for eight counties: Wexford (Gardiner and Ryan, 1964), Limerick (Finch and Ryan, 1966), Carlow (Conry and Ryan, 1967), Kildare (Conry et al., 1970), Clare (Finch et al., 1971), Leitrim (Walsh, 1973), Westmeath (Finch and Gardiner, 1977), and Meath (Finch et al., 1983). Parts of two other counties, west Cork (Conry and Ryan, 1963) and west Donegal (Walsh et al., 1969), were also surveyed. A general soil map of the country was published in 1969, and an improved version, together with a bulletin "Soil Associations of Ireland and Their Land Use Potential" (Gardiner and Radford), appeared in 1980. In 1979, Hammond published a map and bulletin on Irish peatlands. Cruickshank (1970) discussed pedogenesis in Northern Ireland, and Culleton (1977) and Collins (1981) dealt with the significance of glacial geology studies for soil investigations.

THE PRESOIL SURFACE

The glaciations described by Warren (Chap. 3, this volume) and McCabe (Chap. 4, this volume) were responsible for laying down a complex mantle of deposits which formed the parent material of almost 80% of the soils of the island. Apart from a few areas, only the higher mountains and steeper slopes remained drift free.

In their passage the glaciers incorporated material from the underlying bedrock, together with any weathered regolith in their path, and deposited the mixture either as till or sorted sediments (Fig. 1). The grain size of this material was of fundamental significance in determining the pedogenetic pathways of Irish soils. Most rocks are quickly broken down to their terminal grain sizes in the glacier, hence the physical composition of glacial sediments is closely related to the underlying bedrock. The fine-grained Upper Carboniferous shale of Co. Leitrim, north Kerry, Co. Clare, and the Castlecomer Plateau, for example, gave rise to poorly drained, impermeable soils, while on the coarser grained Lower Palaeozoic shale of Counties Wexford and Louth, well-drained soils developed.

Although local rock is usually dominant, glacial sediments are generally polylithological, and this intermixing of rock types leads to considerable physical and chemical variation. One result of this heterogeneity is a complex suite of clay minerals, since the deposits contain not only the primary minerals of igneous origin but also the secondary minerals inherited from sedimentary and metamorphic rocks, for example, shale and schist. Kaolinite, quartz, and 14-Å minerals, mostly chlorites and chlorite–vermiculite intergrades, are usually present, with mica (probably illite) often dominant (Gardiner and O'Callaghan, 1967).

Not all glacial sediments are derived from the underlying bedrock. For example, large tracts of land along the east coast are covered by glaciomarine deposits derived from the bed of the Irish Sea. These extend inland as far as Slane, Navan, Enfield, and Naas (McCabe, 1973; Finch *et al.*, 1983). In Co. Carlow, limestone-dominant drift overlies the Leinster granite for a distance of several kilometres (Conry and Ryan, 1967); in Counties Galway and Mayo, siliceous materials from the Palaeozoic rocks form a broken veneer on the limestones to the east; and the limestone-floored valleys of Munster are partially filled with Old Red Sandstone and shale from the anticlines.

Glaciofluvial deposits, by nature coarse textured, are widespread in the Central Plain and elsewhere. Because of their great variation in physical composition, bedding sequence, and relief, these give rise to a complex soil pattern, often too intricate to separate in the field or to show on the pub-

Fig. 1.   Distribution pattern of major soil parent materials: 1. limestone drift; 2. limestone bedrock and thin drift; 3. sandstone till and bedrock; 4. Lower Palaeozoic shale and slate till with sandstone in places; 5. schist, gneiss, and quartzite till and bedrock; 6. acid igneous and quartzite till and bedrock; 7. basalt till and bedrock; 8. glaciomarine drift; 9. Upper Carboniferous shale.

lished county maps. Conry and Ryan (1967) discussed this problem in relation to the Athy Soil Complex in Co. Carlow, while Collins (1976) documented the spatial variation in a 5.0-m long soil section developed in a stratified, coarse-textured deltaic deposit in west Co. Cork. Different profiles which would place the soil in diverse Great Soil Groups were found side-by-side in the section.

Soil formation commenced on the mineral material composing the upper layer of the lithosphere. It seems likely that in Ireland this layer was no more homogenous vertically than laterally. Stratification, to a greater or lesser extent, is recognizable in many glacial sediments. Also, at the dissolution of the ice sheets, considerable quantities of water would have moved down through the unconsolidated glacial deposits, probably transporting fines to lower depths (Boulton and Dent, 1974). Abundant evidence exists, too, from lateglacial sites of removal of fines from the lithosphere surface (Mitchell, 1951; Colhoun and Mitchell, 1971; Culleton, 1976; Craig, 1978). Collins and O'Dubhain (1980) relate fragipans in the Dublin–Wicklow Mountain flanks to periglacial activity on the basis of the arrangement of their silt fractions. These processes resulted in an upper layer in which considerable alteration in particle size distribution had already taken place before soil formation commenced. In particular, translocation of fines led to coarsening of the upper layer and to the formation of clay-rich layers lower down. This process, rather than massive pedogenetic clay movement, may account for the clay increases in the B horizons of certain Irish soils.

The presoil surface on the lowlands, then, can be pictured as consisting mainly of glacial sediments of variable lithological, physical, and chemical composition. Where the topography was hilly and mountainous, the drift and weathered rock would have been moved downslope through periglacial action, leaving a bare or thinly covered rock surface. In the river valleys, alluvium may have overlain the drift or, more likely, the bedrock.

SOIL FORMING PROCESSES IN IRELAND

Simonson (1959) visualized the major soil forming processes as consisting of losses, gains, translocations, and transformations of constituents within the soil. Soil gains organic matter but loses mineral constitutents such as nitrogen and exchangeable cations, for example, Ca^{2+}, Mg^{2+}, and K^+. Some constituents may be translocated only from upper to lower parts of the soil, notably, iron and aluminium oxides, humus, and fine clay particles. Transformations are constantly occurring in soil: organic matter is broken down by bacteria, and minerals are subject to breakdown and alteration through replacement of certain pivotal elements in the crystal structure. These minerals were formed under equilibrium conditions which do not now exist at the earth's surface. The degree to which they break down or alter depends on the rate of weathering, on the strength of leaching, and on the formation of secondary minerals (Paton, 1978, p. 17). Various studies to date (e.g., Walsh et al., 1957; Gardiner and O'Callaghan, 1967; Conry et al., 1972: Collins et al., 1975; Curtin and Smillie, 1981) indicate that the clay minerals present in Irish soils are mainly inherited from the parent materials and that little secondary pedogenetic alteration has taken place.

Leaching and Podzolization

Central to the soil forming processes under Irish conditions is the contact with, and downward movement of, water through the soil. This occurs where precipitation exceeds evapotranspiration. In Ireland the balance is heavily tipped in favour of the former, and constituents released by pedochemical weathering in the surface layer are moved relatively quickly to lower depths. Some, such as the highly soluble elements Ca, Mg, K, and Na, may be removed entirely in drainage water, while others, such as humus and iron oxides, may be precipitated below the surface.

A fundamental aspect of pedogenesis in Ireland is the progressive acidification of the surface (A) horizon. The rate and depth of acidification is related to the nature of the minerals forming the parent material, to the chemical composition of the soil solution, and to the volume of water and the rate at which it moves downward through the soil. Even on soils developed on limestone, the surface eventually becomes acidic. Man has striven to arrest this process through manuring and marling but particularly through liming and fertilizing in recent years; thus, it is virtually impossible in Ireland to find a lowland soil in its virgin state.

Where the rate of removal greatly exceeds the rate of release of new products through weathering, the lower part of the A horizon may be depleted of most of its weatherable constituents. Iron, aluminium, and humus may be leached downward, leaving a grey- or ash-coloured A2 horizon very different from the overlying brownish A1 horizon. Some of the nonclay constituents removed from the A1 and, particularly, the A2 horizons are deposited at lower depths in the B2 horizon of the soil. Humus and iron and aluminium oxides are among the constituents which accumulate (Table I). B2 horizons, like A2 horizons, develop in areas of high rainfall and excessive drainage and are a relatively common feature of up-

Table I   Differentiated Horizons of Depletion and Accumulation in a Peaty Podzolized Soil Derived from Granite[a]

Horizon	C (%)	Fe_2O_3 (%) Dithionite extractable	Al_2O_3 (%)
02	19.3	—	—
A1	5.3	0.53	0.18
A2	4.7	0.36	0.19
B1	4.2	1.23	1.25
B2irh	7.3	4.16	0.86
B3	3.4	0.97	1.32
C	1.1	0.56	0.44

[a] From O'Dubhain and Collins (1981).

Table II Profile Analyses of Soil Showing Clay Movement and Deposition[a]

	Particle size analysis of mineral fraction							
Horizon	Coarse sand (%)	Fine sand (%)	Silt (%)	Clay (%)	pH	C (%)	Free iron (%)	Carbonates (%)
A11	14	30	34	22	6.0	3.1	1.1	—
A12	18	28	33	21	6.0	1.5	1.2	—
A2	10	43	30	17	6.2	0.7	1.2	—
B2t	2	21	51	26	7.3	0.5	1.4	3.5
C	9	37	38	16	7.8	0.4	0.3	54.6

[a] From Finch *et al.* (1971).

land and coarse-textured soils in Ireland. Indurated iron pans are found in many podzols, generally at moisture-gradient interfaces.

Podzolization, as the above process is called, is not fully understood. It is known that organic compounds, particularly polyphenols, are active complexing agents and can attach constituents such as iron, aluminium, and humus. Polyphenols are derived from higher plants and are most plentiful in vegetation growing on acid soils. Breakdown of organic matter also produces organic acids capable of chelating (complexing) metal ions which then move down the profile in the soil solution. Anderson *et al.* (1982) have shown that aluminium and iron can also be translocated in an inorganic form.

Clay Movement

Another important pedogenetic process in Ireland is movement of fine clay through the profile and its deposition in a lower horizon, often referred to as an argillic or "textural" B horizon (Table II). In the past, clay movement was regarded as part of the podzolization process, especially in the Russian School of soil science. But in more recent years it is considered a soil forming process in its own right. Again, the exact mechanism is not fully understood, but the most likely theory is that after a dry period, fine clay particles become dispersed on rewetting and are washed down the profile; this clay is then deposited on ped faces and pore walls at the wetting fringe. As conditions for water percolation are only likely to prevail after a dry spell, movement and deposition of clay is probably infrequent and sporadic. Hence, textural B horizons are generally not well developed in Irish soils and are often referred to as minimal Bt horizons (Table II).

Clay movement does not usually take place until carbonates have been leached from the solum. However, it would seem from the study profiles of

Ryan and Walsh (1966) and from profile data in the bulletins of the National Soil Survey (*vide infra*) that such movement can take place when most, but not necessarily all, the carbonates have been removed. Those remaining may be in the form of hard limestone gravels in an otherwise noncalcareous matrix (Soil Taxonomy, 1975, p. 19).

Some Irish soils show remarkably large increases in B horizon clay content, not all of which can satisfactorily be accounted for by pedogenetic clay movement. For example, the Mortarstown soil in Co. Kildare showed an increase in clay content from 23% in the A2 to 50% in the B2t horizon, a ratio greater than 2:1 (Collins *et al.*, 1975), and micromorphological examination of this soil showed only slight evidence of clay illuviation. Stratification within the original geological deposit is suspected in many instances.

Gleying

Gleying is a pedogenetic process involving contact with water. This contact may be due to a seasonal high water table, to the presence of seepage caused by stratification in till or bedrock, or to the small sizes of soil pores which retard downward movement of water. Such pores are often associated with high clay or silt content. The absence of oxygen and presence of organic compounds ensure reducing conditions, and released iron is maintained in the ferrous form and may leave the soil environment completely. The characteristic grey colours are due mostly to the absence of iron even though some ferrous compounds impart bluish-green tinges.

Gleying may occur throughout the profile or only in certain parts. Where the water table is high the soil may be gleyed to the surface, but where seasonal fluctuation occurs the soil may be only partially gleyed. In the latter case, mottling will be evident, caused by reprecipitation of reduced iron compounds in local spots where oxygen is available. Grey streaks may develop along cracks and old root channels. Where soil permeability is low, reduction of iron may also take place, leading to grey colours in the upper horizons and sometimes to the development of a peat layer.

Because of high rainfall, the impermeable nature of some parent materials, and topographic effects, some 30% of Irish mineral soils exhibit hydromorphic characteristics (Gardiner and Radford, 1980).

Peat Formation

Surface soils in Ireland contain various amounts of organic matter. In a survey of Irish pastures, Brogan (1966) found that organic matter contents in topsoil range from about 4 to 36%. In certain environmental situations, this organic matter may constitute over 20% of the surface layer and develop to a thickness of more than 45 cm. Where this occurs, such a soil is termed a peat (Hammond, 1979, p. 21).

Each of the three major peat types in Ireland (fen peat, raised bog, and blanket bog) had a different origin, although raised bog seems to have required an underlayer of fen. Fen peats formed in hollows and basins with base-rich ground water, particularly in the Central Plain, and hence are known as minerotrophic peats. Mitchell (1976) pictured the basins and depressions with open water being gradually filled by peat derived from aquatic and semiaquatic plants from at least 9,000 years ago. Where this built up and became firm, trees such as willow (*Salix*), birch (*Betula*), and even pine (*Pinus*) grew and eventually formed a woody fen peat. But as the fen peat grew well above the water table, *Sphagnum* moss, which can survive on rain water, took over the surface and gradually built up a thick dome-shaped layer of *Sphagnum* peat, known as raised peat or, more scientifically, ombrotrophic peat.

FACTORS IMPORTANT IN SOIL FORMATION IN IRELAND

Soil formation results from the influence of climate and vegetation acting on the surface stratum and modified by relief and man. These factors interact in a complex manner over time, and it is not possible to separate their individual effects except in general outline.

Climate

The oceanic nature of the Irish climate and the country-wide variation in rainfall and evapotranspiration have important implications for soil formation. Frequent rainfall and low evapotranspiration ensured a downward movement of constituents released through weathering (Fig. 2). As base status was lowered, more acid-tolerant plant taxa such as *Vaccinium* and *Erica* became dominant, particularly where the forest cover had been removed. These, in turn, accentuated podzolization, which probably increased westward with the increasing rainfall: evapotranspiration ratio. Given similar parent materials, a greater degree of leaching toward the west is evident on permeable soils (as well as the more frequent occurrence of gleys in low-lying positions and on impermeable materials) and also a stronger expression of gleying conditions. Temperature also played a part; colder temperatures are less favourable for plant-decomposing microorganisms and also reduce evapotranspiration rates. Annual mean daily air temperatures in Ireland range from 9°C in the northeast to 10.5°C in the south (Rohan, 1975).

Fig. 2. Mean annual water surplus (mm) for the period 1941–1970, calculated from mean annual rainfall by assuming a potential evapotranspiration of 400 mm per annum (adapted from Doherty and Logue, 1977).

Another important result of Ireland's relatively high rainfall was the formation of peat in waterlogged basins and depressions and the growth of blanket peat where conditions were very wet, for example, along the west coast and in areas of high altitude in other parts of the country.

Vegetation

A forest canopy and understorey, varying in both composition and density, covered the country during the greater part of the postglacial period. Pollen evidence (Mitchell, 1976) indicates that from about 10,000 years ago the canopy was mainly deciduous, except where *Pinus* established itself on dry, acid soils after about 9,250 years bp. From then until removed by man, a forest of tall deciduous woodland dominated by oak (*Quercus*), elm (*Ulmus*), hazel (*Corylus*), and ash (*Fraxinus*) covered the well-drained limestone lowlands. In wetter areas alder (*Alnus*) and *Salix* would have been dominant, while the uplands were probably covered by *Betula* and *Pinus*.

The presence of this tree and shrub canopy resulted in (1) a lower water table, and hence less gleying, and (2) greater evapotranspiration and consequently less water percolating through the soil to cause leaching and podzolization. Ovington (1954) estimated that under coniferous forest, only 30–40% of total rainfall reaches the soil under our climatic conditions, and O'Hare (1972) found that on blanket peat at Glenamoy, Co. Mayo, the depth to the water table in October under *Pinus contorta* was 35.3 cm compared to 9.7 cm under grass. The figures for June were 100 and 49.5, cm respectively.

Irish soils then, have been subject to several phases of varying water regime. Before the full development of forest cover, gleying would have been dominant on less permeable soils, as well as in depressions; the process would then have been arrested by the spread of forest cover but given new impetus by its clearance by man. Phases of deforestation and subsequent regeneration varied both in extent and intensity over time. Obviously, better drained soils would have been deforested first, but these attacks took place over a period in excess of 4000 years and were not continuous. The type of farming also varied between arable and pastoral (O'Connell, 1980; Edwards, Chap. 9, and Woodman, Chap. 11, this volume).

Soil development would have been either accentuated or inhibited by the vegetation cover and the different types of forest litter and residues on the surface. The particular influences of previous vegetation cover in Ireland have been investigated by a number of researchers. From a study of podzolic soils on calcareous gravels in the Midlands, Ryan and Walsh (1966) concluded that vegetation was at least as significant in causing profile differentiation as was variation in texture. Conry *et al.* (1972) considered the role of vegetation crucial in soil formation and showed that podzolization took place under oakwoods in southwestern Ireland. Deforestation and cultivation were not prerequisites for podzolization. They emphasize the podzolizing effect of ground cover and suggest that its effect was accentuated by the presence of the heath taxa *Vaccinium, Calluna,* or *Erica,* in contrast to where an understorey of holly (*Ilex*) predominated. In a detailed study on the development of humus–iron podzols at Ballycastle, Co. Antrim, Cruickshank and Cruickshank (1981) showed that podzolization intensified following deforestation and the development of heath vegetation, although climatic deterioration may also have been a factor.

Decayed vegetation in the form of organic matter or peat exerted a dual role in soil formation, depending on its thickness. Where only an organic or thin peaty surface developed, the organic acids generated enhanced the podzolization process—hence peaty podzols are widespread in Ireland. But when sufficient peat accumulated, it acted as a sponge, retarding downward movement of water; podzolization slowed down and eventually ceased as the peat layer built up. Similarly, clay movement was arrested by peat development (Hammond, 1968; Barry *et al.*, 1973).

Relief

At the national scale there is a close relationship between lithology and relief. The limestone of the Midlands forms a gently undulating plain, mostly below 120 m, while the pre-Carboniferous and Tertiary rocks form a discontinuous, peripheral upland rim. The glacial sediments range from the gently undulating ground moraine of the Golden Vale, through the kame and kettle topography of the Midland gravels and sands, to the classic examples of drumlin topography to the north and northwest.

The influence of relief is reflected in soil changes with degree and length of slope, and with altitude. The lower slopes of mountains and hills are often covered by glacial deposits, while colluvium, scree, or weathered rock may form the parent material of soils upslope or on the crests. Water runoff, seepage, and erosion, as well as the degree of water penetration in the soil, are affected by slope.

An altitudinal sequence found on the Galtees (Lee *et al.*, 1964) is typical of many hill and mountain situations in Ireland with brown podzolics on the lower slopes, followed by podzols and peaty podzols, with lithosols on the steeper slopes and blanket peat at the highest elevations. Some studies on drumlins, in which other factors are relatively constant, demonstrate more directly the effects of relief on soil formation. Lee and Ryan (1965) found a relationship between soil type and degree of slope on a sandstone till drumlin in Co. Cavan. The better drained soils occurred on the steeper slopes, the wetter soils on the gentler slopes, and the wettest soils in flat interdrumlin areas, as well as on the flat drumlin crests. McCabe *et al.* (1977) found even greater variability on a drumlin in the same county derived mainly from Silurian shale and slate tills. Of the 10 soil orders identified at the highest level in Soil Taxonomy (1975), 5 were represented on this drumlin.

Relief also has been a determining factor in peat formation as evidenced by the extent and depth of basin peat in the Irish Midlands.

Man

It is generally assumed that when farming began in Ireland about 5,500 years ago, the early agriculturalists encountered a forested landscape. Successive attacks on these woodlands and the resultant changes in vegetation have been documented by Mitchell (1965, 1976), Smith (1970, 1975), Pilcher and Smith (1979), and Edwards (Chap. 9, this volume). The greater amount of water moving through the soil as a result of deforestation would have enhanced both the gleying and podzolizing processes, thus accentuating local and regional soil variation. In places the water table would have reached much nearer the surface than before, while elsewhere, particularly on sandy soils, the rate of podzolization would have increased. Deforestation, cultivation, and subsequent rainfall must also have led to translocation downslope of fine material from the soil surface. Thus, at Garradreen, Co. Wexford, 1 m of mineral soil overlies a layer of organic matter containing charcoal which gave a ^{14}C date of 1,485 ± 85 years bp (Culleton and Mitchell, 1976), and, on a drumlin in Co. Cavan, McCabe and Collins (1977) found that topsoil depth in fields increased downslope and that fences parallel to the contours acted as barriers to soil movement.

If early farmers assisted soil degradation through deforestation, their descendants now expend much effort in reversing the process through additions of lime and fertiliser. Soils data from Johnstown Castle Research Centre, Co. Wexford, show that within the past 30 years the pH of surface soils in Ireland has risen considerably, particularly on the better soils, as a result of liming (Brogan et al., 1981). Similarly, where land drainage lowers the water table, the gleying process is halted if not reversed (Larney and Collins, 1981). Man has also altered the normal pedogenetic pattern by "marling," i.e., the addition of calcareous clay, sea sand, and seashells to the land, a practice documented by Lucas (1969) and Conry (1971). In some drumlin areas, soil which had been washed downslope was carted upslope, a process known locally as "sliping."

But man's most spectacular effect on soils in modern times is to be seen in the systematic cutting away of many Irish boglands by Bord Na Mona (the Peatlands Board). This will result in many thousands of hectares being left with perhaps less than 50 cm of peat where once the cover was over 10 m. In these areas, farmers of the future will be farming on minerotrophic (fen) peat or on organomineral mixtures rather than on the edges of deep acid ombrotrophic peats. Such dramatic soil changes by man are rare. However, it is believed that over the centuries, and certainly in the last century, much peat has been cut from mountainsides and small basins in the lowlands, and some soils once considered peats are now classed as peaty podzols and peaty gleys. Finch and Ryan (1966) have shown that on the sandstone hills in southeastern Co. Limerick, the surface horizons of former podzols have been modified by cultural practices. As a result, original 0 and A horizons have been homogenized, but the lower portion of the leached A2 horizon is still to be seen in a few places. Similar changes were described from other areas by Conry (1972b).

Time

It is difficult to formulate a time scale for soil formation in Ireland, particularly when even time zero for its initiation has not yet been established. Pollen analyses have shown that grassland predominated in the country during much of the lateglacial period, hence soil changes would not have been extensive up to the beginning of the postglacial. Mitchell (1976) suggested that the present soils began to form only about 10,000 years ago and that earlier soils north of the Southern Irish End-Moraine were destroyed by glaciers and those to the south by periglacial action. Even the limited soil formation of the lateglacial warm phase he considered to have been destroyed during the later cold snap. The limited clay mineral analyses available to date substantiate the view that Irish soils are relatively young.

As Bunting pointed out in 1965 (p. 79) (and little new information has become available since), data on the length of time required to form a unit depth of soil in a given situation are limited. Jenny (1941) quoted time sequence studies which show that on sandy soils in Finland spodic B horizons were present after 500 years but continued to develop and to come closer to the surface over a period of 5,000 years. At the same time the thickness of the A2 horizon increased with advancing age. In a sandy soil in Sweden, Tamm (1920) estimated that podzols required from 1,000 to 1,500 years to develop, but Collins and Coyle (1980) found that where 5–7 cm of acid moss peat debris was added to a mineral soil an eluvial/illuvial sequence developed within 30 years.

At Goodland, Co. Antrim, there is evidence of podzolization before about 5,000 years ago (Case et al., 1969). Lynch (1981), on the basis of radiocarbon dates, suggested that podzols have been developing on the Beara Peninsula since around 4,000 years ago but that the date of commencement and the rate of formation varies widely, depending on other factors such as texture and topography. From a detailed study in Co. Antrim, Cruickshank and Cruickshank (1981) concluded that humus–iron podzols had developed from acid brown earths within the past 2,000 years.

Dates for the initiation of peat vary widely. In a Midland basin, basal peat was dated to around 8,500 years ago, and upslope the peat became progressively younger, reaching 1,700 years bp on the edge of the basin (Hammond, 1968). For blanket peat, Edwards (Chap. 12, this volume) quotes all available dates for basal blanket peat in Ireland, and these range from 4,600 to 800 bp.

TOWARDS A SYNTHESIS

Pedologists have long grappled with the problem of devising a soil classification system which will separate soils into meaningful classes. The prob-

Table III   Dominant Pedogenetic Processes and Their
Associated Great Soil Groups and Subgroups

Dominant pedogenetic process	Great Soil Group	Subgroup
	Lithosol	—
Dynamic equilibrium	Rendzina	—
	Brown earth	Acid High base status
Clay illuviation	Grey brown podzolic	Minimal Well developed Gleyed Degrading
Podzolization	Podzol	Brown podzolic Humus Iron pan (gleyed) Peaty
Gleying	Gley	Surface water Ground water
Peat "ripening"	Peat	Fen Raised Blanket

lem is determining on what basis can they be so grouped as to convey the maximum meaningful amount of information about each group. At the highest level, soils in Ireland are classified into Great Soil Groups and Subgroups based on the presence or absence of certain diagnostic horizons or other characteristics. The distinctions made cannot be based on the processes themselves because new knowledge is certain to change our ideas about the processes. But indications of the processes can be observed, measured, and used as a basis for distinction (Soil Taxonomy, 1975).

Table III lists those processes dominant in Ireland and the resultant Great Soil Groups and Subgroups, while Fig. 3 shows the distribution pattern of the dominant processes. It must be stressed that the map is highly simplified. The actual field situation is never so clear–cut, and a great variety of soils is present within the major areas defined, depending on the influence of each soil forming factor in any particular area. The dominant processes are, quite obviously, associated with a moisture surplus, that is, excess rainfall over evapotranspiration (Fig. 2). Vast areas of blanket peat occur in the west where moisture surplus is greatest. Similarly, podzolization is dominant on elevated areas in the southwest, northwest, and on the Leinster batholith. The influence of calcareous parent material on clay illuviation is evident in the Midlands and south-central Ireland, as is the preponderance of gleying on fine-textured soils, for example, on the Castlecomer and Limerick

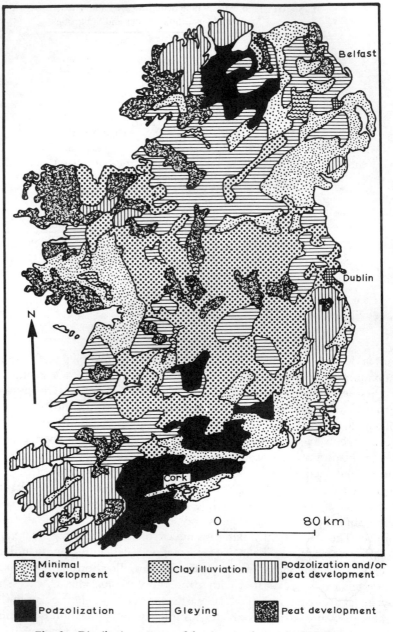

Fig. 3.   Distribution pattern of dominant pedogenetic processes.

Plateaux and on the drumlins of north central Ireland. Thus, although Irish soils are relatively young, they display a variety of morphological characteristics which reflect the significance of the various factors and processes involved in their formation.

The primary category used in soil mapping is the soil series, which comprises soils with similar type and arrangement of horizons, developed from similar parent material, and with similar land-use range. While the variation in properties such as texture within the Great Soil Group may be wide, the allowable variation of properties within a given series is quite narrow; for example, the same series could not contain textures ranging from sandy loam to clay loam. Several soil series may be included in a single Great Soil Group.

Soil formation in Ireland appears to have gone through at least two major phases. The first probably lasted from lateglacial–postglacial transition times through to the Neolithic period when the landscape was still largely forested. The second phase commenced during the Neolithic period when deforestation started, and for which considerable evidence exists for pedogenesis. Whether these changes were the result of man's interference with the woodlands or of a climatic deterioration, or both, is difficult to establish.

Because Irish soils are relatively young and not highly weathered, they are still strongly influenced by parent material. An attempt is made below to trace the major pedogenetic pathways on these parent materials, as they were affected by shifts in other factors and by changes in the materials themselves.

Soil Development on Calcareous Parent Materials

The carbonate content of soil parent material may vary from 0 to almost 100%. Where limestone or limestone drift is of nearly pure carbonate, little colloidal material is left on weathering, hence there is little soil development. Soils formed in these situations consist of a dark-coloured, organic surface layer, the dark colour being due to the calcium–humus complex (rendzinas). Drainage is always free to excessive. These soils occur mainly on the Burren in Co. Clare and in parts of east Counties Galway and Mayo, and their development is generally associated with grassland vegetation (Buol et al., 1973, p. 240). In time they would probably develop into brown earths of high base status or, where depth permits, into grey-brown polzolic soils. But apart from the incorporation of organic matter in the surface horizons, they appear to have changed little in the first 10,000 years of their existence. Under more adverse climatic conditions, podzols (as on the Leitrim Plateau and the Antrim Plateau chalks) and peat (as on Ben Bulben,

Co. Sligo) may develop. On slightly calcareous parent material in some situations, little or no clay movement took place, and brown earths of high base status were formed. In depressional areas and on slowly permeable parent material, gleying conditions would have been dominant since soil development commenced, and gleys, peaty gleys, and peat were formed. The increased soil moisture surplus following deforestation would have led to intensification of this gleying process and to an enlargement of the areas under these soils.

The most significant soil development on calcareous parent materials has undoubtedly been clay movement and deposition, giving rise to grey-brown podzolic soils. This process did not usually commence until all carbonates had been removed from the upper layer. That clay movement had not taken place at Clonsast, Co. Offaly, before 6,000 to 5,000 years bp was shown by Hammond (1968). The process was probably widespread under deciduous forest where wetting and drying conditions prevailed.

Following deforestation and a possible deterioration in climate, these grey-brown podzolic soils may have developed in several ways:

1. Where drainage became impeded the soils remained moist throughout most of the year, clay movement ceased, and gleyed conditions developed. These partially relict soils are now classed as podzolic gleys but might more properly be termed gleyed or peaty gleyed grey-brown podzolics.
2. Where permeability prevented intensive leaching of basic cations and possibly where vegetation consisted of grassland, little change took place and clay movement continued.
3. Where permeability was rapid, the upper layer became acidic, and podzolic features developed, for example, the sandy, excessively drained soils in the Midlands (Ryan and Walsh, 1966). These soils can be classed as "degrading" grey-brown podzolics.

Soil Development on Noncalcareous Parent Materials

Within this group, the soils which show the least amount of development have formed on quartz-rich, resistant rocks (e.g., granite, sandstone, and quartzite). They consist of skeletal, stony soils, predominantly mineral, overlying in most cases solid or shattered bedrock (lithosols). Their main extent is on the Old Red Sandstone uplands of the south and on the granite lowlands of Iar Connacht. As on calcareous parent material, gleying probably always predominated in certain depressional areas and in very fine-textured soils. Under free-draining conditions and podzolizing vegetation, the brown earths were converted to brown podzolics and podzols. In many places peat developed, giving peaty podzols and sometimes deep peat. But where the rate of leaching did not greatly exceed the rate of release of basic cations, brown earths seem to have undergone little significant change.

It seems likely that on noncalcareous parent materials, the dominant soil type under deciduous forest would have been the acid brown earth. Deforestation and possibly climatic deterioration would have led to large-scale changes in soil formation. Where wet conditions predominated, gleys, peaty gleys, and peat would have developed. But free-draining conditions would have allowed greater leaching and caused a disequilibrium between the rate of release and rate of removal of basic cations, thus accentuating podzolization and leading to the formation of brown podzolics, podzols, peaty podzols, or peat. Where a moisture gradient occurred, an impermeable iron pan may have developed, leading to gleying and peat development above the pan.

Acknowledgments

The authors wish to thank their colleagues in the National Soil Survey on whose work this paper is largely based. Our thanks are also due to Dr. J. F. Collins, University College, Dublin, and to Professor G. F. Mitchell, Trinity College, Dublin, for comments on earlier drafts of the paper, and to Mr. J. Lynch, Johnstown Castle Research Centre, for the maps and diagrams.

References

Anderson, H. A., Berrow, M. L., Farmer, V. C., Hepburn, A., Russell, J. D., and Walter, A. D. (1982). *J. Soil Sci.* **33,** 125–136.

Anonymous. (1853). "Instructions to Valuators and Surveyors appointed under the Act 15 and 16 Vict. Cap. 63, for the uniform Valuation of Lands and Tenements in Ireland." Alex Thom, Dublin.

Barry, T. A., Carey, M. L., and Hammond, R. F. (1973). "A survey of cut-over peats and underlying mineral soils." Bord na Mona, Cnoc Dioliun Group, An Foras Talúntais/Bord na Mona, Dublin.

Boulton, G. S., and Dent, D. L. (1974). *Geogr. Annaler* **56A,** 121–133.

Brogan, J. C. (1966). *Ir. J. Agric. Res.* **5,** 169–176.

Brogan, J. C., Kelly, O. D., and O'Keeffe, M. F. (1981). "The Lime Status of Soils of the Republic of Ireland." An Foras Talúntais, Dublin.

Bunting, B. T. (1965). "The Geography of Soil." Hutchinson, London.

Buol, S. W., Hole, F. D., and McCracken, R. J. (1973). "Soil Genesis and Classification." The Iowa State Univ. Press, Ames.

Case, H. J., Dimbleby, G. W., Mitchell, G. F., Morrison, M. E. S., and Proudfoot, V. B. (1969). *J. R. Soc. Antiq. Irel.* **99,** 39–53.

Colhoun, E. A., and Mitchell, G. F. (1971). *Proc. R. Ir. Acad., Sect. B* **71,** 211–244.

Collins, J. F. (1976). *Geoderma* **15,** 143–156.

Collins, J. F. (1981). *Agric. Rec. (1980),* 3–8.

Collins, J. F., and Coyle, E. (1980). *J. Soil Sci.* **31,** 547–558.

Collins, J. F., and O'Dubhain, T. (1980). *Geoderma* **24,** 215–224.

Collins, J. F., Freyne, D. F., and Conry, M. J. (1975). *Proc. R. Ir. Acad., Sect. B* **75,** 585–597.

Conry, M. J. (1971). *J. Soil Sci.* **22,** 401–416.

Conry, M. J. (1972a). *Geoderma* **8,** 139–146.

Conry, M. J. (1972b). *Sci. Proc. R. Dublin Soc.* **3,** 137–157.

Conry, M. J., and Ryan, P. (1963). *In* "Physical Resources of the Area, Section A, West Cork Resource Survey," pp. 1–68. An Foras Talúntais, Dublin.

Conry, M. J. and Ryan, P. (1967). "Soils of County Carlow." An Foras Talúntais, Dublin.

Conry, M. J., Hammond, R. F., and O'Shea, T. (1970). "Soils of Co. Kildare." An Foras Talúntais, Dublin.

Conry, M. J., de Coninck, F., Bouma, J., Cammaerts, C., and Diamond, J. J. (1972). *Proc. R. Ir. Acad., Sect. B* **72,** 359–402.

Craig, A. J. (1978). *J. Ecol.* **66,** 297–324.

Cruickshank, J. G. (1970). *In* "Irish Geographical Studies in Honour of E. E. Evans" (N. Stephens and R. E. Glasscock, eds.), pp. 89–104. The Queen's Univ. of Belfast, Belfast.

Cruickshank, J. G., and Cruickshank, M. M. (1981). *Oikos* **36,** 238–253.

Culleton, E. B. (1976). "Pleistocene Deposits in South Wexford and Their Classification as Soil Parent Materials." Unpublished Ph.D. thesis, Trinity College (Dublin Univ.), Dublin.

Culleton, E. B. (1977). *Ir. J. Agric. Res.* **16,** 171–185.

Culleton, E. B., and Mitchell, G. F. (1976) *J. R. Soc. Antiq. Irel.* **106,** 120–130.

Curtin, D., and Smillie, G. W. (1981). *Clays Clay Minerals* **29,** 277–284.

Diamond, J. J., Ryan, M., and Gardiner, M. J. (1965). "Soils of Athenry Agricultural College, Co. Galway." An Foras Talúntais, Dublin.

Doherty, H. B., and Logue, J. J. (1977). *Climatol. Note* No. 5, Meteorological Service, Dublin.

Drew, D. P. (1983). *J. Hydrol. (Amsterdam)* **61,** 113–124.

Finch, T. F., and Gardiner, M. J. (1977). "Soils of Co. Westmeath." An Foras Talúntais, Dublin.

Finch, T. F., and Ryan, P. (1966). "Soils of County Limerick." An Foras Talúntais, Dublin.

Finch, T. F., Culleton, E. B., and Diamond, J. J. (1971). "Soils of Co. Clare." An Foras Talúntais, Dublin.

Finch, T. F., Gardiner, M. J., Comey, A., and Radford, T. (1983). "Soils of County Meath." An Foras Talúntais, Dublin.

Gallagher, P. H. (1942). *Proc. R. Ir. Acad., Sect. B* **48,** 213–229.

Gallagher, P. H., and Walsh, T. (1942). *Proc. R. Ir. Acad., Sect. B* **47,** 205–249.

Gardiner, M. J., and Culleton, E. B. (1978). *J. Earth Sci. (Dublin)* **1,** 143–150.

Gardiner, M. J., and O'Callaghan, J. P. (1967). *Sci. Proc. R. Dublin Soc.* **2B,** 87–97.

Gardiner, M. J., and Radford, T. (1980). "Soil Associations of Ireland and Their Land Use Potential." An Foras Talúntais, Dublin.

Gardiner, M. J., and Ryan, P. (1964). "Soils of Co. Wexford." An Foras Talúntais, Dublin.

Hammond, R. F. (1968). *Proc. Int. Peat Congr., 3rd,* 109–115.

Hammond, R. F. (1979). "The Peatlands of Ireland." An Foras Talúntais, Dublin.

Jenny, H. (1941). "Factors of Soil Formation." McGraw-Hill, New York.

Kilroe, J. R. (1907). "Soil-Geology of Ireland." H. M. Stationery Office, Dublin.

Larney, F., and Collins, J. F. (1981). *Ir. J. Agric. Res.* **20,** 81–99.

Lee, J., Finch, T. F., and Ryan, P. (1964). *Ir. J. Agric. Res.* **3,** 175–187.

Lee, J., and Ryan, M. (1965). *Ir. J. Agric. Res.* **4,** 1012.

Lucas, A. (1969). *In* "Studies in Folk Life" (J. J. Jenkins, ed.), pp. 183–203. W. Cloves and Sons Ltd., London.

Lynch, A. (1981). "Man and Environment in South-west Ireland, 4000 BC–AD 800, A Study of Man's Impact on the Development of Soil and Vegetation." BAR Brit. Ser. No. 185, Oxford.

MacNiocaill, G. (1971). *Eriu* **22,** 80–86.

McAleese, D. M., and McConaghy, S. (1957–1958). *J. Soil Sci.* **8,** 127–140; *ibid.* **9,** 66–68, 289–297.

McCabe, A. M. (1973). *Proc. R. Ir. Acad., Sect. B* **73,** 355–382.

McCabe, F., and Collins, J. F. (1977). *Ir. Geogr.* **10,** 19–27.

McCabe, F., Collins, J. F., and McAleese, D. M. (1977). *Sci. Proc. R. Dublin Soc.* **5,** 491–502.

Mitchell, G. F. (1951). *Proc. R. Ir. Acad., Sect. B* **53,** 113–206.

Mitchell, G. F. (1965). *Spec. Pap. Geol. Soc. Am.* **84,** 1–16.

Mitchell, G. F. (1976). "The Irish Landscape." Collins, London.

O'Connell, M. (1980). *New Phytol.* **85,** 301–319.

O'Dubhain, T., and Collins, J. F. (1981). *Pedologie* **31,** 81–98.

O'Hare, P. J. (1972). *In* "Research Papers in Meteorology," pp. 126–133. Aberystwyth Symposium.

Ovington, J. D. (1954). *Forestry* **27,** 41–53.

Paton, T. R. (1978). "The Formation of Soil Material." George Allen and Unwin, London.

Pilcher, J. R., and Smith, A. G. (1979). *Philos. Trans. R. Soc. London, Ser. B* **286,** 345–369.

Rohan, P. K. (1975). "The Climate of Ireland." The Stationery Office, Dublin.

Ryan, P., and Walsh, T. (1966). *Proc. R. Ir. Acad., Sect. B* **64,** 465–507.

Simington, R. C., and Wheeler, T. S. (1945). *Studies* **34,** 539–551.

Simonson, R. W. (1959). *Soil Sci. Soc. Am. Proc.* **23,** 152–156.

Smith, A. G. (1970). *In* "Irish Geographical Studies in Honour of E. E. Evans" (N. Stephens and R. E. Glasscock, eds.), pp. 65–88. The Queen's Univ. of Belfast, Belfast.

Smith, A. G. (1975). *In* "The Effect of Man on the Landscape: The Highland Zone" (J. G. Evans, S. Limbrey, and H. Cleere, eds.), pp. 64–74. CBA Res. Rep., No. 11, London.

Soil Taxonomy. (1975). Agric. Handbook No. 436, Soil Conservation Service. U.S. Department of Agriculture, Washington, D.C.

Tamm, O. (1920). *Medd. Statens Skogsforskningsinst. (Swed.)* **17,** 49–300.

Walsh, M. (1973). "Soils of Co. Leitrim." An Foras Talúntais, Dublin.

Walsh, T., McDonnell, P. M., and Ryan, P. (1957). *Agrochimica* **1,** 350–364.

Walsh, M., Ryan, M., and van der Schaaf, S. (1969). *In* "Soils and Other Physical Resources. Part 1. West Donegal Resource Survey," pp. 1–79. An Foras Talúntais, Dublin.

CHAPTER **8**

QUATERNARY VEGETATION CYCLES

William A. Watts
Trinity College
Dublin, Ireland

INTRODUCTION

Research into Ireland's Quaternary vegetation history began in the 1930s when Professor Knud Jessen of the University of Copenhagen was invited to develop a programme of pollen-analytical research into Irish bogs and lake deposits by the Royal Irish Academy's Committee for Quaternary Research in Ireland. The Committee included such distinguished scholars as Robert Lloyd Praeger and Anthony Farrington. Jessen's work resulted in the first identification of Irish lateglacial deposits at Ballybetagh, Co. Dublin (Jessen and Farrington, 1938), and, after the interruption of the war years, he produced a classic publication on Ireland's lateglacial and postglacial vegetation history (Jessen, 1949). He subsequently identified as interglacial a remarkable deposit rich in macrofossils at Gort, Co. Galway (Jessen et al., 1959). Jessen's pioneering work, and that of his pupil G. F. Mitchell, laid the foundations for Quaternary botany in Ireland. By present-day standards Jessen's pollen studies were technically primitive, but his profound interest in and knowledge of plant macrofossils has exercised a long-term influence on the subject. His stratigraphic studies of sediment profiles are also exemplary. More recently, emphasis has moved to much more detailed and exacting pollen analysis, and from bog to lake sediments. The development of radiocarbon dating has provided a most important tool for exact correlation. Examples of these developments are presented in this chapter.

Strong climatic fluctuation on relatively short time scales is one of the most characteristic features of the later Quaternary. In studies of plant fossils, climatic inferences may be made from the known collective tolerances of the assemblage of species and their geographic ranges. Climatic inference is particularly important in the lateglacial period which is very rich in short-term events, but study of interglacial deposits has led also to the identification of cycles of climatic change (Iversen, 1958) with a duration of some 10,000 years. Recurrent "warm" interglacial periods are characteristic of the Quaternary (Shackleton and Opdyke, 1976). The Tertiary also displays climatic change but on much longer time scales and with less extreme expression.

TERTIARY AND INTERGLACIAL FLORAS

Tertiary and Quaternary floras older than 13,000 years are rather poorly known because of a lack of reported deposits. The older Tertiary has been reviewed (Boulter, 1980; Wilkinson et al., 1980), but the Neogene is known from only one site, Hollymount, in the valley of the River Barrow near Carlow (Hayes, 1978; Boulter, 1980). The thick sediments (20 m) at Hollymount occupy a sinkhole in Carboniferous limestone that is sealed by till. It is likely that other such sites remain to be discovered. Boulter (1980) used form-genus names only, but identified the flora as having a strong Taxodiaceous swamp/woodland component. Hayes (1978) identified pollen referable to *Pinus* (pine), *Quercus* (oak), *Corylus* and *Myrica* (hazel and bogmyrtle), and Ericaceae (heaths) as the predominant elements in the Hollymount pollen flora. Pollen of *Taxodium* type (swamp cypress), *Symplocos* (sweetleaf—mainly tropical today), *Tsuga* (hemlock), *Sciadopitys* (Japanese umbrella pine), *Liquidambar* (sweet gum), and Palmae type (palms) also occurs. The flora resembles Miocene or earliest Pliocene assemblages from northwest Europe described by van der Hammen et al. (1971). The richly diverse but essentially modern tree flora with representation of swamp species suggests a warm temperate to subtropical climate.

Elsewhere, in Britain (West, 1980) and in the Netherlands (van der Hammen et al., 1971), the floristic transition by local extinction from the Tertiary to the Quaternary can be followed, but Ireland lacks sites to document either the end of the Tertiary or the earlier interglacial stages. It is likely that many sites remain to be found during construction or drilling operations, but the extensive glaciation of Ireland during the later Quaternary may have destroyed many deposits by ice action or buried them under till so that they are less abundant than in East Anglia, for example (West, 1980), where glaciation was not so widespread and severe.

Interglacial phases are periods of time during the Quaternary in which the climate and biotic communities became comparable to those of today in the same region, although the biota often differed greatly in detail of both the species present and the order in which they immigrated after glaciation. The glacial periods, which alternated with the interglacials, are characterized by generally colder climates than today. They are very diverse in events, varying in northwest Europe from the presence of ice sheets, to ice-free but very cold climates, and to warmer "interstadial" climates with diverse plants and animals, but still never as warm as today. The number of interglacial periods is uncertain and is probably underestimated from land records. West (1980) records seven interglacial phases from East Anglia, but more have been documented from marine cores (Shackleton and Opdyke, 1976).

Interglacial phases display a cyclic development of vegetation, analyzed in principle by Iversen (1958). In northwest Europe, according to Iversen's scheme, the end of a glaciation is followed by invasion of pioneer vegetation of herbs and shrubs on poorly developed soil (protocratic stage). This is followed by broadleaved forest and the development of brown forest soils (mesocratic stage). Then, in a more oceanic climate related to rising sea levels and a transgressive sea, forest is diversified, and bogs and heaths develop as soils become podzolized (telocratic stage). At the end of the cycle, trees, shrubs, and herbs tolerant of colder climates replace more demanding species (cryocratic stage), and a new glaciation begins with a transition in which frost phenomena disturb the soil surface until ice sheets form. The erosion caused by glacial ice tends to destroy interglacial peats and lake deposits completely, but where these are preserved by chance in protected localities, their tops are often truncated.

In Ireland only six sites are known to preserve a large portion of an interglacial cycle (Fig. 1). Of these Gort, Co. Galway, Burren, Co. Mayo (not the well-known limestone region of Co. Clare), and Benburb, Co. Tyrone, are exposed in stream sections. Kilbeg, Co. Waterford, and Kildromin and Baggotstown, both in Co. Limerick, are known only from well excavations and boreholes. Other reported sites, for instance Fenit, Co. Kerry (Mitchell, 1970; Warren, 1981), are less complete.

All of the Irish sites have features which suggest that they are of the same age, although there is some variety in detail. The common features are seen in a pollen diagram from Baggotstown (Fig. 2; Watts, 1964). Strictly, an orthodox pollen diagram is not available from Baggotstown. The diagram is a reconstruction based on a series of pollen samples taken from very diverse types of sediment available from the spoil heaps of a well dug at the site, and on descriptions of the well diggers. The spectra are organized in what appears to be a logical succession closely similar to a pollen diagram from a core from the nearby site of Kildromin (Fig. 1; Watts, 1967) which

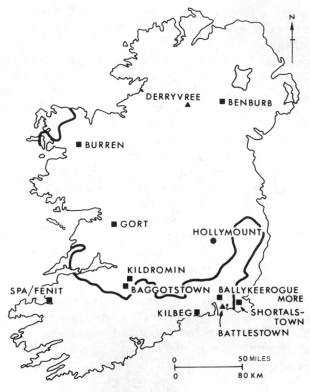

Fig. 1.   Location map for Tertiary (●), interglacial (■), and interstadial (▲) sites referred to in the text. The dark line marks the position of the Southern Irish End-Moraine of the last glaciation.

lacks the pioneer stage present at Baggotstown and is not so rich in macro-fossils.

The Baggotstown sequence can be divided into stages to which Iversen's system can be applied:

1. Pioneer stage (protocratic).
    a. Creeping willow (*Salix repens*) with grass and herbs.
    b. Shrub flora of juniper (*Juniperus communis*) and sea buckthorn (*Hippophae rhamnoides*).
    c. Woodland or forest of tree birch (*Betula* spp.).
2. High forest stage (mesocratic). Forest of pine (*Pinus sylvestris*), birch, and oak (*Quercus* spp.) with abundant ivy (*Hedera helix*) at first; other trees rare.
3. Atlantic-climate forest stage (telocratic). Forest with fir (*Abies alba*), spruce (*Picea abies*), alder (*Alnus glutinosa*), yew (*Taxus baccata*), and

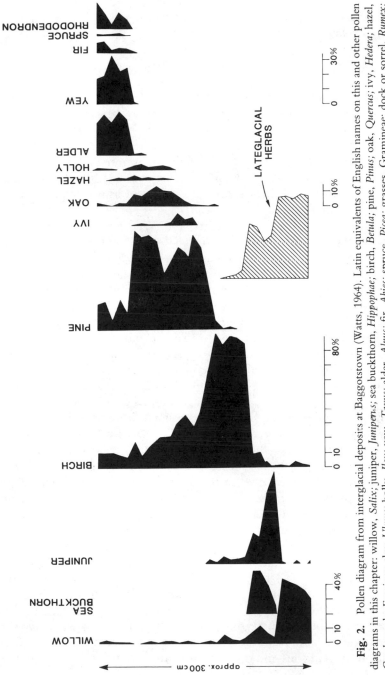

Fig. 2. Pollen diagram from interglacial deposits at Baggotstown (Watts, 1964). Latin equivalents of English names on this and other pollen diagrams in this chapter: willow, *Salix*; juniper, *Juniperus*; sea buckthorn, *Hippophaë*; birch, *Betula*; pine, *Pinus*; oak, *Quercus*; ivy, *Hedera*; hazel, *Corylus*; ash, *Fraxinus*; elm, *Ulmus*; holly, *Ilex*; yew, *Taxus*; alder, *Alnus*; spruce, *Picea*; fir, *Abies*; grasses, Gramineae; dock or sorrel, *Rumex*; sedge, Cyperaceae; bracken, *Pteridium*; plantain, *Plantago lanceolata*; cereals, Cerealia. (Redrawn from original data.)

heaths or bogs with rhododendron (*Rhododendron ponticum*) and heaths (*Calluna, Erica, Bruckenthalia, Daboecia*).

4. Stage with tolerant northern trees and herbs (cryocratic). Not preserved at Baggotstown or any other Irish site.

The cycle differs from the lateglacial and postglacial in (1) the abundance of *Hippophae* in the protocratic stage, (2) the very low diversity of broadleaved trees in the mesocratic stage [the relatively low percentages of *Quercus* and the scarcity of elm (*Ulmus*) and hazel (*Corylus*) are in striking contrast to the postglacial where hazel is much the most common pollen type on most soils], and (3) *Taxus* is not so conspicuously abundant in the postglacial, while *Picea, Abies,* and *Rhododendron* do not occur as native plants after the last glaciation.

The sequences at Kildromin (Watts, 1967) and Kilbeg (Watts, 1959) are similar to those at Baggotstown. Gort (Jessen *et al.*, 1959) differs in the apparent absence of an early tree birch phase and in the very high percentages of yew, of which there are two peaks. Gort, like Baggotstown, has a well-developed protocratic flora with *Hippophae,* and a remarkable frequency of *Buxus* (box), also absent from the modern flora of Ireland, in the telocratic phase. Of the remaining sites, Benburb (Boulter and Mitchell, 1977) has a fir–spruce–yew flora, so that the exposed section represents a telocratic phase like those at the other sites. The section, which is exposed in a stream cutting, has not been bored to the base. At Burren, part of an interglacial cycle, including both mesocratic and telocratic phases, is exposed in a 2-m-high section. Interglacial deposits occur in at least two other localities in neighbouring stream sections but are evidently disturbed by overlying till. At Ballykeerogue More, Co. Wexford (W. A. Watts, unpublished observations), the situation is even more extreme, for the sediments are grossly distorted and involuted so that there is no reliable sequence. This suggests caution in the interpretation of interglacial deposits from boreholes, where distortion caused by ice pressure may bring about transport of blocks of sediment with possible duplication, upending of sediments, and other undetectable problems in interpretation.

The similarity of floristic development between the known sites suggests that they all belong to one interglacial stage, named the Gortian after the first site investigated (Jessen *et al.*, 1959). The interglacial is remarkable because of the very rich flora revealed by detailed pollen and macrofossil studies in the telocratic stage. There appears to have been extensive development of rhododendron heath. Rhododendron is not now a native shrub, but it is widely naturalized in western Ireland where it is very invasive of woodland on acid soils and on peaty soils that are not permanently waterlogged. Today, its natural occurrence is very local in Portugal. It is abundant on the Black Sea coasts of Bulgaria and, especially, Turkey (Godwin, 1975). It is known as a fossil in interglacial sites from western continental Europe, but not from Britain. In addition, the heaths *Erica mackaiana*, *Erica* cf. *ciliaris*, *Daboecia cantabrica*, and *Bruckenthalia spiculifolia* each occur at several of the sites. The first two are confined to the Roundstone area of Co. Galway now, and *Daboecia* occurs more widely in the same region. All have their main distribution today from southwest France to Portugal but are found especially in the Cantabrian Mountains of Spain where they occur in heathy woodland. *Bruckenthalia* is now confined to southeast Europe. Its ecology is similar to that of *Calluna vulgaris* (common heather). Reference of Gortian fossils to *Erica scoparia* var. *macrosperma* (Jessen *et al.*, 1959) appears to be a misidentification of *Bruckenthalia* (Phillips, 1976). Spores of Hymenophyllaceae (filmy ferns) are also recorded (Jessen *et al.*, 1959). They are epiphytic on tree trunks and branches, on boulders in oceanic areas, and, in an extreme case (*Trichomanes*, the Killarney fern), in the spray zone of waterfalls. The family provides an indication of acid soils and high rainfall. The telocratic flora of the Gortian permits an unambiguous climatic interpretation. The assemblage of species suggests an oceanic high-rainfall climate, similar to that of southwest Ireland today, affecting all or most of the island.

Among the wealth of Gortian aquatics, *Azolla filiculoides* (water fern), a species confined to western North America today, is known from Kilbeg, Kildromin, and Baggotstown. It is known from interglacial deposits older than the last (Eemian) interglacial in Europe but not from the Eemian itself, although many deposits have been investigated in northwest Europe. *Brasenia schreberi* (purple waterlily) is absent from Europe now, but its pollen and seeds are widespread in European interglacial deposits, Eemian and pre-Eemian. It is recorded from Gort, as are seeds of *Nymphoides* cf. *cordata* (Watts, 1971), a North American species first recorded as new to science and tentatively named *Menyanthes microsperma* by Jessen *et al.* (1959). *Eriocaulon septangulare*, a common species of eastern North America which also occurs in western Ireland and, very locally, in Scotland, is found fossil at Gort and at Kilbeg. The finds prove the geological antiquity of the species in Europe.

The possibility that extinct species occur in the Gortian should not be dismissed, although most Quaternary plant candidates for this distinction have proved to be living species now occurring in geographically remote regions. An unknown pollen grain (illustrated and discussed by Phillips, 1976) occurs with some frequency at several Gortian sites. Phillips suggests that it may belong to the Oleaceae (olive family), within which it is similar to the genus *Phillyrea,* an evergreen shrub of Mediterranean maquis. A new investigation of the Gort site would be timely, because its great wealth in macrofossils may throw light on this and other problems. One of these is the record of *Fagus* (beech) at Gort, a single cupule unaccompanied by pollen, which may be a contaminant, for beech plantations, in spite of Jessen's statement to the contrary, occur close to the site.

There remains the question of the age of the Gortian deposits. On many grounds they appear to be of the same age as the Hoxnian (penultimate interglacial) of Britain (West, 1980). Common features of the Gortian and Hoxnian are (1) the high percentages of *Hippophae* in the protocratic stage, (2) the presence of *Azolla,* known from Hoxnian but not from Eemian deposits in Europe (West, 1980), and (3) the abundance of *Abies* in the telocratic phase. In contrast, Ipswichian (Eemian, last interglacial) deposits in Britain have a telocratic phase dominated by *Carpinus* (hornbeam) from which *Abies* is absent (West, 1980). Abundant *Hippophae* and presence of *Azolla* are unknown in the Ipswichian. It should be conceded that the Irish sequences differ from the British in the presence of *Rhododendron* and in the relatively small role played by broadleaved trees in the mesocratic stage. It is assumed that the differences between Ireland and Britain are explained by a broad region of distinctively oceanic woodland in which much of Ireland's vegetation lay and contrasted with that of Britain, just as the woodland of the Killarney area contrasts with that of eastern Ireland today.

Warren (1979) suggests that stratigraphic considerations must place the Gortian in the "last interglacial." In favour of this argument is the absence otherwise of last interglacial deposits which should, in principle, be much easier to find than deposits of the penultimate interglacial because they would have been less subject to glacial erosion. In unglaciated areas, they should lie close to the surface, covered, as in Denmark (Jessen and Milthers, 1928) by solifluction deposits only. It can also be observed, with justice, that several sites (Baggotstown, Kildromin, Burren, Benburb) lie close to the surface in areas undisputably glaciated during the most recent glaciation. Kilbeg, however, which is deeply buried by till, lies well outside the prominent Southern Irish End-Moraine of the last glaciation (Fig. 1; Mitchell, 1976; Charlesworth, 1928). The glacial deposits south of the End-Moraine clearly represent an older glacial episode because of the evidence they preserve of widespread solifluction and of the occurrence of ice-wedge casts and fossil pingos (Mitchell, 1973, 1976) which are largely absent north of the End-Moraine (see further discussion in Warren, 1979, Warren, Chap. 3, and Lewis, Chap. 5, this volume).

The view adopted here is that the "objective stratigraphic approach," which Warren (1979, p. 329) claims must place the Gortian within the last interglacial, is not sufficiently well founded to upset the correlation of Gortian with Hoxnian which the floristic evidence clearly supports. In the present very imperfect state of knowledge of Ireland's Quaternary, it may not be possible to arrive at secure conclusions about some of the questions raised. It seems better to indicate the present "state of play" and to leave advancement of the argument to further research or new discoveries. The main matters requiring consideration appear to be the following:

1. Southern Ireland south of the Southern Irish End-Moraine (Fig. 1) has landscape forms (ice-wedge casts, solifluction phenomena, and fossil pingos) which are infrequent or unknown in the "younger" landscape further north. This suggests that the tills south of the End-Moraine are of greater age than Weichselian. According to Warren, they should be placed early in the last glaciation. More traditional views (Mitchell, 1976) place them in the second-to-last major glaciation, separated from the last glaciation by a period of interglacial climate from which no certain deposits have survived.

2. The "Gortian" deposits are probably all of the same age because of the similarity of their floras. The possibility that deposits of more than one age are represented cannot be dismissed but appears unlikely. The cycle of floristic development is much more like the Hoxnian than the last interglacial (Ipswichian, Eemian) of Britain. The presence of abundant *Abies* is especially critical in the comparison. If it were shown that the Hoxnian itself is incorrectly dated (see Bristow and Cox, 1973), then the age assigned to the Gortian would also be incorrect. However, this possibility has not received serious consideration (West, 1980).

3. It is undoubtedly puzzling that several Gortian deposits lie close to the surface in areas glaciated during the last glaciation. The Kilbeg deposit, however, lies well to the south of the End-Moraine and is buried by till. If all the deposits are of the same age, then all lie stratigraphically below tills formed by glaciations older than that which formed the End-Moraine.

4. The tills at the surface to the south of the End-Moraine may be of older Weichselian or of pre-Eemian age. Objective choice between these two requires conclusive field evidence. Clearly, the discovery of new interglacial deposits containing a previously unknown vegetation cycle would help to resolve the issue. Finds of a *Carpinus*-dominated flora without *Abies* would be evidence for a classic Eemian cycle in Ireland and would confirm the correlation of Gortian with Hoxnian.

In summary, we need both more field research on the glacial stratigraphy and more study of interglacial deposits, especially in exposures where the relationship to the glacial stratigraphy can be recorded satisfactorily. Without more information, it is doubtful that the argument can advance to a satisfactory conclusion.

INTERSTADIAL FLORAS

Few interstadial deposits are known in Ireland, and their stratigraphic position is usually uncertain. The most significant deposit is recorded from Derryvree, Co. Fermanagh (Fig. 1; Colhoun et al., 1972), where a road cut exposed organic silts and fine sands under a drumlin. They were dated to 30,500 $\pm^{1170}_{1030}$ radiocarbon years bp and contained a rich flora of seeds and mosses in addition to molluscs (Pisidium nitidum) and diverse beetles. Pollen was largely of grasses and sedges. The ecological interpretation is open tundra in a periglacial climate. Derryvree is important because it provides unambiguous evidence for a relatively young nonglacial interval which preceded the glaciation that deposited drumlins widely over north-central Ireland.

In southern Ireland, enclosed basins occur widely outside the Southern Irish End-Moraine. None has yet been found to contain interglacial deposits, but at Battlestown, Co. Wexford (Craig, 1973), a species-poor flora and fauna occurs at the base of a 3-m silt profile in such a basin. The flora suggests open and perhaps unstable vegetation in a severe and limiting environment. The flora is undated because of the low organic content of the sediments, but it could belong in any cold phase of the later Quaternary. It might precede the advance of ice to the End-Moraine, and the overlying silts could be contemporaneous with the maximum extension of ice, but this can be no more than speculation.

At Kilbeg, what appears to be an interstadial deposit with an arctic flora overlies the interglacial deposit unconformably. If Kilbeg represents the penultimate interglacial, the deposit lies early in the glacial period between Gortian and Eemian (Munsterian Glaciation sensu Mitchell, 1976).

LATEGLACIAL VEGETATION AND CLIMATE
(13,000–10,000 bp)

Introduction

The first investigation of lateglacial sediments in the British Isles was carried out by Jessen and Farrington (1938) at Ballybetagh, Co. Dublin. They showed that organic lake mud which contained remains of giant Irish deer (Megaloceros giganteus) was sealed by stony inorganic sediments believed to be formed by solifluction. The interpretation was that a period of relatively temperate climate was brought to an end by a phase of arctic cold. Comparison was made with the lateglacial stratigraphy of Denmark at the classic Allerød site. Subsequently, Jessen (1949) developed a general scheme of lateglacial events in Ireland (Table I) and correlated them with events in Denmark.

Table I Jessen's Scheme for the Lateglacial in Ireland

Ireland	Characteristics	Denmark
Younger *Salix herbacea* period (Zone III)	Tundralike vegetation with sub-arctic heaths, solifluction	Younger *Dryas* period
Lateglacial birch period (Zone II)	Birch copses, open herb-rich vegetation	Allerød period
Older *Salix herbacea* period (Zone I)	Tundralike vegetation, solifluction	Older *Dryas* period

Jessen's simple tripartite scheme with a "warm" Allerød interstadial sandwiched between two cold phases with inorganic sediments was generally accepted until relatively recently when more detailed pollen analysis revealed that the lateglacial was more complex than had been realized. In particular, Jessen's failure to identify juniper, an abundant and important component of the vegetation during part of the lateglacial, was a serious weakness. His relatively primitive pollen analysis, however, was balanced by the meticulous attention he gave to the determination of plant macrofossils which proved to be a rich and invaluable source of information for ecological and phytogeographical interpretation. Jessen's tradition was developed by Mitchell (1953, 1954) who studied numerous lateglacial sites and greatly extended the range of plant macrofossil determinations. He also demonstrated (Mitchell, 1941, 1976; Mitchell and Parkes, 1949) that giant deer and reindeer (*Rangifer tarandus*) were characteristic animals of the Allerød interstadial, inhabiting mainly regions with limestone soils. They seem to have become extinct at the transition from the Allerød to the Younger *Salix herbacea* period.

With the application of more detailed pollen analysis, it became clear that Jessen's scheme was no longer adequate. Changes in the vegetation cover did not necessarily correlate exactly with changes in sediment type (Watts, 1963), and there were too many distinct events in the vegetation history to be fitted into three units (Singh, 1970; Smith, 1970). More recently, concern with terminology and classification (Mangerud *et al.*, 1974; Mitchell *et al.*, 1973) has made analysis and description of the period from 13,000 to 10,000 bp more complex, because it would require the use of chronozones with boundaries defined in conventional radiocarbon years (Mangerud *et al.*, 1974) or of local nomenclature based on reference sections (Mitchell *et al.*, 1973). Watson and Wright (1980, p. 153) argue that formal chronostratigraphic units "are unrealistic for application to late-Quaternary history, which involves events that are demonstrably time-transgressive owing to measurable lags between the causative force and the geologic response." This view is adopted here. Each site is treated independently, and the use of

Table II Nomenclature for Subdivisions of the Lateglacial

Mangerud et al. (1974)		Mitchell et al. (1973)		Present paper	
Flandrian	10,000 bp–present	Littletonian	Postglacial	10,000 bp–present	
Late Weichselian				*Artemisia* Phase	
Younger *Dryas*	11,000–10,000	Nahanagan Stadial		(Younger *Dryas*)	10,900–10,000
Allerød	11,800–11,000	Woodgrange Interstadial		Grass Phase	11,800–10,900
Older *Dryas*	12,000–11,800	Late Devensian — Lateglacial		Erosion Phase	~12,000–11,800
Bølling	13,000–12,000			First Juniper Phase	12,400–12,000
?	?–13,000			*Rumex* Phase	13,000–12,400
			Last glaciation		?–13,000

a general scheme of pollen zones is abandoned. Correlation between sites is made by radiocarbon dates. The validity of this approach has come to be recognized in the most recent literature (Birks, 1982; Mangerud *et al.,* 1982). This acknowledges the differences that exist between sites in the details of the flora present at any one time, and in the timing of change from one type of vegetation to another. A rather loose narrative of events within a time framework provided by radiocarbon dating appears to be a better way to represent what really happened than a rigid structure of zones. Simplicity of nomenclature, widely understood, for ease of international communication is also preferred to detailed local nomenclature (Table II).

The Sequence of Vegetational Events

It is certain that the climate and vegetation of the phases of the lateglacial were complex, with much variation between sites, depending on soil type and elevation. The subject has most recently been reviewed by Watts (1977, 1980), to which the reader is referred for more extensive discussion.

Fig. 3.   Location of lateglacial and postglacial sites referred to in the text.

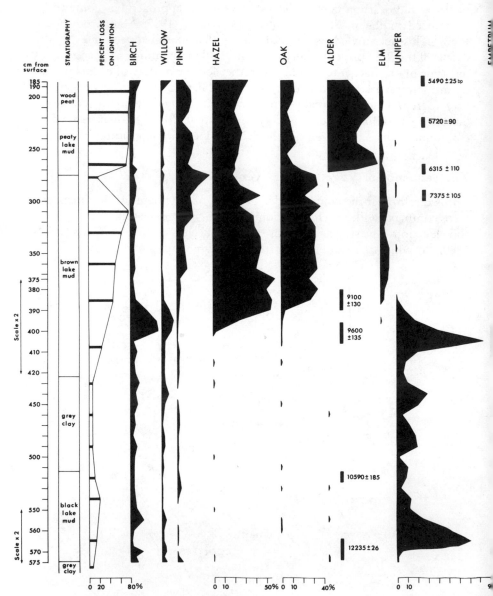

Fig. 4.   Pollen diagram of the lateglacial and early postglac

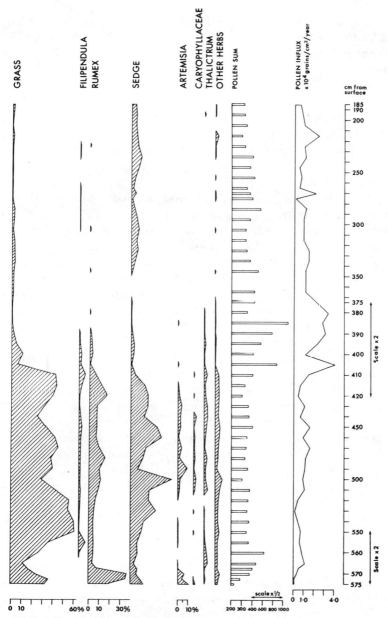

at Belle Lake, Co. Waterford (Drafted from data in Craig, 1973.)

Pollen diagrams from Woodgrange (Singh, 1970), Roddans Port (Morrison and Stephens, 1965), Ballybetagh and Dunshaughlin (Watts, 1977), and Coolteen and Belle Lake (Craig, 1973, 1978) in eastern Ireland have many common characteristics, and a common sequence of events may be discerned (Watts, 1977, p. 276). Initially pollen is sparse, grass is the most frequent type, and secondary pollen of trees is a component of the total pollen count. This phase, best represented at Woodgrange, records the very first plant colonization after ice of the last glaciation had melted about 13,000 years ago. The vegetation, which is poorly known, probably consisted of very sparse, scattered herbs in a landscape in which bare soil still predominated.

In the next phase, recognizable throughout Ireland, pollen of *Rumex acetosa/acetosella* type (sorrel) was abundant together with *Salix herbacea* (dwarf willow), leaves of which are often found. This was a time of rapidly increasing biological productivity and closing of the vegetation cover. The type of vegetation is difficult to define. The abundance of *Rumex,* up to 50% of the pollen sum at some sites, is almost certainly due to its large pollen production rather than to any necessary predominance in the landscape. Probably there was a greatly increased cover of grasses, sedges, and other herbs. Dwarf willow and *Rumex* suggest late snow patch vegetation. This phase lasted from approximately 13,000 to about 12,400 bp.

The *Rumex* Phase is succeeded by a major peak in juniper pollen, well seen at Belle Lake (Fig. 4). In the east, juniper was very abundant, but this species does not reach the same percentage values in the west, where it was common nevertheless. In the west, *Empetrum* (crowberry), which expanded at about the same time as juniper or a little earlier, was also common. In the First Juniper Phase (Table II) there was probably very extensive juniper scrub in eastern Ireland, with juniper/crowberry scrub or heath in the west. At Coolteen (Craig, 1978) the influx of juniper pollen to the lake basin reached very high values, suggesting both free flowering and real abundance. The juniper peak ends quite abruptly, and, especially in the Dublin region, there is a small but distinct peak of sea buckthorn, a shrub of unstable gravelly and sandy soil. Sea buckthorn is no longer a native plant in Ireland, but fossil leaf hairs confirm that it was present at Dunshaughlin, Ballybetagh, and Roddans Port. At Belle Lake, juniper remained common through the rest of the lateglacial, but at other sites, such as Coolteen, it became rare after its first peak and disappeared completely in the *Artemisia* (Younger *Dryas*) Phase.

After the peak in sea buckthorn, a grass-dominated phase lasted from about 11,800 to 10,900 bp. This is Jessen's "Lateglacial Birch period." In fact birch was abundant only at the Long Range (Watts, 1963) where it was apparently a purely local occurrence. Elsewhere grass predominated, and the landscape must have been open, with a stable cover of grass and herbaceous communities and with local birch copses. *Helianthemum* (rock-rose) was frequent in the Grass Phase. The surprising failure of birch to form woodland has been discussed by Craig (1978). The presence of *Betula pubescens* (tree birch) is proved at several sites by macrofossils, but it was not abundant, either because of heavy browsing by giant deer, or, more probably, because it was close to its climatic limit.

The Grass Phase ends about 10,900 bp, with a peak of crucifer pollen at some sites, perhaps marking the breakup of a stable herb cover with increasing soil erosion. A major change of vegetation then followed to dwarf willow, *Artemisia,* Caryophyllaceae, and diverse herbs. Pollen is often sparse and difficult to count, and the minerogenic, often stony, sediments indicate soil instability. The flora is consistent with a diversity of arctic plants with low percentage cover. It may reasonably be called tundra vegetation. Jessen named this phase the Younger *Salix herbacea* period because of the characteristic abundance of dwarf willow leaves, but "Younger *Dryas* period" is preferred here because it is more widely understood, although *Dryas* leaves are much less frequent than dwarf willow in Ireland. After the end of the Younger *Dryas* period a rapid expansion of grass with *Rumex,* then a second peak of juniper, then abundant tree birch and willow usher in the postglacial.

There is considerable regional and local variation in lateglacial vegetation. The behaviour of juniper, already mentioned, is a striking example. Perhaps the most distinctive feature is the abundance of crowberry at Roundstone (Jessen, 1949) and Glenveagh (Telford, in Watts, 1977) which shows that heaths replaced other kinds of vegetation in infertile areas now characterized by blanket peats. *Betula nana* (dwarf birch, now extinct in Ireland) and *Dryas octopetala* occur frequently as macrofossils, but almost invariably in limestone regions. Thus, there is no record of dwarf birch at Ballybetagh, a nonlimestone site with an exceptionally rich macroflora (Jessen and Farrington, 1938; L. C. Cwynar, unpublished observations). *Saxifraga hypnoides* (mossy saxifrage) was abundant at Dunshaughlin, Co. Meath, but not at Ballybetagh, Co. Dublin where *Saxifraga oppositifolia* (purple saxifrage) leaves are extremely abundant in the Younger *Dryas* period (Watts, 1977). In brief, the vegetation showed as it does now, diversity in its response to soil type, elevation, and exposure, and many distinct plant associations must have occurred.

Lithostratigraphy and Climate

The lithostratigraphy of the lateglacial is also variable from site to site. At many localities the total thickness of sediment is less than 100 cm, but some very thick sequences are known. The deepest part of the Coolteen basin (Craig, 1978) has up to 800 cm of lateglacial sediment; Ballybetagh has 400 cm and Dunshaughlin, 300 cm. Such rapid accumulation is scarcely encountered elsewhere in northwest Europe (Pennington, 1977). It means that Ireland is particularly favoured for detailed stratigraphic study of the lateglacial. The oldest lateglacial sediments are usually fine silts and clays with no countable pollen, or pollen at a very low concentration. These are rock flour deposits of decaying ice sheets or erosional sediments washed into basins from unvegetated mineral soil surfaces. Sometimes thicknesses of several metres of such material are encountered.

Clays and silts are followed by progressively more organic deposits, often calcareous. The Coolteen profile (Fig. 5) shows black organic muds with up to 50% loss on ignition values early in the lateglacial, correspond-

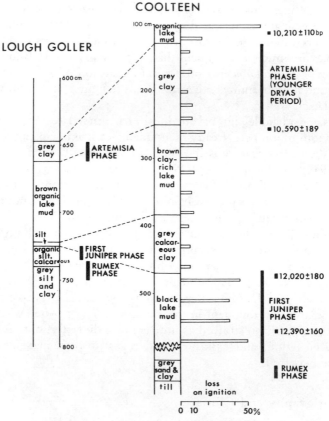

Fig. 5. Lateglacial sediment profiles from Coolteen (Craig, 1973) and Lough Goller (Watts, 1963) with loss-on-ignition curve from Coolteen.

ing with the first phase of juniper abundance. This may have been the time with the warmest climate in the lateglacial. The organic or carbonate content of sediments was high, and diverse aquatic plants such as *Ceratophyllum demersum* (hornwort) and several *Potamogeton* (pondweed) species flourished. The algal flora was unusually diverse at Roddans Port (Morrison and Stephens, 1965) and at Dunshaughlin (Watts, 1977). These and other lines of evidence, reviewed by Watts (1977) and Craig (1978), suggest that the First Juniper Phase had an exceptionally favourable climate, not equalled until the postglacial.

Coope (1977) has presented evidence from fossil beetles to show that a warm interstadial (the Windermere Interstadial) took place in northwest Britain in the closing stages of the last (Devensian) glaciation before 13,000 bp but later than 14,000 bp. The climate is considered to have been as warm in summer as it is today with winter temperatures a little lower than at present. A cooling took place about 12,200 bp, and a period of over 1,000 years (the Allerød oscillation) followed, during which summers were about 3°C cooler than during the thermal maximum. The dates proposed by Coope are somewhat older than those suggested here for Ireland, but the similarity of the analysis of the climatic sequence from both plant and beetle evidence is very striking and at variance with the accepted view from continental Europe (Iversen, 1954), which tends to equate the Bølling and Allerød climates and does not distinguish an early warm phase so clearly.

Until this point the vegetation sequence is that of an interglacial cycle (Iversen, 1958), and one would expect invasions of more demanding pioneering trees and shrubs, such as tree birches, tree willows, and aspen, to follow. However, a brief period of erosion between about 12,000 and 11,800 bp appears to have broken the cycle. At many sites, especially on the west coast (Glenveagh, Lough Goller, Gortlecka, and Muckross; W. A. Watts, unpublished data), a conspicuous layer of silt is found after an earlier period of juniper expansion. In the Dublin area, sea buckthorn, a plant of unstable soils, appeared briefly, and juniper itself fell to low values once more (Watts, 1977). At Coolteen (Fig. 5) high organic productivity and high juniper pollen concentrations ended simultaneously. Climatically, whatever change took place caused soil movement and a decline in juniper. The latter may have been caused by a lowering of both temperature and precipitation, for juniper is killed by wind in the subarctic unless protected by a snow cover (Iversen, 1954). The end of the First Juniper Phase therefore shows a serious climatic deterioration and erosion of the upland, marked by silt movement into lake basins. It is maintained that this event is similar in climatic character to the Younger *Dryas* event, although smaller in scale, and probably had the same causes.

The Grass Phase occupied about 1000 years. At Dunshaughlin the sediments became more inorganic and the early diversity of algal species was lost. At Coolteen (Fig. 5) the grey calcareous clay of the Erosion Phase gave way to brown clay-rich lake mud with a higher organic content, but still less than half that of the Juniper Phase. The received opinion is that the Grass Phase (in part, Jessen's Zone II or Allerød) was the warmest part of the lateglacial. This seems unlikely to be true. It is more probable that it was a relatively cold time with moderate erosion or deflation from herb-dominated vegetation that cannot have formed a complete cover. It may have been steppelike in character with trees or shrubs confined to protected places. Thus the environment for giant deer and reindeer may have been tundralike grassland, well suited to cold-adapted grazing and browsing mammals.

At the end of the Grass Phase severe erosion began once more, marking the beginning of the Younger *Dryas* period. The sediments are predominantly inorganic and may contain stones. Leaves of dwarf willow are often abundant. This is consistent with solifluction and late snow beds. The distinctive pollen flora and often sparse pollen grains suggest an incomplete vegetation cover of arctic species. Mitchell (1973) has suggested that pingos, which are abundant in southern Ireland outside the Southern Irish End-Moraine, were active during the Younger *Dryas*. Investigations up to the present support that view, but there are still too few site studies to confirm it strongly. At Lough Nahanagan, a mountain lake south of Dublin (Colhoun and Synge, 1980; Watts, 1977), there is clear evidence that a glacier formed in the Younger *Dryas* at an elevation of 450 m and moved from the back wall of the cirque to form a substantial moraine. This incorporates blocks of organic sediment dated to 11,600 ± 260 bp with an early lateglacial pollen assemblage. The depression of the snowline to permit a new glacial advance at Lough Nahanagan represents a considerable lowering, calculated at 7.2°C, of the average annual temperature in comparison with today.

Open system pingos, apparently the type found remnant in Ireland and Wales (Mitchell, 1973; Watson, 1971), occur at present in the discontinuous permafrost zone of central Alaska, with mean annual temperatures as high as −1 to −2°C (Washburn, 1973). If the pingos were active in the Younger *Dryas*, at least discontinuous permafrost was present. The advance of cirque glaciers and the evidence for widespread solifluction point to a very cold climate, which has not been quantified satisfactorily as yet in Ireland. None of the phenomena observed requires high precipitation. They would be consistent with much lower precipitation than at present. Ruddiman and McIntyre (1981) observe that the eastern North Atlantic Ocean, though very cold, was not frozen over in the Younger *Dryas*. This would favour transport of moisture from the oceans to form precipitation, so that it is unlikely that the cold phase approached aridity.

The causes of the Younger *Dryas* climatic deterioration have been clarified by recent studies of ocean cores (Ruddiman and McIntyre, 1981). Cores north of 45° in the North Atlantic are virtually barren of foraminiferal and coccolith assemblages from 16,000 to 13,000 bp. Previously, they were dominated by a single polar water species. The interpretation is that oceanic productivity fell steeply because fresh meltwater from decaying ice sheets and the presence of icebergs made the surface waters of the oceans an unfavourable environment for foraminifera. After 13,000 bp the North Atlantic became productive once more and supported a more diverse subpolar fauna. The amount of meltwater entering the ocean and iceberg activity are assumed to have been reduced. Between 11,000 and 10,000 bp, an exclusively polar fauna appeared in the cores once more. A fall in ocean surface temperature of 10°C is assumed. The favoured explanation (Ruddiman and McIntyre, 1981) is that the disintegration of an ice shelf over the Arctic Ocean and the northern Norwegian Sea caused the southward dispersal of large tabular icebergs, as originally proposed by Mercer (1969).

Whatever the cause, the Younger *Dryas* period on land correlates with the new southward advance of the polar front. Figure 6 shows the position of the polar front in the North Atlantic at the height of the last glaciation, during deglaciation, and during the Younger *Dryas* period. The reappearance of cold oceanic water off the west coast of Ireland was directly responsible for the Younger *Dryas* climatic deterioration. The earlier deterioration in the Erosion Phase and the maintenance of a cold climate during the Grass Phase suggest that the polar front had reversed its retreat already by 12,000 bp and may have readvanced slowly over the next 1,000 years with a further rapid and severe deterioration at the beginning of the Younger *Dryas* period. In general, authors have tended to blur the distinctions between the subdivisions of the lateglacial. It is now clear that an early warm period (the First Juniper Phase) was succeeded by a colder Grass Phase and then by a very much colder Younger *Dryas* period. Studied with careful attention to stratigraphy, it becomes clear that the major units have quite distinctive floras, each with distinct ecological and climatic implications that still require quantification.

In Ireland, a sea level fall of 100 m at the maximum of glaciation would not expose a large land area to the west (Watts, 1977, his Fig. 5) so that the climatic impact of cold polar water on the western seaboard would have been very direct and sudden. For this reason Ireland's vegetation is more responsive to lateglacial climatic change than are other regions of Europe, because the response weakens with increasing distance from the ocean, until the effect of the Younger *Dryas* advance of the polar front is slight to undetectable in central and eastern Europe (Watts, 1980). Ruddiman and McIntyre (1981) suggest that melting was vigorous after 16,000 bp. Sea level was already rising during the lateglacial. It is noteworthy that pol-

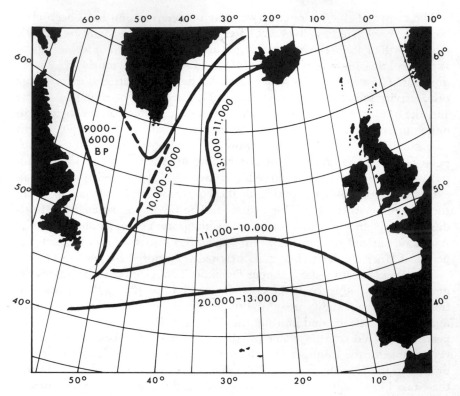

Fig. 6.   Position of the polar front at the maximum of the last glaciation, during deglaciation, and during the Younger *Dryas* readvance according to ocean core data (after Ruddiman and McIntyre, 1981).

leniferous lateglacial freshwater sediments of "Allerød" age lie some 55 m below the sea surface on the floor of Bantry Bay (Stillman, 1968).

THE EARLIER POSTGLACIAL (10,000–5,000 bp)

If we regard the lateglacial as an aborted interglacial cycle, then the true postglacial cycle began 10,000 years ago when polar water finally withdrew from the west coast at the end of the Younger *Dryas*. As warmer oceanic water replaced polar, the temperature rose rapidly. The rapidity with which types of pioneering vegetation succeeded one another suggests that the climate was out of phase with the vegetation, which lagged behind the rise in temperature. Species such as juniper, birch, and hazel either had to build up their numbers from small populations already present in Ireland or had to migrate to Ireland from distant sources in which they had survived the climate of the Younger *Dryas*. Although not proved, it is likely that the climate was already suitable for forest trees in a very short time after the Younger *Dryas*.

The modern Irish flora originates in the Younger *Dryas,* for many species present earlier in the lateglacial, including tree birch, may have become extinct and had to reimmigrate. The first response of the flora to warming is expressed as an increase in grass and *Rumex,* followed by a brief reappearance of crowberry and a peak of *Filipendula* (meadowsweet). A major expansion of juniper followed. It became much more abundant at most sites than in the lateglacial (Watts, 1977; Craig, 1978; see simplified pollen diagram from Belle Lake in Fig. 4) and must have formed continuous scrub over wide regions. Juniper, after a brief period of dominance, was invaded in its turn by tree birch, tree willow, and aspen (*Populus tremula*), which formed woodland and may have had a closed canopy locally or more widely. At this time, substantial diversification of the flora was already taking place, for example, the shrub *Viburnum opulus* (guelder rose) was already present in birch woodland in the Burren region (Watts, 1984). Birch woodland was then invaded by hazel. For a brief period hazel was enormously abundant, reaching 90% of the pollen sum at some sites in eastern and central Ireland. It is difficult to envisage the vegetation. There may have been almost continuous hazel scrub or low woodland throughout Ireland as can be seen in limestone regions of the west of the country today. Hazel may have competed successfully with the taller birches and tree willows because of its shade tolerance and suppression of the seedlings of its competitors. It is probable, however, that hazel produced pollen very freely in full light or light shade, and its real abundance in the vegetation was less than the pollen diagrams suggest, although still very great. It is noteworthy that hazel continued to have the largest pollen percentages among trees and shrubs in pollen diagrams from most of Ireland for the rest of the postglacial.

In the nomenclature of Jessen (1949), the period from the end of the Younger *Dryas* to the first invasions of birch woodland by hazel is Pre-Boreal (his pollen zones 4 and 5), and the peak of hazel marks the beginning of the Boreal (his zone 6). For reasons already stated above, it is preferred to retain a narrative account of events with a chronology based on radiocarbon dates, rather than to use the traditional zones and nomenclature. This procedure preserves the variability and time transgressiveness of events from site to site.

The hazel peak may be dated to about 9,000 bp. Pine, oak, and elm were invading by that date. In some places, especially in the west, pine arrived first, even before hazel in the Burren region (Watts, 1984). At many sites in the fertile Midlands and east, elm was the first high forest tree to arrive; less frequently (Belle Lake, Fig. 4), oak was the leader. At all sites, by shortly after 9,000 bp, after a period of mutual adjustment, high forest with oak, elm, and pine was present. Usually, all three species were present at any site, but pine was by far the most common tree, and elm infrequent, on rocky and infertile sites in the west (e.g., Glenveagh, Telford, 1977), while elm and oak were predominant in the fertile regions. Elm was more common on limestone soils, and oak tended to be more common on non-limestone soils in the southeast (Belle Lake) and in the northeast (Singh and Smith, 1973). The pollen flora, reflecting the dominance of a few high forest species, is low in diversity. Hazel, birch, and willow continued frequent, and holly and ivy were both present.

After a period of 2,000 years or more, alder (*Alnus glutinosa*) invaded Ireland and became widespread. A date of about 7,000 bp may be assigned to the arrival of alder, but it appears at different times at different sites, and its progress may have been time transgressive (Smith and Pilcher, 1973). It probably came to dominate lake shores, alluvial deposits on river-banks, and other sites where high tolerance of waterlogged or intermittently flooded soils gave it an advantage. Its arrival marked the passage from a landscape marked by high forest and open lakes or fens (mesocratic stage), to one in which swamping of surfaces and the transition to bog formation was beginning to take place. The period of abundant alder from approximately 7,000 to about 5,100 bp (the Atlantic period of Jessen, 1949) is thought by many to have been the warmest part of the postglacial cycle. The evidence for this view in Irish lake or bog deposits is quite slight, and more convincing evidence is available from other countries. The extinction of *Najas marina,* a submerged aquatic plant of eutrophic waters which was present in Ireland from the beginning of the postglacial until about 4,000 bp, may suggest that the earliest part of the postglacial had warmer summer temperatures than the later part, for the species survives farther south and in more continental regions today. The suggestion can only be very tentative, however, because other possible causes of extinction may have been at work.

The earliest man in Ireland is dated to 8,960–8,440 bp in the northeast (Woodman, 1978). He was Mesolithic in culture, a hunter–fisherman, and used fine flint points called microliths. Subsequently, the Mesolithic Larnian culture (Mitchell, 1976) used much coarser flint or chert flakes. This apparently simple culture of fishermen and food-collectors survived until it merged with or gave origin to the first Neolithic farming cultures. The date of the first agriculture is not certain, but it seems possible that farming was practised in Ireland by about 5,845 bp (Lynch, 1981; Groenman-van Waateringe, 1983; Edwards and Hirons, 1984). There seems little unequivocal evidence that Mesolithic man managed his environment to any great extent, although there has been some speculative literature on the subject (Smith, 1970, 1981; and see Edwards, Chap 9, this volume). As soon as the Neolithic population became large enough, forest clearance necessarily followed, with dramatic effects on the high forest of the first half of the postglacial.

THE LATER POSTGLACIAL
(5,000 bp to the Present)

At about 5,100 bp, pollen of elm declines precipitously in pollen diagrams throughout Ireland and elsewhere in northwest Europe. The elm fall is most marked in fertile limestone regions of the Central Plain, and is less conspicuous or even not recognizable in nonlimestone regions where the species was never common. At Scragh Bog, Co. Westmeath (O'Connell, 1980), elm fell from 28% to 3% of the pollen sum over a vertical interval of 8 cm which, although undated, may represent less than 200 years. The elm fall coincided with an increase in hazel, plantain, grass, and cereals.

It is widely assumed (Mitchell, 1976; Pilcher and Smith, 1979; O'Connell, 1980) that the elm fall was caused by Neolithic agriculture, and the case has been persuasively and imaginatively argued that man would have caused forest clearance ("landnam") by ring barking, burning and felling, by use of twigs and coppice shoots for fodder, and even by eating the inner bark at times of famine. The topic is discussed elsewhere in this volume (Edwards, Chap. 9, this volume), and it is not appropriate to explore it in detail here. However, an alternative, natural, explanation of the elm fall is possible. This assumes that Neolithic people were already present in northwest Europe before 5,100 bp, for which there is some evidence in Ireland (e.g., Pilcher and Smith, 1979; Göransson, 1981; Daniel, 1981; Edwards and Hirons, 1984). The elm fall can be seen as a natural catastrophe, caused by disease (Watts, 1961; Groenman-van Waateringe, 1983), which man was able to exploit. We have no direct evidence for disease, although the modern destruction of the elm population by Dutch elm disease is suggestive. A similar fall in *Tsuga canadensis* (hemlock) in North America has been attributed to disease as its most likely cause (Davis, 1981). The significance of the latter observation is in showing that a tree population can "crash" in the absence of Neolithic farming. The arguments against the elm fall being caused by man are, briefly, (1) the universality and severity of the fall throughout Ireland, apparently in one short time period (Smith and Willis, 1962) sandwiched between periods when elm was abundant and during which man was also present. The timing and duration of the elm fall are not exactly known because of the error inherent in radiocarbon dates, but it could have been as little as a decade or two (the probable answer, in the author's opinion) or as long as several centuries (see Edwards, Chap. 12, this volume). Exact dates may become available in the future with the aid of annually laminated lake sediments. (2) The second argument is the absence of evidence that man was equally widespread, that he used elm for any purpose, or that his archaeological remains are concentrated in areas where elm was abundant. It seems possible that elm died extensively because of disease, developed resistance, and staged a recovery and that the spread of Neolithic man through the forests may have been made easier by its temporary removal.

From about 4,500 bp onward, ash (*Fraxinus*) became abundant for the first time in association with the recovering elm population. Ash may well have been favoured by the elm fall and by prehistoric forest clearance, especially between 4,500 and 2,000 bp as it is an opportunistic "weedy" tree. It has been observed (Mitchell, 1956, 1965) that ash and elm were abundant at the same times and that their periods of joint abundance alternated with periods of rarity when grass, plantain, bracken, and cereal pollen gave evidence of forest clearance by man. These features can be seen in the pollen diagram from Redbog, Co. Louth (Fig. 7). Mitchell (1965) has argued that this should not be interpreted as a series of local histories of forest clearance and reinvasion of cleared land by secondary woodland. Instead the pollen diagrams may be a record of the total population pressure on woodland over very large regions. Periods of expansion of ash and elm would reflect low total population, and periods with abundance of herbs would result from clearing of forest by larger farming populations. This interesting hypothesis, testable by radiocarbon dating, that the ash–elm peaks are synchronous at several sites, has not yet been given the critical examination it deserves.

In the Midlands both ash and elm fell to low levels from which they did not recover in the early centuries AD (Mitchell, 1956; Watts, 1961). From that time onward major forest trees were less frequent than before in pollen diagrams. By Medieval times forest had largely disappeared from the Central Plain, but large stands of oak forest survived in the southwest and northeast, which were finally cleared in the seventeenth century (McCracken, 1971). Elsewhere secondary woodland, scrub, and weedy pasture prevailed until modern times.

Outside the more fertile regions, pine was an important tree until its general extinction between 4,000 and 1,500 bp (Watts, 1984). Extinction took place about 3,200 bp in northwest Ireland, but pine had already disappeared somewhat earlier at Glenree in Co. Mayo (Herity, 1981). In contrast, it survived until the early centuries AD at Gortlecka in Co. Clare, at the

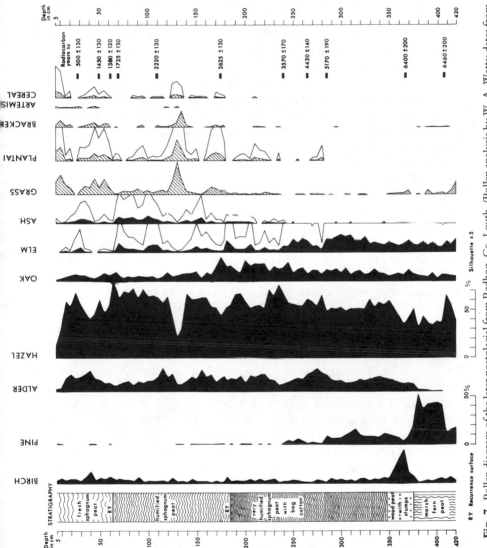

Fig. 7. Pollen diagram of the later postglacial from Redbog, Co. Louth. (Pollen analysis by W. A. Watts; dates from McAulay and Watts, 1961.)

Long Range (Watts, 1984) and in the raised bog at Clonsast in the Central Plain (Mitchell, 1956). The reasons for its extinction, given its survival in ecologically similar areas of northwest Scotland, remain unexplained (see Edwards, Chaps. 9 and 12, this volume). Yew, an underrecorded species because its pollen has only recently been recognized, appears to have been frequent in the later postglacial. It is especially common at Muckross (Vokes, 1966) and at Gortlecka and Rinn na Mona in the Burren (Watts, 1984), where it expanded strongly on limestone pavement after the elm fall. The time of arrival of yew, apparently in the early postglacial at Muckross and after the elm fall in the Burren, merits further investigation. Its ecological role is not yet understood, but it may be suspected that future pollen analysis will prove it to have been an abundant tree right up to modern times, as its high frequency in place names suggests.

Iversen (1958) emphasizes the importance of soil development in his concept of interglacial cycles. Hammond (1968) documented the history of soil profile development in mineral soils sealed by progressive growth of raised bog in the Central Plain. The profiles show initially loss of carbonates, then clay movement and podzolization over a series of several thousand years. Essentially, carbonate-rich boulder clays of the last glaciation with little profile development early in the postglacial progressively became podzols with downward loss of clay by 1,500 bp. The numerous fossil soils sealed by raised bogs provide a perhaps unique opportunity to expand this type of study in Ireland (Cruickshank and Cruickshank, 1981).

Development of raised bogs in the Central Plain is illustrated by stratigraphy at Redbog (Fig. 7). At most sites, after an early period of open water or fen vegetation, acid peats with *Sphagnum* (bog moss) were deposited. The older peats are rather humified, and the plant material, although well preserved at some levels, has largely disintegrated to form a dark-coloured dense peat. The youngest peat is usually light coloured, less dense, and contains very well-preserved plant debris with little humification. Horizons where there is an abrupt change in sediment type, suggesting that the bog had temporarily ceased growth, are known as recurrence surfaces (Fig. 7). These attracted considerable attention in the 1950s (e.g., Mitchell, 1956), in the belief that they must express regional climatic fluctuations of value in correlating profiles. The current more widely held view is that recurrence surfaces are expressions of the growth dynamics of individual bogs and lack regional significance (Walker and Walker, 1961; Casparie, 1972).

Raised bogs are certainly, and the blanket bogs of western and mountainous Ireland are largely, of natural origin. Blanket and montane bogs are not discussed here because of lack of space (see Edwards, Chaps. 9 and 12, this volume), but in a remarkable site at Glenree, Co. Mayo, Herity (1981) showed that thick blanket bog developed over mineral soil which was clearly cultivated as fields before 3,300 bp.

CONCLUSIONS

In conclusion, the vegetation succession of the postglacial, its soil and peatland development, and its history of sea level movement (Mitchell, 1976) are similar in principle to the fundamental interglacial cycle described by Iversen. The postglacial is in all essential respects like other interglacial periods and may be regarded as one.

Several aspects of the postglacial have been referred to where interesting problems are still unresolved—the possible assessment of population size changes from pollen diagrams, the conflicting explanations of the elm fall, and the unique opportunities for the study of soil profile development. Because of constraints of space, the analysis of change in the postglacial climate has not been seriously attempted, although this may become increasingly possible through tree-ring studies (Pilcher, 1973). As yet in Ireland, few pollen influx studies have been attempted (Craig, 1978; Hirons, 1983) and pollen surface samples little used (O'Connell, 1980). Also, we have no studies on annually banded lake sediments. The enigmas of the dating of Irish interglacial deposits will not be resolved until further site discoveries are made. The find of even one site with a hornbeam-dominated flora referable to the last interglacial would place all arguments about stratigraphy in a clear perspective. In an island so rich in lakes, bogs, and archaeological sites, the opportunities for innovative studies remain very great and little exploited.

References

Birks, H. J. B. (1982). *Striae* **16,** 99–105.

Boulter, M. C. (1980). *J. Earth Sci. (Dublin)* **3,** 1–11.

Boulter, M., and Mitchell, W. I. (1977). *Ir. Nat. J.* **19,** 2–3.

Bristow, C. R., and Cox, F. C. (1973). *J. Geol. Soc. London* **129,** 1–37.

Casparie, W. A. (1972). "Bog Development in South Eastern Drenthe (The Netherlands)." W. Junk, Publishers, The Hague.

Charlesworth, J. K. (1928). *Q. J. Geol. Soc. London* **84,** 293–392.

Colhoun, E. A., and Synge, F. M. (1980). *Proc. R. Ir. Acad., Sect. B* **80,** 25–45.

Colhoun, E. A., Dickson, J. H., McCabe, A. M., and Shotton, F. W. (1972). *Proc. R. Soc. London, Ser. B* **180,** 273–292.

Coope, G. R. (1977). *Philos. Trans. R. Soc. London, Ser. B* **280,** 313–340.

Craig, A. J. (1973). "Studies on the Ecological History of South-East Ireland, Using Pollen Influx Analysis and Other Methods." Unpublished Ph.D. thesis, Trinity College (Univ. of Dublin), Dublin.

Craig, A. J. (1978). *J. Ecol.* **66,** 297–324.

Cruickshank, J. G., and Cruickshank, M. M. (1981). *Oikos* **36,** 238–253.

Daniel, G. (1981). *Antiquity* **55,** 83.

Davis, M. B. (1981). *Proc. Int. Palynol. Conf. 4th Lucknow (1976–1977)* **3,** 216–228.

Edwards, K. J., and Hirons, K. R. (1984). *J. Archaeol. Sci.* **11,** 71–80.

Godwin, H. (1975). "History of the British Flora," 2nd edn. Cambridge Univ. Press, Cambridge.

Göransson, H. (1981). In "The Carrowmore Excavations, Excavation Season 1981" (G. Burenhult, ed.), pp. 180–195. Stockholm Archaeol. Rep., No. 8. Inst. of Archaeol., Univ. of Stockholm, Stockholm.

Groenman-van Waateringe, W. (1983). In "Landscape Archaeology in Ireland" (T. Reeves-Smyth and F. Hamond, eds.), pp. 217–232. BAR Brit. Ser. No. 116, Oxford.

Hammen, T. van der, Wijmstra, T. A., and Zagwign, W. H. (1971). In "The Late Cenozoic Glacial Ages" (K. K. Turekian, ed.), pp. 391–424. Yale Univ. Press, New Haven, Connecticut.

Hammond, R. F. (1968). Proc. Int. Peat Congr., 3rd 109–115.

Hayes, F. L. (1978). "Palynological Studies in the South-East United States, Bermuda, and South-East Ireland." Unpublished M.Sc. thesis, Trinity College (Univ. of Dublin), Dublin.

Herity, M. (1981). Pop. Archaeol. (1981) 36–37.

Hirons, K. R. (1983). In "Landscape Archaeology in Ireland" (T. Reeves-Smyth and F. Hamond, eds.), pp. 95–117. BAR Brit. Ser. No. 116, Oxford.

Iversen, J. (1954). Dan. Geol. Unders., [Afh.], Raekke 2 80, 87–119.

Iversen, J. (1958). Uppsala Univ. Årsskr. 6, 210–215.

Jessen, K. (1949). Proc. R. Ir. Acad., Sect B 52, 85–290.

Jessen, K., and Farrington, A. (1938). Proc. R. Ir. Acad. Sect. B 44, 205–260.

Jessen, K., and Milthers, V. (1928). Dan. Geol. Unders., [Afh.], Raekke 2 48, 1–379.

Jessen, K., Andersen, S. T., and Farrington, A. (1959). Proc. R. Ir. Acad. Sect. B 60, 1–77.

Lynch, A. (1981). "Man and Environment in South-West Ireland." BAR Brit. Ser. No. 85, Oxford.

Mangerud, J., Andersen, S. T., Berglund, B. E., and Donner, J. J. (1974). Boreas 3, 109–128.

Mangerud, J., Birks, H. J. B., and Jäger, K.-D. (1982). Striae 16, 1–6.

McAulay, I. R., and Watts, W. A. (1961). Radiocarbon 3, 26–38.

McCracken, E. (1971). "The Irish Woods Since Tudor Times." David and Charles, Newton Abbot, England.

Mercer, J. H. (1969). Arct. Alp. Res. 1, 227–234.

Mitchell, G. F. (1941). Proc. R. Ir. Acad. Sect. B. 46, 183–188.

Mitchell, G. F. (1953). Proc. R. Ir. Acad., Sect. B. 52, 225–281.

Mitchell, G. F. (1954). Dan. Geol. Unders., [Afh.], Raekke 2 80, 73–86.

Mitchell, G. F. (1956). Proc. R. Ir. Acad. Sect. B 57, 185–251.

Mitchell, G. F. (1965). Spec. Pap. Geol. Soc. Am. 84, 1–16.

Mitchell, G. F. (1970). Proc. R. Ir. Acad. Sect. B 70, 141–162.

Mitchell, G. F. (1973). Proc. R. Ir. Acad., Sect. B 73, 269–282.

Mitchell, G. F. (1976). "The Irish Landscape." Collins, London.

Mitchell, G. F., and Parkes, H. M. (1949). Proc. R. Ir. Acad., Sect. B 52, 291–314.

Mitchell, G. F., Penny, L. F., Shotton, F. W., and West, R. G. (1973). Geol. Soc. London Spec. Rep. 4, 1–99.

Morrison, M. E. S., and Stephens, N. (1965). Philos. Trans. R. Soc. London, Ser. B. 249, 221–255.

O'Connell, M. (1980). New Phytol. 85, 301–319.

Pennington, W. (1977). Philos. Trans. R. Soc. London, Ser. B 280, 247–271.

Phillips, L. (1976). Philos. Trans. R. Soc. London, Ser. B 275, 215–286.

Pilcher, J. R. (1973). Tree-Ring Bull. 33, 1–5.

Pilcher, J. R., and Smith, A. G. (1979). Philos. Trans. R. Soc. London, Ser. B 286, 345–369.

Ruddiman, W. F., and McIntyre, A. (1981). Palaeogeogr., Palaeoclimatol., Palaeoecol. 35, 145–214.

Shackleton, N. J., and Opdyke, N. D. (1976). *Mem. Geol. Soc. Am* **145,** 449–464.

Singh, G. (1970). *Proc. R. Ir. Acad., Sect. B* **69,** 189–216.

Singh, G., and Smith, A. G. (1973). *Proc. R. Ir. Acad., Sect. B* **73,** 1–51.

Smith, A. G. (1970). In "Studies in the Vegetational History of the British Isles" (D. Walker and R. G. West, eds.), pp. 81–96. Cambridge Univ. Press, Cambridge.

Smith, A. G. (1981). *Proc. Int. Palynol. Conf. 4th Lucknow (1976–1977)*

Smith, A. G., and Pilcher, J. R. (1973). *New Phytol.* **72,** 903–914.

Smith, A. G., and Willis, E. H. (1962). *Ulster J. Archaeol.* **24–25,** 16–24.

Stillman, C. J. (1968). *Sci. Proc. R. Dublin Soc.* **A3,** 125–127.

Telford, M. B. (1977). "Glenveagh National Park: The Past and Present Vegetation." Unpublished Ph.D. thesis, Trinity College (Univ. of Dublin), Dublin.

Vokes, E. (1966). "The Late and Postglacial Vegetational History of Killarney, Co. Kerry in South-West Ireland." Unpublished M.Sc. thesis, Trinity College (Univ. of Dublin), Dublin.

Walker, D., and Walker, P. M. (1961). *J. Ecol.* **49,** 169–185.

Warren, W. P. (1979). *Geol. Surv. Irel. Bull.* **2,** 315–332.

Warren, W. P. (1981). *Biul. Peryglac.* **28,** 241–248.

Washburn, A. L. (1973). "Periglacial Processes and Environments." Edward Arnold, London.

Watson, E. (1971). *Geol. J.* **7,** 381–392.

Watson, R. A., and Wright, H. E., Jr. (1980). *Boreas* **9,** 153–163.

Watts, W. A. (1959). *Proc. R. Ir. Acad. Sect. B* **60,** 79–134.

Watts, W. A. (1961). *Proc. Linn. Soc. London* **172,** 33–38.

Watts, W. A. (1963). *Ir. Geogr.* **4,** 367–376.

Watts, W. A. (1964). *Proc. R. Ir. Acad. Sect. B* **63,** 167–189.

Watts, W. A. (1967). *Proc. R. Ir. Acad., Sect. B* **65,** 339–348.

Watts, W. A. (1971). *New Phytol.* **70,** 435–436.

Watts, W. A. (1977). *Philos. Trans. R. Soc. London, Ser. B* **280,** 273–293.

Watts, W. A. (1980). In "Studies in the Lateglacial of North-West Europe" (J. J. Lowe, J. M. Gray, and J. E. Robinson, eds.), pp. 1–22. Pergamon, Oxford.

Watts, W. A. (1984). In "Lake Sediments and Environmental History" (E. Y. Haworth and J. W. G. Lund, eds.), pp. 359–376. Leicester Univ. Press, Leicester.

West, R. G. (1980). *New Phytol.* **85,** 571–622.

Wilkinson, G. C., Bazley, R. A. B., and Boulter, M. C. (1980). *J. Geol. Soc. London* **137,** 65–75.

Woodman, P. C. (1978). "The Mesolithic in Ireland: Hunter–Gatherers in an Insular Environment." BAR Brit. Ser. No 58, Oxford.

CHAPTER 9

THE ANTHROPOGENIC FACTOR IN VEGETATIONAL HISTORY

Kevin J. Edwards
Department of Geography
University of Birmingham
Birmingham, England

INTRODUCTION

The end of the lateglacial (Midlandian Stage) in Ireland around 10,000 years ago signalled the coming of the warm postglacial stage (the Littletonian) in which we live today. A probable rapid rise in summer temperatures (Osborne, 1976) to levels 1 or 2°C warmer than the present would appear to have characterized the period known throughout Europe as the Climatic Optimum (~6,000–4,000 radiocarbon years bp). The great Irish naturalist Robert Lloyd Praeger was able to capitalize on the Scottish researches of Jamieson (1865) and Robertson (1877) to demonstrate the postglacial warm period from molluscan remains in the estuarine clays of Belfast Lough (Praeger, 1888, 1896). This increased warmth, combined with the more sluggish development of soils and the attendant spread of vegetation communities, provided the physical backdrop against which the earliest unambiguous presence of man, during the Irish Mesolithic, must be viewed.

Evidence for Mesolithic peoples takes us back to around 7,000 bc (~8,950 radiocarbon years bp) with the dated site at Mount Sandel (Woodman, 1978, and Chap. 11, this volume) in the Bann Valley (Fig. 1). By this time the flora of Ireland had passed through various recognizably distinct stages as in the rest of Europe, although its diversity was then, as now, poorer

187

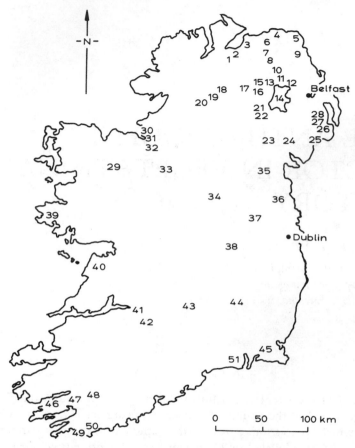

Fig. 1. Location of sites mentioned in the text. Map numbers refer to the following sites: 1, Mullaboy Td.; 2, Ballygroll Td.; 3, Gortcorbies Bog; 4, Whitepark; 5, Goodland Td.; 6, Mount Sandel; 7, Cannons Lough; 8, Fallahogy; 9, Parkmore; 10, Newferry; 11, Ballyscullion; 12, Sluggan Moss; 13, Toome Bay; 14, Lough Neagh; 15, Slieve Gallion; 16, Ballynagilly; 17, Beaghmore; 18, Lough Catherine; 19, Derryandoran; 20, Meenadoan; 21, Killymaddy Lough; 22, Weir's Lough; 23, Gosford Castle; 24, Carrivmoragh; 25, Dundrum; 26, Ballydugan; 27, Magheralagan; 28, Woodgrange; 29, Cloonlara; 30, Carrowmore; 31, Cloverhill Lough; 32, Ballygawley Lough; 33, Corlona; 34, Scragh Bog; 35, Redbog; 36, Newgrange; 37, Agher; 38, Clonsast; 39, Dolan; 40, Carron Depression; 41, Dromsallagh; 42, Lough Gur; 43, Littleton Bog; 44, Leigh; 45, Garradreen; 46, Cashelkeelty; 47, Dromatouk; 48, Maughanasilly; 49, Lough Ine; 50, Ballyally Lough; 51, Belle Lake.

than elsewhere. At present, for example, Ireland has only some 67% of the total British flora (Webb, 1943). This disparity is perhaps due to the restricted range of habitats as well as the early disappearance of a land bridge between Great Britain and Ireland (Praeger, 1934, 1939). The latter event possibly occurred before 9000 bp (Tooley, 1978; Synge, Chap. 6, this volume). Although such an early date might go some way toward explaining the notable faunal absences (van Wijngaarden-Bakker, 1974; Stuart and van Wijngaarden-Bakker, Chap. 10, this volume), it would create difficulties in explaining the later migrations of trees. It might be necessary, for example, to invoke the idea of plant refugia; the long–distance transport of seeds by wind, birds, water, or floating vegetation; or the temporary existence of a later postglacial land bridge.

The pattern of past vegetation communities in Ireland can be reconstructed from the microfossil evidence of pollen and the macroremains of plants contained within deposits such as peat, lake sediments, and soils. Given a satisfactory stratigraphic context, preferably with secure dating evidence, it is possible to recover plant remains and from them to infer past vegetation as well as the environments in which the plants lived. Pollen analysis (palynology) has been the most widely used technique, and it was applied in Ireland from an early stage by a Swede, Gunnar Erdtman, who visited in 1924 and 1926 (Erdtman 1924, 1927, 1928). This was followed up by the visit of a Dane, Knud Jessen, in 1934 and 1935, which encouraged G. Frank Mitchell in his own studies (Jessen, 1949; Mitchell, 1976). Subsequent work has come principally from two centres—Trinity College Dublin, where Frank Mitchell and William Watts have researched all periods of the Quaternary, and The Queen's University of Belfast, where Alan Smith (now of Cardiff), Jonathan Pilcher, and associates have carried out extensive and detailed work on the postglacial vegetational history of the north.

The absence of polleniferous deposits coeval with stratified artifacts or the existence of sites where the relationships could be tentatively proved, although without the support of detailed palynological information, is not uncommon. Perhaps one of the unappreciated facts among modern workers is that in three publications alone (Erdtman, 1928; Mitchell, 1945; Jessen, 1949), more than 40 polleniferous sites in which human artifacts were stratified or to which finds could be correlated were reported. Mitchell's major (1956) paper on the raised bogs of Ireland presented the first set of Irish pollen diagrams in which clear signs of anthropogenic activity were evident.

The vegetational landscape which existed prior to the conspicuous depradations of Neolithic peoples can be inferred from Fig. 2 and the diagrams from Belle Lake and Redbog (Figs. 4 and 7 in Watts, Chap. 8, this volume). The open aspect of the lateglacial (late Midlandian) landscape, dominated by

DEPTH (CM)

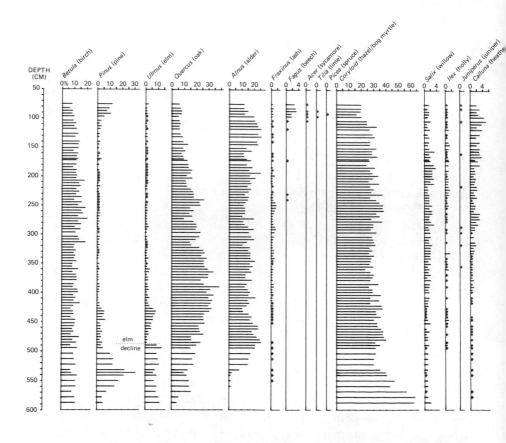

Fig. 2. Relative pollen and spore diagram from Lough Catherine II,

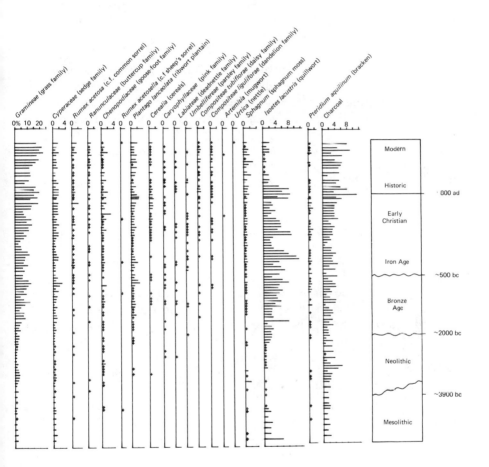

Co. Tyrone (selected taxa only). Plus indicates <1% present.

grass–sedge–mugwort assemblages of Nahanagan Stadial times (Mitchell, 1976; Watts, 1977) gave way to juniper–willow shrubland. This was succeeded by birch woodland around 9,500 bp as a prelude to the dominant hazel-pine-oak-elm elements of the Irish woodland of Boreal times (~9,500 bp onwards). Mesolithic peoples would have been familiar with such a woodland cover and the later spread of alder in wetter areas after the transition to the possibly wetter Atlantic period around 7,000 bp. It is important to point out that such a picture thus presented is very much simplified, and, for some areas, the pollen record suggests that various of these woodland types were only minor constituents in the localities sampled. By the same token, many hundreds of other taxa were present, and the landscape of Ireland was not one uninterrupted forest but was punctuated, in wetter areas particularly, by the great raised bogs which formed in the basins created in the glacial landscape.

In examining the impact of man on the vegetation of Ireland it is perhaps most convenient to discuss the process within the tentative chronological framework of cultural groups in prehistory and early historic times.

THE MESOLITHIC (~7,000–3,300 bc)

The evidence for Mesolithic peoples in the northeast especially (see Woodman, Chap. 11, this volume) but also in the east-central area of the island constitutes the earliest undisputed evidence for prehistoric communities in Ireland. It is to the rich "Bann flake" area of the north that one must also look for the clearest evidence of possible Mesolithic impact on the Irish landscape.

Toome Bay lies at the northwestern end of Lough Neagh and northward from it flows the Lower Bann River, via Lough Beg, to the sea near Coleraine. Immediately to the north of Lough Beg is the abandoned village of Newferry. The flint-bearing deposits at Newferry and Toome Bay have been the focus of considerable archaeological attention (Whelan, 1934, 1936, 1938; Movius, 1936, 1942; Woodman, 1977, 1978). At Toome Bay, Gunnar Erdtman (1928) carried out pollen analyses of the peat and diatomaceous earths deposits which contained unspecified flint implements. He was able to show only that the deposit "would be termed (Late) Atlantic–Subboreal. That is in close agreement with the archaeological records" (Erdtman 1928, p. 163). In an excavation of 1932, C. Blake Whelan found that the "late Upper Palaeolithic Facies" (Whelan, 1936) was associated with a peat layer which Erdtman placed tentatively at the Boreal–Atlantic transition (Whelan, 1933). Knud Jessen visited the site in 1934, and his investigations suggest that the "Toome Bay Culture" (compare with the Early Larnian

Culture of Movius, 1942) may be "almost the same as that of the Boreal hazel maximum" (Jessen, 1949, p. 121). The emphasis on age assignation rather than upon aspects of the contemporary environment did not change until G. F. Mitchell investigated the site in 1951 (Mitchell, 1955). Apart from charcoal, hazelnuts, pierced hazel wood, and worked pine, the pollen spectra from the archaeological horizons show a Boreal period fossil assemblage with hazel (*Corylus*) declining and a possible abrupt post-Mesolithic rise in alder (*Alnus*) pollen frequencies. Mitchell does not ascribe the vegetational changes described here to the activities of man, and there are no indications in the pollen diagrams that the known use of pine wood or the resources of hazel had any broad impact on the vegetational cover. The rise in alder could account for the relative decline in hazel pollen values, and such changes can be explained by the ecological and sedimentological changes which themselves would be contingent upon climatic or wider physiographic factors. It is of interest to note, however, that after a reexamination of polleniferous deposits at nearby Newferry, Smith speculated that Mesolithic man may have effected environmental changes, although he was not prepared to state that early man was anything more than "a minor local factor in these processes" (Smith, 1981a, p. 256).

The site at Newferry was most recently excavated by Peter Woodman (1977), following upon earlier archaeological and palynological investigations (Jessen, 1936, 1949; Movius, 1936; Smith and Collins, 1971). The earlier work had demonstrated that the Bann cultural material was primarily of Mesolithic age, but there were doubts (Smith, 1970; Smith and Collins, 1971) as to the likelihood of human interference with the Mesolithic woodland cover. Some of the most recent analyses from Newferry (especially Monolith II: Smith, 1981a, 1984) are based upon a site 200 m distant from the original one (Smith and Collins, 1971). The new site is thought to be closer to the former flood plain margin of the River Bann and to have been capable of supporting dry-land vegetation. It is the propensity of the site to produce a more local pollen record, which may explain its distinctive features (Fig. 3). The Newferry Monolith II diagram reveals hazel-dominant basal pollen assemblages with a secondary *Corylus* peak close to the point where *Alnus* values expand (denoting the generally recognized Boreal–Atlantic transition). Smith considered that these high hazel pollen frequencies may be attributable to human activity since, following work in Brazil and the United States (Rawitscher, 1945), it may be the case that hazel can withstand fires by virtue of its strong rooting system and can spring up readily from burnt stumps (Smith, 1970, 1981a). This view has been questioned with regard to the European *Corylus avellana* (Rackham, 1980), while the presence of hazel pollen peaks in deposits distant from known Mesolithic activity in Ireland (Jessen, 1949; Mitchell, 1956) and

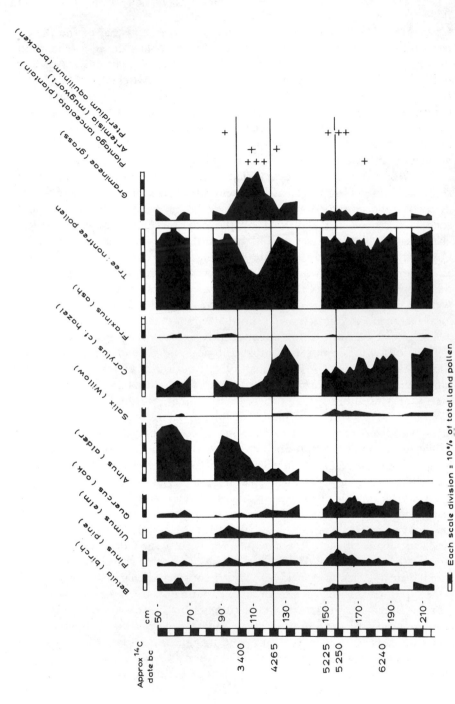

Fig. 3. Relative pollen diagram from Newferry Monolith II, Co. Antrim (selected taxa only) (after Smith, 1981a). Plus indicates <1%. The

■ Each scale division = 10% of total land pollen

■ Relative pollen diagram ■□ Tree : nontree pollen

Scotland (Edwards, 1982; Edwards and Ralston, forthcoming) should also prompt caution. Nevertheless, other features in the Monolith II diagram from Newferry encouraged Smith to speculate on early Mesolithic impact. At the point where the *Alnus* curve rises, pine (*Pinus*) and birch (*Betula*) decline, and there are temporary maxima for ash (*Fraxinus*) and willow (*Salix*) pollen frequencies. Pine and birch charcoal had been recovered during an earlier excavation at Newferry, while *Fraxinus* and *Salix* spp. are light-demanding seral taxa which like to colonize clearings. Such events as these persuaded Smith to the almost inescapable conclusion that they represent "the effects of man in locally opening up the forest cover but culminating in hazel scrub which was eventually removed" (Smith, 1981a, p. 254). This "small-scale activity," which could conceivably be large scale if the events described were more distant from the sampling locality than considered by Smith (see discussions in Edwards, 1979a, 1982), is dated at around 5,400 bc. This episode is succeeded by a major grass (Gramineae) pollen peak which is accompanied by such open-land indicators as ribwort plantain (*Plantago lanceolata*) and mugwort (*Artemisia*). This phase is of considerable duration (more than 500 radiocarbon years) and is interpreted to represent a local forest clearing perhaps kept open by continued grazing. Such grazing, especially by deer, may have occurred in those seasons of the year when the Mesolithic population had deserted the site for pursuits other than fishing. The grass peak comes to an end at around the time of the elm decline at Newferry, ~3,400 bc, by which time the Irish Neolithic was almost certainly current.

The pollen diagrams from elsewhere in Ireland indicate no sign of indisputable Mesolithic tampering with the environment (but see discussions on early agriculture below). This may be a result of many factors. Early impact could have frequently been of a minor nature, pollen sampling sites may be located too far from areas of impact to detect signs of anthropogenic activity, or the pollen diagrams themselves may be insufficiently detailed to clearly reflect short-term ecological changes.

THE NEOLITHIC (~3,900–2,000 bc)

Early Agriculture

Until relatively recently, the decline in elm (*Ulmus*) pollen frequencies (with a concomitant rise in herbaceous taxa) in northwest European pollen diagrams was taken to signify the beginnings of Neolithic agriculture around 3,000 bc (Smith and Pilcher, 1973). Ireland was no exception to this rule (Jessen, 1949; Mitchell, 1956; Smith, 1961a; Watts, 1961), and it was suggested that the occasional small peaks in cultural-type pollen prior to the elm decline at such sites as Ballyscullion and Newferry in Co. Antrim were a response to late Mesolithic activity (Smith, 1970; Smith and Collins, 1971), or they remained enigmatic [as at Leigh, Co. Tipperary, and Agher, Co. Meath (Mitchell, 1956), Parkmore, Co. Antrim (Morrison, 1959), and Scragh Bog, Co. Westmeath (O'Connell, 1980)].

A fresh look at the available palynological and archaeological data would encourage the view that early Neolithic activity is demonstrable prior to the elm decline (Groenman-van Waateringe, 1983; Edwards and Hirons, 1984). Below the level of the elm decline at the Newferry site discussed above, "a single large grass pollen grain, almost certainly a cereal grain was encountered" (Smith and Collins, 1971, p. 18). The possible cereal grain is associated with a slight depression in elm pollen frequencies. At Ballynagilly, also in Co. Tyrone, a cereal-type pollen grain together with fluctuations in the values for such taxa as *Betula*, *Corylus*, and *Pinus*, was found at a level dated to *3,800 bc (* denotes a date estimated from a ^{14}C depth–time sequence) and preceding the elm decline at the site dated to *3,270 bc (Pilcher and Smith, 1979). Analyses at yet a third site in Co. Tyrone, Weir's Lough, revealed four cereal pollen grains dating from *3,670 bc, fluctuations in tree pollen values, and expansions in herbaceous taxa, prior to an elm decline of 3,345 ± 85 bc (Hirons, 1984). At Dolan in Co. Galway, a single cereal-type pollen grain was discovered "at the end of the Atlantic Period" (Teunissen and Teunissen-Van Oorschot, 1980, p. 294), but there are no radiocarbon dates for the peat profile. In the southwest of Ireland in Co. Kerry, Lynch (1981) identified one *Hordeum*-type (cf. barley) and two *Triticum*-type (cf. wheat) pollen grains in the Cashelkeelty I peat monolith. The cereal grains are dated to between 3,895 ± 100 bc and *3,420 bc, while the elm decline is dated to 2,965 ± 95 bc. The author finds the evidence for pre–elm decline cultivation less convincing at two other sites which have produced cereal pollen grains at about the elm decline—Carron Depression in Co. Clare (Crabtree, 1982) and Ballygawley Lough in Co. Galway (Göransson, 1981). Groenman-van Waateringe (1983), however, considered that these two sites together with another nine may provide evidence for widespread Neolithic agriculture in Ireland in the pre–elm decline period. To these Irish examples may be added further instances from Britain and the European mainland (Edwards and Hirons, 1984), and such palynological data should be seen to underline caution in referring immediately pre–elm decline pollen changes to Mesolithic disturbance (Smith, 1975, 1981b; Simmons *et al.*, 1981). As a corollory, of course, this early agriculture "may have been effected by indigenous Mesolithic peoples who had acquired the techniques and materials of cereal cultivation" (Edwards and Hirons, 1984, p. 78; see also Dennell, 1983).

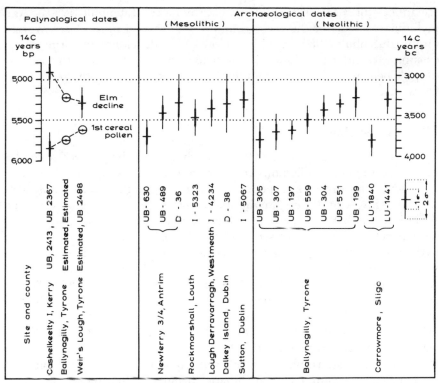

Fig. 4. Radiocarbon (^{14}C) dates from palynological and archaeological contexts for the period of early agriculture in Ireland.

Archaeologists have provided hints for a possible pioneer phase of Neolithic colonization (e.g., Smith, 1974; Coles, 1976; Bradley, 1978). Archaeological evidence for Neolithic activity in pre–elm decline times is, in fact, available from Ireland (Fig. 4). At Ballynagilly, there are 14 ^{14}C dates from Neolithic contexts of which 7 precede the elm decline estimation of *3,270 bc (Pilcher and Smith, 1979). The oldest date from Ballynagilly is 3,795 ± 90 bc for a hearth and ash pit, and this is indistinguishable from the proposed earliest cereal pollen date of *3,800 bc at the site. At Carrowmore in Co. Sligo, there are 2 early ^{14}C dates for megalithic constructions, the oldest being 3,800 ± 85 bc (Burenhult, 1980a; and see critiques in Caulfield, 1983, and Woodman, Chap. 11, this volume). The temporal overlap with Mesolithic archaeological contexts is demonstrated in Fig. 4. As was noted above, the palynological evidence does not preclude the possibility of Mesolithic agriculture. The archaeological data do not go far in supporting this possibility, although there is the negative evidence provided by the aceramic nature of the material from Carrowmore (Burenhult, 1980a). As a preliminary model of the socioeconomic background of the Carrowmore tombs, Burenhult (1980b, p. 5) has suggested "a development within a pre-existing Mesolithic population." Caulfield (1983, p. 206) warns that against this, "There is not a single mesolithic-type artefact from the tombs" or known from the Sligo area.

The Elm Decline

Within 1–2 centuries of 3,000 radiocarbon years bc (~3,800 calendar years BC) the pollen diagrams from Ireland, like those from Britain and elsewhere in northwest Europe, exhibit a decrease in the pollen frequencies for elm. This "elm decline" traditionally marks the boundary between the Atlantic and Sub-Boreal periods and has attracted extensive comment since the pioneering work of Iversen (1941) in Denmark. Various explanations have been advanced for the apparent reduction in elm pollen representation which all revolve around concepts of climatic change, disease, and the activities of man (Troels-Smith, 1960; Smith, 1961a, 1970, 1981b; Watts, 1961, and Chap. 8, this volume; Ten Hove, 1968; Godwin, 1975). The fact that the elm decline is the most synchronous boundary in northwest European pollen diagrams (Hibbert *et al.*, 1971; Smith and Pilcher, 1973; Edwards, Chap. 12, this volume) might encourage the view that some kind of all-embracing "catastrophic" explanation would be a most likely contender. This would favour a sudden climatic change affecting the sensitive elm or the rapid spread of an elm disease, perhaps analogous to the recent decimation of elms in Europe and elsewhere under attack from Dutch elm disease (see Jones, 1981). If man were to be responsible, where he might be involved in such activities as tree felling (the elm favours light, base-rich soils suitable for arable cultivation), the collection of leaves and twigs for fodder (which would reduce flowering and pollen production), or even "ring-barking" by domestic animals, then this would necessitate the presence of Neolithic man close to every pollen site at about the same time.

All three possible major causes for the *Ulmus* decline, singly or in combination, have had their adherents although it would probably be true to say that anthropogenic explanations have received most acceptability. Apart from negative aspects of the nonanthropogenic explanations (Ten Hove, 1968; Smith, 1981b), the role of man is reinforced by the presence of pollen types consistent with agricultural activity at and around the level of the elm decline. Such cultural indicators include expansions in the representation of pastoral, arable, or open-land taxa (e.g., grass, plantain, fat-hen, mugwort) and, most convincingly, the presence of cereal pollen grains. Additional support for the impact of Neolithic communities comes from the widespread archaeological evidence for Neolithic occupation and agriculture at this time (Renfrew, 1974; Herity and Eogan, 1977).

A relatively recent statement on the elm decline and its association with elm disease has come from Groenman-van Waateringe (1983). She noted the lack of direct evidence for a disease explanation—no general outbreak of the elm bark beetle *Scolytus scolytus* is found in the palaeoecological record, nor is there palaeomycological evidence for the relevant fungus or spores of *Ceratocystis ulmi*. It is also acknowledged that the evidence for man in the pollen record at the time of the elm decline is pronounced. But Groenman-van Waateringe saw special significance in the existence of pre–elm decline agriculture. Neolithic man was versatile, and it seemed to her inconceivable that experienced early farmers suddenly discovered, for example, that elm leaf fodder was best suited to his stock-keeping practices in all areas. The spread of an elm disease would provide convenient forest openings suitable for agriculture on the best soils. An incomplete recovery of elm may indicate areas where erosion or leaching altered local conditions, whereas in those areas where elm reestablished itself there may have been only a small farming population. Groenmann-van Waateringe has brought much detailed and convincing evidence to bear on her arguments. Reference can be made to the exciting find of the elm bark beetle *Scolytus scolytus* just before the elm decline at West Heath Spa, Hampstead Heath, London (M. Girling, personal communication; Girling and Greig, forthcoming). Rackham (1980) and Huntley and Birks (1983) see the coincident spread of Neolithic farmers and a pathogen, together with the failure of beech (*Fagus*) and hornbeam (*Carpinus*) to migrate into gaps left by the dead elms in Northern Europe, as probable necessary conditions "required to explain all aspects of the spectacular mid-Holocene decline of elm pollen" (Huntley and Birks, 1983, p. 415). There is also the suggestion that the hemlock (*Tsuga*) decline seen around 2,850 bc in North American pollen diagrams has a disease explanation (Davis, 1981, where it is also compared with the historic chestnut blight of the twentieth century in the same continent).

Relative and Absolute Pollen Studies

Most published Irish pollen diagrams covering the period of the elm decline have been based on traditional sampling methods which represent each pollen taxon as a proportion of some fixed pollen sum. Thus, with various modifications, the elm pollen values are presented as a percentage of total land pollen (e.g., Smith and Pilcher, 1972; Pilcher, 1973; O'Connell, 1980; Lynch, 1981) or of total tree pollen (Mitchell, 1954, 1956; Morrison, 1959; Smith and Willis, 1962; Buzer, 1980), or perhaps both in different diagrams (Smith, 1961b; Pilcher, 1969). This method of calculation presents one major difficulty—the interdependence of the variables means that a reduction in percentage of one taxon will cause increases in the values of others, although these other taxa may themselves have decreased their real contribution to the pollen rain of a site. To overcome this deficiency, pollen data calculated on an absolute concentration (grains/cm^3) or "influx" accumulation rate basis (grains/cm^2/year) enable the pollen taxa to be represented as independent variables (Davis and Deevey, 1964; Pennington, 1973, 1975; Beckett and Hibbert, 1976; Craig, 1978).

The few absolute diagrams from the anthropogenic levels of British postglacial pollen diagrams suggest that the level of the elm decline is frequently accompanied by reductions in tree taxa other than elm (e.g., Sims, 1973; Pennington, 1975; Beckett and Hibbert, 1976; Edwards, 1978a). The importance of these sites is the demonstration that the percentage reductions in *Ulmus* pollen rates at the elm decline can sometimes be explained as statistical artifacts of the orthodox relative pollen method. This further means that there may not always have been a selective decline of elm woodland and that elm trees may not have grown in discrete stands as has been suggested for various parts of Ireland (see Mitchell, 1956, 1976; Morrison, 1959).

Support for the reduced presence of other taxa at the elm decline in Irish diagrams comes from many sites, where relative pollen diagrams suggest the decline in one or two taxa along with elm, for instance, pine together with hazel at Ballynagilly (Pilcher and Smith, 1979), pine and birch at Beaghmore (Pilcher, 1969), pine, alder, and hazel at Redbog (Mitchell, 1976), pine, birch, and oak at Littleton Bog (Mitchell, 1965), and birch and oak at Scragh Bog (O'Connell, 1980). However, absolute pollen studies underway at Lough Catherine in west Co. Tyrone (Edwards and Larmour, in preparation) provide an indication of events for one area in the north of Ireland where the elm decline really did seem to involve a reduction in *Ulmus* alone. In nearby east Co. Tyrone, research at Killymaddy Lough similarly suggests prolonged absolute pollen falls for *Ulmus* at the elm decline, with no evidence for the reduction in the pollen of other tree species, while herb pollen influx values increased (Hirons, 1983, 1984).

Patterns of Clearance Activity

The expression "clearance phase" seeks to describe the removal of woodland for agricultural and settlement purposes. Clearance was first suggested for Ireland by Mitchell (1951), and a particularly well-dated example has been demonstrated from Fallahogy, near Kilrea in Co. Londonderry (Smith and Willis, 1962), where an episode of inferred clearance, farming, and regeneration lasted about 340 radiocarbon years. Even greater longevity is reported from three other sites in the north of Ireland at Ballynagilly, Beaghmore, and Ballyscullion, where Pilcher *et al.* (1971) discerned initial clearance and arable farming stages lasting 100–400 ^{14}C years, succeeding pastoral episodes of 150–200 years duration, and forest regeneration in 50–100 years (but see discussion below). The Cashelkeelty pollen diagram from Co. Kerry (Lynch, 1981) displays two extensive clearance phases of 350–400 ^{14}C years duration. Both of these predated the elm decline, and the earlier, beginning around 3,895 bc, featured palynological evidence for wheat (*Triticum*) and barley (*Hordeum*) cultivation. These extensive phases of activity, if taken together with similar evidence from Scotland (Edwards, 1978a, 1979a) and northwest England (Pennington, 1975), suggest that pollen diagram clearance phases are not the short "landnam" episodes hypothesized by Iversen (1941, 1956). The resolution of pollen diagrams is not, in most cases, sufficient to detect individual events of short duration (Edwards, 1979a; but see Garbett, 1981). Considering that a 1-cm thick slice of lake sediment could represent 30 or more years of deposition time, it is clear that the pollen diagrams represent an aggregated pollen record which would make precise detection of the pattern of vegetational impact very difficult to discern. The composite activity displayed by a clearance phase may be the result of a single community at work, perhaps involved in shifting agriculture around a settlement; alternatively, it may result from many separate groups of farmers entering and leaving an area over a prolonged period. Extreme "clearance" longevity may even result from the postoccupational grazing of former cleared areas by wild animals (Smith, 1970; Buckland and Edwards, 1984). Many permutations of activity could result in the registration of similar palynological phenomena (Edwards, 1979a, 1982)—the likelihood of equifinality in the pollen record [i.e., the existence of the same end result via different causes (Haines-Young and Petch, 1982)] is very strong. This should not stop speculation as to the causes we perceive in the pollen record—indeed such conjecture is surely mandatory—but one should bear in mind the plurality of hypotheses (Chamberlin, 1890) which could account for apparent cultural data.

Although charcoal is found in the Neolithic and/or later levels of such sites as Ballynagilly (Pilcher, 1975; Pilcher and Smith, 1979) and Lough Catherine (Edwards and Larmour, in preparation; Fig. 2), there is no certainty that it was used for woodland clearance although this could be a valid reason for its presence. Lightning, "muir burn," domestic burning, or the burning of axe-felled woodland are just as likely to be responsible for charcoal (Edwards, 1979a; Edwards and Ralston, forthcoming), and the lack of its reported consistent presence in pollen preparations from many parts of Ireland suggests perhaps that woodland was generally felled by axe, possibly after initial ring-barking to kill the tree. Slash-and–burn agriculture has, however, been proposed by Lynch (1981), but such inferences must for the moment remain conjectural.

What patterns, if any, are to be found in the Neolithic age records of the Irish vegetation? The general picture is one of a definite decline in elm accompanied sometimes by declines in other tree types, dependent on the geographical area. Pine and to a lesser extent oak seem to share in this reduced woodland cover. At the same time, ash was able to colonize openings in woodland together with the opportunistic spread of birch and hazel. All of these woodland types, except for elm, are found as worked wood in Neolithic contexts from Irish archaeological sites (Coles et al., 1978). To the expansion, natural or otherwise, of such nonarboreal plant types as plantain, grass, mugwort (*Artemisia*), goosefoot (Chenopodiaceae), and bracken (*Pteridium aquilinum*) can also be added that of the cultivars in arable agriculture. The pollen evidence suggests the presence of such genera as *Triticum* [Cashelkeelty (Lynch, 1981), Newgrange (Groenman-van Waateringe and Pals, 1982), Scragh Bog (O'Connell, 1980), Ballygawley Lough (Göransson, 1981)], and *Hordeum* [Cashelkeelty, Ballygawley Lough, and Dolan (Tennissen and Tennissen-van Oorschot, 1980), Lough Catherine (Edwards and Larmour, in preparation)]. The remains of seeds suggest, more specifically, einkorn (*Triticum monococcum*) and emmer (*T. dicoccum*) at Dunnly cairn and naked barley (*Hordeum vulgare*) at Whitepark, both in Co. Antrim (Jessen and Helbaek, 1944).

A general pattern of Neolithic clearance activity for Britain and Ireland has been proposed by Whittle (1978). His thesis is that "there seems to have been a phase around the middle of the third millenium bc when many clearances were abandoned and woodland or forest allowed to regenerate after the initial clearings had been effected from the fourth millenium onwards, and before renewed activity at the end of the third millenium bc" (Whittle, 1978, p. 34). The evidence has been criticized on palynological and temporal grounds elsewhere (Edwards, 1979a). Spatial relationships between the sampling sites and the Neolithic communities which were involved make inferences very uncertain. Whittle's thesis may be correct, but much more information is required for its validation.

Patterns of land use change might best be demonstrated for mid-Ulster, for which many detailed pollen diagrams are available. In the earlier discussion on clearance phase longevity, mention was made of the three sites of Ballynagilly, Beaghmore, and Ballyscullion (Pilcher et al., 1971). The changes inferred for these sites followed a sequence of "arable" cultivation at the elm decline (stage A) followed by "pastoral" farming (stage B) and then "regenerating forest" (stage C). The Ballyscullion site did not, however, produce cereal pollen grains at the elm decline—an absence ascribed to the poor dispersal of cereal pollen. The descriptive system of Pilcher et al.

(1971) is followed in its essentials by Pilcher and Larmour (1982), where the Co. Tyrone sites of Ballynagilly, Beaghmore, Derryandoran, Meenadoan, and Slieve Gallion are compared (their Fig. 6). Concentrating, for the moment, on the elm decline and post–elm decline discussion of Pilcher and Larmour (1982), it may be noted that all sites feature an "arable or mixed farming" stage. This is misleading for Meenadoan at least, because the pollen diagram shows no cereal-type pollen ("Gramineae > 40 μm") until about 1,350 bc, the elm decline section of the diagram shows no expansions of arable-type weed taxa, and the authors talk of "the absence of an early crop growing stage" (Pilcher and Larmour, 1982, p. 292). It is suggested that the late ^{14}C dates for the elm declines at Meenadoan (2,860 ± 125 bc) and Derryandoran (2,900 ± 90 bc), perhaps reflect a combination of a westerly situation and altitude (see Kirk, 1974–1975). This explanation is doubtful, however, because the nearby site of Lough Catherine VI has an elm decline date of 3,240 ± 95 bc and lies at an altitude of 60 m, only 14 m lower than Derryandoran. The easterly site of Slieve Gallion, where the elm decline is also relatively late (2,945 ± 65 bc), certainly lies at high altitude (427 m) when compared to Meenadoan (213 m). But then the lower site of Sluggan in the accessible Bann Valley is located at an altitude of 52 m and has an elm decline date of 3,015 ± 75 bc, which is later than that at many other mid–Ulster sites (Fig. 4; Edwards, Chap. 12, this volume). It is, of course, possible that the elm decline is a synchronous event (Smith and Pilcher, 1973; Edwards, Chap. 12, this volume) but that errors inherent in the radiocarbon dating method or problems associated with dating different types of material (e.g., peat with intruding younger roots or lake muds containing old carbon derived from inwashed soils) may obscure the true picture.

Figure 5 portrays schematically the inferred early to late Neolithic local land use pattern along an L-shaped transect of Co. Tyrone sites. Arable agriculture is indicated only if the pollen diagrams show the pollen of either cereals or a suite of taxa frequently associated with arable land in Ireland (e.g., Artemisia, Rumex, Umbelliferae, Chenopodiaceae, and Cruciferae). This represents a very cautious view of the data. On the other hand, pastoral activity is inferred particularly from expansions in Gramineae, Plantago lanceolata (ribwort plantain), and Pteridium aquilinum; this may be incorrect because these pastoral-type indicators may simply have colonized woodland openings created by dead elms or cleared by man and subsequently abandoned. The elm decline is taken to be a synchronous datum for all sites in order to emphasize land use variations following this first probable major impact on the vegetation of the Neolithic period. The possible loss of temporal fidelity is perhaps unfortunate, but the exercise is not intended to demonstrate strict time comparisons.

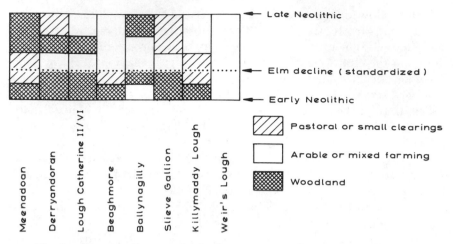

Fig. 5.   Schematic diagram of early clearance events of Co. Tyrone pollen sites.

The tentative reinterpretation of the local land use patterns in Fig. 5 differs markedly from those of the individual diagram constructors. The Ballynagilly column, for example, not only includes the proposed pre–elm decline cereal cultivation (Edwards and Hirons, 1984) but differs from Pilcher et al. (1971) in the present author's failure to detect a separate pastoral phase (stage B) after initial arable activity. The full long profile pollen diagrams from Ballynagilly (Pilcher and Smith, 1979) reveal that a cereal pollen grain appears during the pastoral stage of Pilcher et al. (1971). A mixed agricultural economy is also suggested by the presence of cereal pollen grains in the post–elm decline Gramineae- and *Plantago lanceolata*-rich spectra of the Ballynagilly monoliths A and E diagrams. These may correspond chronologically with the stage B pastoral episodes (Pilcher et al., 1971), but Pilcher and Smith (1979) indicated stratigraphic evidence of possible site erosion which makes precise temporal interpretation hazardous. For Meenadoan, the present author found some evidence for immediate pre– and post–elm decline clearance or pastoral activity but none for late Neolithic arable or mixed farming. Pilcher and Larmour (1982) seem to have based their reconstructions for Neolithic activity at the site particularly on changes in pollen frequencies for elm, pine, and hazel. Pilcher and Larmour, as is the case for other workers at the other sites, may be quite correct in their land use inferences; however, the apparent change in woodland could be of a more regional nature or may indicate, for instance, changes resulting from plant competition. The lack of local nontree pollen indicators at many of the sites underscores the need to be cautious (Edwards, 1979a; Behre, 1981). It seems to this author that apart from the fact that the easterly sites have been more open for a longer time, no clear sequential pattern of land use change is visible in Fig. 5—even the adjacent stream-linked sites of Weir's Lough and Killymaddy Lough differ quite markedly, although their patterns of woodland change are similar (Hirons, 1984). There is clearly a need for more objectivity to be introduced into palynological inferences concerning anthropogenic change (see Edwards, 1979a; Maguire, 1983; Maguire *et al.*, 1983).

Soil Erosion

The changes observable in the pollen diagrams pose obvious problems as to the changing configurations of trees in the woodland of Ireland. What is clear, however, is that the apparent depression of arboreal taxa around 3,000 bc marks a "floating" datum which anticipates the eventual relegation of woodland to a minor position in the vegetational landscape. The apparent reduction in various tree taxa around the time of the elm decline, assuming a real if not monocausal relationship with the frequently accompanying signs of agricultural activity in the pollen record, probably marks the major presence of man. A resulting impact can also be inferred from the sedimentary record. At Lough Catherine, the accumulation rate of the lake muds rises from an estimated 4.9 cm per calendar year at the elm decline to 5.9 cm per calendar year by the end of the Neolithic, after which values decrease again (Fig. 6). Since lake sediments are largely a record of inwashed soils (Mackereth, 1966), the fact that this increase in sedimentation rates coincides with the first presence of cereal (*Hordeum*-type) pollen and an expansion of grass pollen values provides circumstantial evidence for linking agricultural activity with soil erosion. This is supported by increased magnetic susceptibility concentration values precisely at the elm decline (Fig. 6; see Thompson and Edwards, 1982). The magnetic properties of lake sediments can be closely related to the ferrimagnetic mineral content of soils in lake basins (Thompson *et al.*, 1976). An inwash of soils rich in magnetic minerals, prompted by agricultural activity (but see Dearing and Flowers, 1982), would therefore lead to increases in the magnetic susceptibility of sedimentary detritus deposited within the lake. Sediment chemistry can also provide analogous evidence for erosion (Mackereth, 1966), although the values for potassium in the Lough Catherine profile (Fig. 6) appear to provide a less sensitive minimal indication of erosion in the early Neolithic. It should be noted that the origin of the materials providing evidence for environmental change (e.g., pollen, magnetic and potassium-rich soils) are likely to vary spatially within the lake catchment.

Evidence for erosion during the Neolithic may also be found elsewhere in Co. Tyrone at Ballynagilly in sample monoliths A, C, and E (Pilcher and

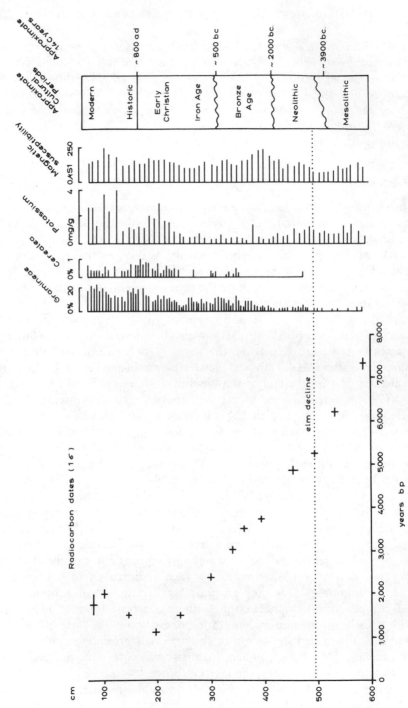

Fig. 6. Selected palaeoenvironmental indicators from the Lough Catherine II site, Co. Tyrone.

Smith, 1979) and at Weir's Lough and Killymaddy Lough (Hirons, 1983, 1984). Additional indications of Neolithic age erosion may be provided by the clay layer containing cereal-type pollen in the Carrivmoragh peat monolith in Co. Down and by the subpeat clay and twig layer from Gosford Castle Forest, Co. Armagh (see notes to UB-870 and UB-874 ^{14}C dates in Pearson and Pilcher, 1975; Holland, 1975).

THE BRONZE AGE (~2,000–500 bc)

The close of Neolithic times sometimes saw a recovery of the forest cover, but within about 500 years any cover had effectively disappeared from the greater part of Ireland. The present treeless nature of the island in many areas can be seen to stem from Bronze Age times. The causes would appear to be a result of the actions of man and climate operating independently and perhaps, at times, in concert.

In the north of Ireland, the pollen diagram from the lowland raised bog at Sluggan Moss in Co. Antrim (Smith, 1975) shows Beaker/Early Bronze Age declines in tree pollen from 85% of the total down to about 45%, followed by a regeneration–clearance–regeneration pattern in the "Later" Bronze Age, with woodland pollen totals stabilizing at around 80%. Conversely, the upland valley bog at Gortcorbies in Co. Londonderry (Smith, 1975) produced evidence of major long-term reductions in tree pollen. The representation of arboreal taxa fell from around 80% at about 2,000 bc, with intermittent clearances, to a low of around 25% in the "Later" Bronze Age (1,000 bc), before partial recovery to 60% of the total pollen sum before the start of the Iron Age. The clearance phases of Gortcorbies appeared to be on a more massive scale than was the case at Sluggan Moss, although such an inference would have to be qualified by the unknown distance of vegetational impact to the sampling sites. At both sites grass, plantains, and heaths play major roles in the nonarboreal pollen component and suggest not only the hand of man but also the coincident spread of blanket peat or soil impoverishment.

Although similarities exist between the Bronze Age pollen records at Sluggan Moss and Littleton Bog and between Gortcorbies and the Co. Tyrone sites of Beaghmore and Ballynagilly (Smith, 1975), contrasts are not difficult to find. At Cannons Lough, the arboreal pollen values undergo a progressive reduction which had begun at the elm decline, and this is perhaps due to the pressures of agriculture in an area where the sandy soils were already rather poor (Smith, 1961b, 1964). Similar progressive declines in woodland pollen representation, with sporadic partial regeneration, can be seen at Lough Catherine and Parkmore in the north (Morrison, 1959) and at Cashelkeelty (Lynch, 1981) and perhaps Lough Ine in the south (Buzer, 1980). The pollen diagram from Ballyally Lough, Co. Cork, may show a single major Bronze Age regeneration of tree taxa following a general Neolithic reduction (Buzer, 1980).

The differences displayed in pollen diagrams from these sites are perhaps not surprising given the likely variations over different parts of the island in such factors as population density, agricultural practice, vegetation, and soil types. Against such variability, however, can frequently be seen the signs of two widespread phenomena—a marked decline in the frequences of pine pollen and the spread of blanket peat.

The Decline of Pine

Erdtman (1928, p. 165) noted examples of bogs in various parts of Ireland where *Pinus* appeared to be confined to peaty areas during the Sub-Boreal and "entirely disappeared" during the Sub-Atlantic. Jessen (1949) also considered that in the Sub-Boreal pine practically disappeared from the Irish woods although it probably continued to exist on the bog surfaces and around their edges. He further noted that the "petering out of the pine-curve and the often almost total lack of pine pollen in the upper levels of the diagrams is hardly known from any other localities in northwest Europe" (Jessen 1949, p. 256). Mitchell (1956) thought that the importance of pine varied considerably from one locality to another and that it may have survived into the Medieval period around some of the large raised bogs of the Central Plain. He also saw the *Pinus* decline as being essentially diachronous and therefore impossible to use for chronological zoning purposes. Evaluating radiocarbon dates for the final *Pinus* decline in the north of Ireland, Smith and Pilcher (1973) showed that the means of the ^{14}C determinations fell within the period 300–400 years either side of 4,000 bp and suggested that while the decline "does not appear to be synchronous" (p. 910) it also "appears to be one of the least diachronous horizons" in the postglacial (p. 911). The "bunching" of radiocarbon dates around 4,000 bp cannot be attributed to atmospheric variations in ^{14}C levels since the dendrochronological calibration curve for the period in question is smooth rather than "wiggly" (Pearson et al., 1977; Lynch, 1981). Pilcher and Smith (1979) noted a final decline in pine pollen values at Ballynagilly at about 2,150 bc. They also suggested that there is little evidence to indicate that the *Pinus* decline had an anthropogenic cause. Tinsley and Grigson (1981) stated that in Ireland pine was extensively exploited for fuel and timber, and they regarded the spread of available ^{14}C dates as a metachronous feature. It is certain that in various localities pine was exploited by man, but it was also being pushed out, like other taxa, by the deteriorating soils and the spread and regrowth of peat (see Birks, 1975).

Soil Deterioration and the Spread of Blanket Peat

The deterioration of soils and the growth of blanket peat are sometimes seen as part of a pedogenetic continuum whereby time, climate, and agriculture, collectively or individually to a varying extent, exert their influence (Watson, 1939; Goddard, 1971; Mitchell, 1972; Ball, 1975; Moore, 1975; Smith, 1975). Preagricultural climatic deterioration and peat development are known throughout northwest Europe and are most typically associated with the expansion of basin peats and the spread of the wet-loving alder during the Atlantic period. Leached podzol soil profiles are known from the Neolithic period (Proudfoot, 1958; Case et al., 1969) through to the present day (Dimbleby, 1962; Romans and Robertson, 1975; Culleton and Gardiner, Chap. 7, this volume). An interesting example of this occurs in the Dundrum sand dune system in Co. Down (Edwards, 1978b; Cruickshank, 1980; see Edwards and Ralston, 1980). The podzolic soils of the older dune sands have been severely leached and have a Calluna-dominated vegetation cover. The thin soil covering of the fresher seaward dunes is less acid in reaction, with a high shell fragment content, and has a ground flora in which grasses and Liguliflorae are especially common. Pollen spectra from one palaeosol group within the dunes seem to demonstrate similar conditions pertaining in the past. A probable Neolithic age fossil soil produced a Calluna–Rosaceae–Artemisia pollen assemblage. An overlying probable Bronze Age humic horizon, separated from the Neolithic layer by sand and also overlain by sand, contained a Gramineae–Liguliflorae–Tubiflorae–Cruciferae microfossil assemblage. The evidence from Dundrum suggests that a podzol would be the likely end point of pedogenesis in the area, and this is arrived at via intermediate stages. The arrest of the process would result, not unsurprisingly, from burial by sand or perhaps even grazing activity.

The renewed development of peat during the Irish Bronze Age, especially, has been noted in the literature (Smith, 1975; Mitchell, 1976). It is supposed that man's role may have been within a sequence such as (a) woodland clearance, (b) cultivation and impoverishment of soils, (c) leaching aided by decreased woodland cover, (d) iron pan formation, impeded drainage, and higher water tables, (e) peat initiation. If such a process takes place in areas where altitude, exposure, and higher rainfall are a backdrop, then it is difficult to prove that man was not merely accelerating a process which might otherwise have happened quite naturally in the absence of anthropogenic stress over a longer period of time. The deterioration of soils and spread of blanket peat certainly coincided with the presence of man over much of Ireland. There is no doubt that the wooded landscape of the island was dramatically reduced, and the visual dominance of the arboreal taxa would have been replaced by heath, rush, and *Sphagnum* communities, contributing to their accumulating peaty parent materials.

Figure 5 in Edwards, Chap. 12 (this volume), displays the basal radiocarbon dates for blanket peat deposits in Ireland. It can be seen that while most of the dates fall within the Bronze Age, they vary from 2,650 ±65 bc for a sample from Killarney to ad 1,150 ± 30 for Cashelkeelty II, both in Co. Kerry. Few of these determinations are associated with sites for which comprehensive subpeat surveys have been carried out (Tallis, 1964). It is possible, therefore, that they date the accumulation of peat in a depression or on the inclined slope of the land surface—both of which are locations that might lead to "atypical" dating of general peat initiation (Edwards and Hirons, 1982). An example of this problem is illustrated by a date of about 100 bc at Cashelkeelty I and ad 1,150 ± 30 at Cashelkeelty II, sites on the same mountain ledge only 100 m apart (Lynch, 1981). This discrepancy was explained as a function of varying distance from the focus of peat spread. A similar case was met in the north of Ireland on an Iron Age site in Co. Londonderry (Edwards, 1983). The excavator Brian Williams considered that the blanket peat at the site overlay the same podzolized soil surface. The basal peat at the lower Mullaboy Townland site is dated at 80 ± 45 bc, whereas the neighboring Ballygroll Townland site only 365 m further up-slope produced a basal date of ad 445 ± 35. The Ballygroll pollen spectra contained higher frequencies of cultural indicators (e.g., *Hordeum, Plantago lanceolata,* and *Pteridium aquilinum*) and lower values for *Calluna vulgaris* pollen than were found at Mullaboy. Indeed the discrepancies between the pollen assemblages support the idea of noncontemporaneity suggested by the radiocarbon dates. Some additional confusion arises, however, in view of the fact that soil underlying the Mullaboy Townland peat is dated at a more recent ad 1,075 ± 40. If the peat dates are reliable, then peat initiation at Mullaboy was perhaps earlier than that at Ballygroll, because either the latter hilltop site had superior soil drainage and was shedding water for a longer period or the lower site at Mullaboy possibly occupied a topographic depression which was susceptible to earlier peat growth. A dramatic difference in basal peat dates comes from Goodland Townland in Co. Antrim (Case *et al.,* 1969); here, radiocarbon dates of 2,200 ± 200 bc and ad 570 ± 150 were obtained from samples only 7 m apart. Such examples from different parts of Ireland act as a warning against overgeneralization, and together with the temporal variation evident in Fig. 5 (of Edwards, Chap. 12, this volume) should deter overreliance on climatic deterioration as a monocausal explanation for blanket peat growth in Ireland. At the same time, the archaeological, pedological, and palaeoecological evidence does not categorically point to alternative predominant causes but may suggest that many factors contributed to the end result seen over so much of Ireland today.

THE IRON AGE AND EARLY CHRISTIAN
PERIODS (~500 bc–ad 800)

While a deterioration in climate has long been considered a possible feature of Bronze Age Europe, traditionally less doubt has been cast on a supposed increase in wetness and cooling of temperatures during the "Sub-Atlantic" period. The beginning of this phase is usually taken to correspond approximately with the Iron Age in Britain (Godwin, 1975), but in Ireland the lower limit of the Sub-Atlantic was defined by the fall in pine pollen values around 2,000 bc (Jessen, 1949), although Mitchell (1956) abandoned the distinction between the Sub-Atlantic and the preceding Sub-Boreal. Evidence for wetness in the Irish Iron Age has generally been taken from peat-based contexts. Dates for blanket peat initiation within the Iron Age have been discussed above. Another frequently cited indicator of wetness is the recurrence surface, the junction between overlying fresh unhumified *Sphagnum* peat and underlying darker humified peat. Irish peat bogs frequently contain more than one recurrence surface (Mitchell, 1956), and, as is the case elsewhere in Europe, it is doubtful whether a single major *Grenzhorizont* (Weber, 1900), assignable to the supposedly Iron Age RYIII surface (Granlund, 1932), actually exists (Jessen, 1949). The probable local significance of many recurrence surfaces, a function of bog hydrology, has been discussed in the literature (Van Zeist, 1954; Mitchell, 1956), although examples of apparent synchroneity can be found (Dickinson, 1975; Aaby, 1976). While many surfaces prove to be of Iron Age date (Turner, 1981), it is not difficult to produce examples of earlier ones (Seddon, 1967), as is also the case for wooden trackways across peat bogs (Coles and Hibbert, 1968; Tinsley and Grigson, 1981). In Ireland the major recurrence horizon at Fallahogy, Co. Londonderry (Smith, 1958), was dated to 253 ± 120 bc while that at Clonsast, Co. Offaly, predates a pine stump dated to ad 330 ± 130. The *Corylus* trackway at Clonsast (Mitchell, 1956) has a radiocarbon date of ad 30 ± 130, while the trackway in the raised bog at Corlona in Co. Leitrim (Tohall *et al.*, 1955; Mitchell, 1956) was dated to 1,445 ± 170 bc. The associated stratigraphic evidence at Corlona was insufficient to prove that the bog surface was becoming wetter at the time of construction. A tentative suggestion of increased wetness in the Iron Age came from limnic evidence at Lough Gur, Co. Limerick (Mitchell, 1954). Reinterpreting the apparent disturbance of early zone VIIIb sediments, Mitchell (1956) intimated that the cause might lie with springs in the lake basin, thrusting upward with increased vigour and disturbing the muds and their contained pollen.

Conditions in Ireland during the Iron Age were certainly suitable for the regeneration of raised bogs as well as for the continuing spread of blanket peat. After the Late Bronze Age Dowris Period with its rich material culture and active agriculture, there appears to have been a lull in such activity. The pollen diagrams frequently suggest a general reduction (but not absence) of agriculture, and such taxa as hazel, ash, oak, and elm increase their representation. This sequence is clearly shown at Redbog, Co. Louth (Mitchell, 1976; Watts, Chap. 8, this volume) and at Littleton Bog, Co. Tipperary (Mitchell, 1965). For both sites, the situation changes dramatically when cereals and other anthropogenic indicators expand, and elm, ash, and hazel undergo rapid declines. This renewed bout of agricultural activity has been dated to ad 225 ±130 at Redbog. Mitchell (1965) equated this with the Early Christian period, which he placed before the traditional start of Christianity in Ireland as marked by the arrival of St. Patrick in AD 432. The apparent Iron Age passivity in agriculture was tentatively explained by Mitchell (1976, p. 162) as "the culmination of a long-continuing and widespread exhaustion of the soil, rather than a drastic social upheaval brought about by military conquest." The early Christian resurgence was attributed especially to the spread of monasticism and such contemporaneous husbandry techniques as the use of the plough share and coulter and plough teams (Hencken, 1950; Lucas, 1972, 1973).

The pattern of Iron Age decline and Early Christian revival of agricultural activity was also found at other sites investigated by Mitchell, such as Dromsallagh in Co. Limerick, Leigh in Co. Tipperary, and Clonsast in Co. Offaly (Mitchell, 1956). In southwest Ireland, Lynch (1981) found that of the three sites covering the period under discussion, Maughanasilly produced no definite signs of human activity, while pastoral activity is postulated after 70 bc at Dromatouk and mixed agriculture seems to have taken place at Cashelkeelty around ad 180, preceded by a pastoral phase at about 100 bc. Although the Cashelkeelty site produced evidence of increased arable activity after about ad 590 ± 85 with the first appearance of *Avena* (oat) pollen, the intensification of agriculture noted by Mitchell for the Early Christian period in central Ireland is less pronounced in the southwest. In Connemara, Teunissen and Teunissen-Van Oorschot (1980) recorded a fluctuating but steadily increasing influence on vegetation through the Iron Age. The forest area decreases, but they did note a short phase of secondary woodland expansion (*Ulmus* and *Fraxinus*) before a major increase in such herbaceous taxa as *Plantago lanceolata, Rumex, Artemisia, Urtica,* Compositae, and Cerealia, including *Secale* (rye), in the Early Christian period. At Scragh Bog, Co. Westmeath, O'Connell (1980) placed his Early Christian zone boundary at a level where ash and elm both decline, grass and cereal values reach high levels, and *Artemisia* and *Secale* (accompanied by *Avena, Hordeum,* and *Triticum*) make their first appearance. The earlier portions of the Scragh Bog site also suggest, however, that mixed agriculture was also of importance throughout the Iron Age in the local area. Further north, data from Beaghmore, Co. Tyrone (Pilcher, 1969), indicate some agriculture up to 100 bc, with large-scale woodland clearance between 100 bc and ad 350, and after the regeneration of hazel scrub there was a second major clearance from ad 350 to 1275. The Gortcorbies site in Co. Londonderry (Smith, 1975) includes an Iron Age clearance episode at 105 ± 65 bc, with an apparently more drastic clearance episode beginning at a level dated to ad 565 ± 65.

Discrepancies with the Mitchell (1956, 1976) model of vegetation change for this time can also be seen at other sites such as Cannons Lough, Co. Londonderry (Smith, 1961b), and the Lecale, Co. Down, sites of Ballydugan, Magheralagan, and Woodgrange (Singh and Smith, 1973). The interdrumlin basin site at Ballydugan is of particular interest because at the zone VIII/IX boundary (denoting the supposed start of the Early Christian period), tree pollen values drop catastrophically from more than 80% of the total terrestrial pollen sum down to around 20%, a level from which they never recovered. This woodland decline is accompanied not only by major expansions in the pollen of *Plantago lanceolata, P. major,* Gramineae, *Artemisia, Pteridium aquilinum,* Liguliflorae, and Cerealia but also by the deposition of silt upon the organic sediments of the basin. Agricultural impact was sufficiently severe to lead to the removal of material from the adjacent drumlin slopes. It is unfortunate that no radiocarbon dates exist for the Ballydugan site, while comparisons with the dated section at nearby Woodgrange are probably unhelpful in view of a likely hiatus in the Woodgrange sequence (Singh and Smith, 1973). At Cloverhill Lough, Co. Sligo, Göransson (1980) suggested that an inwashed soil, attributable to early zone IX erosion, prevented coring of the full sediment profile.

At Lough Catherine, the depth–time sequence of radiocarbon dates is dramatically disturbed after ad 805 ±95. From this date the ^{14}C determinations are reversed (Fig. 6). This phenomenon can be seen at other times and sites in Ireland such as after about ad 100 at Lough Neagh (O'Sullivan *et al.,* 1973) and after about ad 1315 and ad 1510 at Weir's Lough and Killymaddy Lough, respectively (Hirons, 1984). The inversion of dates would generally be interpreted as a result of the inwash of soils containing old carbon following agriculturally induced erosion (Edwards and Rowntree, 1980). At

Lough Catherine, the reversed dates coincide with high values for cereal and grass pollen, and they follow a prolonged period of likely erosion as indicated by the high values for potassium and magnetic susceptibility (Fig. 6). There is a strong likelihood that the ^{14}C dates preceding the reversed sequence are also falsely old. Although this highlights a problem in the interpretation of radiocarbon dates obtained from lake sediments [partly overcome by palaeomagnetic measurements (Thompson and Edwards, 1982; Hirons *et al.*, Chap. 12, this volume)], it also indicates their potential usefulness in demonstrating environmental impact.

Further signs of impact resulting from agricultural activity in the Early Christian period may be found elsewhere in Ireland. At Cloonlara in Co. Mayo, Jessen (1949, p. 161) recorded "Brown Sphagnum-peat with some content of sand." Mitchell (1956) placed this feature in his zone IX and remarked (p. 216) that "the sand in the peat may have been blown or washed down from the tilled fields on the slopes around the bog." Drew (1982), working in the karstic area of the Burren, Co. Clare, found evidence of former soils overlying the limestone terrain. Although these may have disappeared over a 3,000- to 4,000-year period, severe degradation in "Iron Age–Dark Age times" is favoured, perhaps in response to farming activity (see Crabtree, 1982; Plunkett Dillon, 1983; and Watts, Chap. 8, this volume). At Garradreen in Co. Wexford, Culleton and Mitchell (1976) found organic silt and charcoal overlain by clay loam at the bottom of a drainage ditch. The charcoal was dated to ad 500 ± 85, and the basal organic deposit was considered to represent topsoil material which had been eroded from cultivated upslope areas and subsequently covered by largely inorganic colluvium. The Garradreen site is some 600 m from a small univallate rath. The rath or ringfort would appear in many cases to have been a farm. The Lecale sites are also close to raths, of which tens of thousands survive in Ireland (Aalen, 1978). Many raths are known to be of Early Christian date (Proudfoot, 1961; Collins, 1968), and their wide provenance would be consistent with the extensive agricultural activity of the Early Christian and perhaps later periods.

THE HISTORIC AND MODERN PERIODS
(~AD 800 to the Present)

The Early Christian period in Ireland gave way in different areas to new cultural influences, be they Viking (after about AD 800) Anglo–Norman (after about AD 1150) or Tudor Plantations (after about AD 1550). These influences generally led to a continued destruction of woodland which was hastened by the demands of coopers, shipwrights, house builders, and iron producers. These times, so important for the social history of Ireland, lie outside the remit of this chapter. There is a frequent cessation of peat growth at bog sites and a swamping effect of local pollen as arboreal taxa decrease to low values. The continued presence of cereal pollen, especially at lake sites, attests to the importance of a mixed agricultural economy up to the present day. A marker horizon in some pollen diagrams is provided by the appearance of the pollen of planted trees from about AD 1700 onward— the zone X Afforestation period of Mitchell (1956). At Lough Catherine, for example, pollen spectra from AD 1750 onward are distinguished by the presence of pine, beech, lime, sycamore, and spruce. These taxa reflect the landscaping and woodland plantation of the surrounding Baronscourt Estate and can be related to the available historical evidence (McClintock, 1973; Thompson and Edwards, 1982). In general, the vegetational history of the recent past is best investigated from documentary sources (e.g., Praeger, 1934; McCracken, 1971; Mitchell, 1976).

CONCLUSIONS

Evidence accruing from palaeoecological and environmentally related archaeological research in Ireland underlines the long-term interrelationships between man and vegetation. For the Mesolithic there are several palynological sites which suggest tampering with the vegetation, although a lack of clear evidence for such early impact may be due to sampling problems rather than real absence. There is an embarrassment of evidence for impact during the succeeding Neolithic, although, as always, problems of interpretation provide a tantalizing rather than certain picture of events coinciding with the introduction of agriculture. Cereal cultivation certainly began prior to the elm decline. It may have been started by pioneering Neolithic farmers, but there remains the possibility that indigenous Mesolithic peoples may also have been involved. It is still not possible to assign a definite cause to the elm decline, although *Ulmus,* along with other woodland taxa, suffered reductions in early Neolithic times. The longevity of clearance phases in pollen diagrams warns against elevating speculation to truth—many combinations of folk movement and farming practice would produce similar palaeoecological records. The role of fire is still uncertain, and the evidence does not permit categorical statements as to its precise use, beyond cremation and domestic heating, throughout prehistory.

The Bronze Age continued to see patterns of clearance and regeneration together with two widely recognized phenomena, a marked decline in pine and the spread of blanket peat. The pine decline around 2,000 bc may well be a response to both anthropogenic exploitation and a deterioration in the pedogenic environment. The latter is associated with the appearance of Ireland's mantle of blanket peat, whose initiation may be linked, somewhat uncertainly, with the pressures of agriculture.

The Iron Age provided evidence for the continued favouring of peatland spread and a likely lull, but not absence, of agricultural activity in some areas from which pollen diagrams have been produced. There was for many localities a surge of woodland clearance and agriculture in the Early Christian period, and this can be seen to have produced detrimental effects on the soil cover with probable agriculturally induced erosion. The activities of man during the historic and modern periods served to further reduce Ireland's woodland cover to its present relict state.

Acknowledgments

I owe various debts to former colleagues from the Palaeoecology Centre of the Queen's University of Belfast for their assistance: Drs. Kenneth Hirons and Jonathan Pilcher discussed various aspects of the text; data from Lough Catherine and Derryandoran represents joint work with Robert Larmour, while that from Weir's and Killymaddy Loughs is based on Kenneth Hirons's doctoral thesis; ^{14}C dates from Lough Catherine and Derryandoran were provided by Dr. Gordon Pearson and assistants. I also thank Dr. Maureen Girling of the Department of the Environment, London, for permission to mention the find of *Scolytus scolytus*.

References

Aaby, B. (1976). *Nature (London)* **263,** 281–284.
Aalen, F. H. A. (1978). "Man and the Landscape in Ireland." Academic Press, London.
Ball, D. F. (1975). *In* "The Effect of Man on the Landscape: the Highland Zone" (J. G. Evans, S. Limbrey, and H. Cleere, eds.), pp. 20–27. CBA Res. Rep. No. 11, London.
Beckett, S. C., and Hibbert, F. A. (1976). *Somerset Levels Pap.* **2,** 24–27.
Behre, K.-E. (1981). *Pollen Spores* **23,** 225–245.
Birks, H. H. (1975). *Philos. Trans. R. Soc. London, Ser. B* **270,** 181–226.
Bradley, R. (1978). "The Prehistoric Settlement of Britain." Routledge & Kegan Paul, London.
Buckland, P. C., and Edwards, K. J. (1984). *J. Biogeogr.* **11,** 243–249.
Burenhult, G. (ed.) (1980a). "The Archaeological Excavation at Carrowmore, Co. Sligo, Ireland: Excavation Seasons 1977–1979." Theses and Papers in North-European Archaeology, No. 9, Inst. of Archaeol., Univ. of Stockholm, Stockholm.
Burenhult, G. (ed.) (1980b). "The Carrowmore Excavations, Excavation Season 1980." Stockholm Archaeol. Reps., No. 7. Inst. of Archaeol., Univ. of Stockholm, Stockholm.
Buzer, J. S. (1980). *New Phytol.* **86,** 93–108.
Case, H. J., Dimbleby, G. W., Mitchell, G. F., Morrison, M. E. S., and Proudfoot, V. B. (1969). *J. R. Soc. Antiq. Irel.* **99,** 39–53.
Caulfield, S. (1983). *In* "Landscape Archaeology in Ireland" (T. Reeves-Smyth and F. Hamond, eds.), pp. 195–215. BAR Brit. Ser. No. 116, Oxford.
Chamberlin, T. C. (1890, reprinted 1965). *Science* **148,** 754–759.

Coles, J. M. (1976). *In* "Acculturation and Continuity in Atlantic Europe" (S. J. De Laet, ed.), pp. 59–66. Diss. Archaeol. Gandenses, No. 16, Ghent.

Coles, J. M., and Hibbert, F. A. (1968). *Proc. Prehist. Soc.* **34,** 238–258.

Coles, J. M., Heal, S. V. E., and Orme, B. J. (1978). *Proc. Prehist. Soc.* **44,** 1–45.

Collins, A. E. P. (1968). *Ulster J. Archaeol.* **31,** 53–58.

Crabtree, K. (1982). *In* "Archaeological Aspects of Woodland Ecology" (M. Bell and S. Limbrey, eds.), pp. 105–113. BAR Int. Ser. No. 146, Oxford.

Craig, A. J. (1978). *J. Ecol.* **66,** 297–324.

Cruickshank, J. G. (1980). *Ir. Nat. J.* **20,** 21–31.

Culleton, E. B., and Mitchell, G. F. (1976). *J. R. Soc. Antiq. Irel.* **106,** 120–123.

Davis, M. B. (1981). *Proc. Int. Palynol. Conf. 4th Lucknow (1976–1977)* **3,** 216–228.

Davis, M. B., and Deevey, E. S. (1964). *Science* **145,** 1293–1295.

Dearing, J. A., and Flowers, R. J. (1982). *Limnol. Oceanogr.* **27,** 969–975.

Dennell, R. (1983). "European Economic Prehistory: A New Approach." Academic Press, London.

Dickinson, W. (1975). *J. Ecol.* **63,** 913–935.

Dimbleby, G. W. (1962). "The Development of British Heathlands and Their Soils." Oxford Forestry Memoir No. 23, Oxford.

Drew, D. P. (1982). *In* "Archaeological Aspects of Woodland Ecology" (M. Bell and S. Limbrey, eds.), pp. 115–127. BAR Int. Ser. No. 146, Oxford.

Edwards, K. J. (1978a). "Palaeoenvironmental and Archaeological Investigations in the Howe of Cromar, Grampian Region, Scotland." Unpublished Ph.D. thesis, Univ. of Aberdeen, Aberdeen.

Edwards, K. J. (1978b). *Murlough N. N. R. Scient. Rep. (National Trust)* **4,** 20–21.

Edwards, K. J. (1979a). *J. Archaeol. Sci.* **6,** 255–270.

Edwards, K. J. (1979b). *Scott. Archaeol. Forum* **9,** 27–42.

Edwards, K. J. (1982). *In* "Archaeological Aspects of Woodland Ecology" (M. Bell and S. Limbrey, eds.), pp. 5–22. BAR Int. Ser. No. 146, Oxford.

Edwards, K. J. (1983). *Ulster J. Archaeol.* **44–45** (1981–1982), 43–46.

Edwards, K. J., and Hirons, K. R. (1982). *Quat. Newsl.* **36,** 32–37.

Edwards, K. J., and Hirons, K. R. (1984). *J. Archaeol. Sci.* **11,** 71–80.

Edwards, K. J., and Ralston, I. (1980). *Proc. Soc. Antiq. Scot. (1977–1978),* **109,** 202–210.

Edwards, K. J., and Ralston, I. B. M. (forthcoming). *Proc. Soc. Antiq. Scot.*

Edwards, K. J., and Rowntree, K. M. (1980). *In* "Timescales in Geomorphology" (R. A. Cullingford, D. A. Davidson, and J. Lewin, eds.), pp. 207–223. Wiley, Chichester.

Erdtman, G. (1924). *Sven. Bot. Tidsk.* **18,** 451–459.

Erdtman, G. (1927). *Ir. Nat. J.* **1,** 1–4.

Erdtman, G. (1928). *Geol. For. Stockholm Forh.* **50,** 123–192.

Garbett, G. G. (1981). *New Phytol.* **88,** 573–585.

Girling, M. A., and Greig, J. R. A. (forthcoming). *J. Archaeol. Sci.*

Goddard, A. (1971). "Studies of the Vegetational Changes Associated with Initiation of Blanket Peat Accumulation in North-East Ireland." Unpublished Ph.D. thesis, The Queen's Univ. of Belfast, Belfast.

Godwin, H. (1975). "The History of the British Flora," 2nd Edn. Cambridge Univ. Press, Cambridge.

Göransson, H. (1980). *In* "The Carrowmore Excavations, Excavation Season 1980" (G. Burenhult, ed.), pp. 125–139. Stockholm Archaeol. Reps., No 7, Inst. of Archaeol., Univ. of Stockholm, Stockholm.

Göransson, H. (1981). *In* "The Carrowmore Excavations, Excavation Season 1981" (G. Burenhult, ed.), pp. 180–195. Stockholm Archaeol. Reps., No. 8, Inst. of Archaeol., Univ. of Stockholm, Stockholm.

Granlund, E. (1932). *Sven. Geol. Unders. Afh. Ser. C* **26**, No. 373, 1–193.

Groenman-van Waateringe, W. (1983). *In* "Landscape Archaeology in Ireland" (T. Reeves-Smyth and F. Hamond, eds.), pp. 217–232. BAR Brit. Ser. No. 116, Oxford.

Groenman-van Waateringe, W., and Pals, J. P. (1982). *In* "Newgrange: Archaeology, Art and Legend" (M. J. Kelly, ed.), pp. 219–223. Thames and Hudson, London.

Haines-Young, R. H., and Petch, J. H. (1982). Discussion Papers in Geography, No. 18, Univ. of Salford, Salford.

Hencken, H. O'N. (1950). *Proc. R. Ir. Acad., Sect. C* **47**, 1–76.

Herity, M., and Eogan, G. (1977). "Ireland in Prehistory." Routledge & Kegan Paul, London.

Hibbert, F. A., Switsur, V. R., and West, R. G. (1971). *Proc. R. Soc. London* **B117**, 161–176.

Hirons, K. R. (1983). *In* "Landscape Archaeology in Ireland" (T. Reeves-Smyth and F. Hamond, eds.), pp. 95–117. BAR Brit. Ser. No. 116, Oxford.

Hirons, K. R. (1984). "Palaeoenvironmental Investigations in East Co. Tyrone, Northern Ireland." Unpublished Ph.D. thesis, The Queen's Univ. of Belfast, Belfast.

Holland, S. M. (1975). "Pollen Analytical Study Concerning Settlements and Early Ecology in Co. Down, Northern Ireland." Unpublished Ph.D. thesis, The Queen's Univ. of Belfast, Belfast.

Huntley, B., and Birks, H. J. B. (1983). "An Atlas of Past and Present Pollen Maps of Europe: 0–13000 Years Ago." Cambridge Univ. Press, Cambridge.

Iversen, J. (1941). *Dan. Geol. Unders, [Afh.], Raekke 2* **66**, 1–68.

Iversen, J. (1956). *Sci. Am.* **194**, 36–41.

Jamieson, T. F. (1865). *Q. J. Geol. Soc. London* **21**, 161–203.

Jessen, K. (1936). *Proc. R. Ir. Acad. Sect. C* **43**, 31–37.

Jessen, K. (1949). *Proc. R. Ir. Acad., Sect. B* **52**, 85–290.

Jessen, K., and Helbaek, H. (1944). *Kongr. Danske Vidensk. Selsk.* **3**, 1–68.

Jones, P. (1981). *Trans. Inst. Br. Geogr. New Ser.* **6**, 324–336.

Kirk, S. M. (1974–1975). *Ulster J. Archaeol.* **36–37**, 99–100.

Lucas, A. T. (1972). *Tools Tillage* **2**, 52–62.

Lucas, A. T. (1973). *Tools Tillage* **2**, 67–83.

Lynch, A. (1981). "Man and Environment in South-West Ireland." BAR Brit. Ser. No. 85, Oxford.

Mackereth, F. J. H. (1966). *Philos. Trans. R. Soc. London, Ser. B* **250**, 165–213.

Maguire, D. J. (1983). *In* "Integrating the Subsistence Economy" (M. Jones, ed.), pp. 5–18. BAR Int. Ser. No. 181, Oxford.

Maguire, D. J., Ralph, N., and Fleming, A. (1983). *In* "Integrating the Subsistence Economy" (M. Jones, ed.), pp. 57–105. BAR Int. Ser. No. 181, Oxford.

McClintock, M. (1973). "The Pollen Analysis of Lacustrine Sediments from Lough Catherine Linked with Possible Human Activity over the Last Four Centuries." Unpublished B.A. thesis, New Univ. Ulster, Ulster.

McCracken, E. (1971). "The Irish Woods since Tudor Times." David and Charles, Newton Abbot, England.

Mitchell, G. F. (1945). *Proc. R. Ir. Acad. Sect. C* **50**, 1–19.

Mitchell, G. F. (1951). *Proc. R. Ir. Acad. Sect. B* **53**, 11–206.

Mitchell, G. F. (1954). *Proc. R. Ir. Acad. Sect. C* **56**, 481–488.

Mitchell, G. F. (1955). *Ulster J. Archaeol.* **18**, 1–16.

Mitchell, G. F. (1956). *Proc. R. Ir. Acad. Sect. B* **57**, 185–251.

Mitchell, G. F. (1965). *J. R. Soc. Antiq. Irel.* **95**, 121–132.

Mitchell, G. F. (1972). *24th Int. Geol. Congr. Symp.* **1**, 59–68.

Mitchell, G. F. (1976). "The Irish Landscape." Collins, London.

Moore, P. D. (1975). *Nature (London)* **250,** 439–441.

Morrison, M. E. S. (1959). *Botaniska Notiser* **112,** 185–204.

Movius, H. L. (1936). *Proc. R. Ir. Acad., Sect. C* **43,** 17–40.

Movius, H. L. (1942). "The Irish Stone Age: Its Chronology, Development and Relationships." Cambridge Univ. Press, Cambridge.

O'Connell, M. (1980). *New Phytol.* **85,** 301–319.

Osborne, P. J. (1976). *World Archaeol.* **8,** 150–158.

O'Sullivan, P. E., Oldfield, F., and Battarbee, R. W. (1973). *In* "Quaternary Plant Ecology"
(H. J. B. Birks and R. G. West, eds.), pp. 267–278. Blackwell, Oxford.

Pearson, G. W., and Pilcher, J. R. (1975). *Radiocarbon* **17,** 226–238.

Pearson, G. W., Pilcher, J. R., Baillie, M. G. L., and Hillam, J. (1977). *Nature (London)* **270,**
25–28.

Pennington, W. (1973). *In* "Quaternary Plant Ecology" (H. J. B. Birks and R. G. West, eds.),
pp. 79–104. Blackwell, Oxford.

Pennington, W. (1975). *In* "The Effect of Man on the Landscape: the Highland Zone" (J. G.
Evans, S. Limbrey, and H. Cleere, eds.), pp. 74–86. CBA Res. Rep. No. 11, London.

Pilcher, J. R. (1969). *Ulster J. Archaeol.* **32,** 73–91.

Pilcher, J. R. (1973). *New Phytol.* **72,** 681–689.

Pilcher, J. R. (1975). *Ir. Archaeol. Res. Forum* **11,** 1–6.

Pilcher, J. R., and Larmour, R. (1982). *Proc. R. Ir. Acad., Sect. B* **82,** 277–295.

Pilcher, J. R., and Smith, A. G. (1979). *Philos. Trans. R. Soc. London, Ser. B* **286,** 345–369.

Pilcher, J. R., Smith, A. G., Pearson, G. W., and Crowder, A. (1971). *Science* **172,** 560–562.

Plunkett Dillon, E. (1983). *In* "Landscape Archaeology in Ireland" (T. Reeves-Smyth and F.
Hamond, eds.), pp. 81–94. BAR Brit. Ser. No. 116, Oxford.

Praeger, R. L. (1888). *Proc. Belfast Nat. Field Club* **2,** Appendix 2, 29–51.

Praeger, R. L. (1896). *Proc. R. Ir. Acad.* **4,** 30–54.

Praeger, R. L. (1934). "The Botanist in Ireland." Hodges, Figgis & Co., Dublin.

Praeger, R. L. (1939). *Proc. Linn. Soc. London* **151,** 192–213.

Proudfoot, V. B. (1958). *J. Soil Sci.* **9,** 186–198

Proudfoot, V. B. (1961). *Medieval Archaeol.* **5,** 94–121.

Rackham, O. (1980). "Ancient Woodland: Its History, Vegetation and Uses in England."
Arnold, London.

Rawitscher, F. (1945). *Nature (London)* **156,** 302–303.

Renfrew, C. (ed.) (1974). "British Prehistory, A New Outline." Duckworth, London.

Robertson, D. (1877). *Trans. Geol. Soc. Glasgow* **5,** 192–200.

Romans, J. C. C., and Robertson, L. (1975). *In* "The Effect of Man on the Landscape: the
Highland Zone" (J. G. Evans, S. Limbrey, and H. Cleere, eds.), pp. 37–39. CBA Res.
Rep. No. 11, London.

Seddon, B. (1967). *In* "Weather and Agriculture" (J. A. Taylor, ed.), pp. 173–185. Pergamon,
Oxford.

Simmons, I. G., Dimbleby, G. W., and Grigson, C. (1981). *In* "The Environment in British
Prehistory" (I. G. Simmons and M. J. Tooley, eds.), pp. 82–124. Duckworth, London.

Sims, R. E. (1973). *In* "Quaternary Plant Ecology" (H. J. B. Birks and R. G. West, eds.), pp.
223–236. Blackwell, Oxford.

Singh, G., and Smith, A. G. (1973). *Proc. R. Ir. Acad. Sect. B* **73,** 1–51.

Smith, A. G. (1958). *Proc. R. Ir. Acad. Sect. B* **59,** 329–343.

Smith, A. G. (1961a). *Proc. Linn. Soc. London* **172,** 38–49.

Smith, A. G. (1961b). *Proc. R. Ir. Acad., Sect. B* **61,** 369–383.

Smith, A. G. (1964). *Rep. VI Int. Congr. Quat. Warsaw 1961* **2,** 461–471.

Smith, A. G. (1970). *In* "Studies in the Vegetational History of the British Isles: Essays in Honour of Harry Godwin" (D. Walker and R. G. West, eds.), pp. 81–96. Cambridge Univ. Press, London.

Smith, I. F. (1974). *In* "British Prehistory: A New Outline" (C. Renfrew, ed.), pp. 100–136. Duckworth, London.

Smith, A. G. (1975). *In* "The Effect of Man on the Landscape: The Highland Zone" (J. G. Evans, S. Limbrey, and H. Cleere, eds.), pp. 64–74. CBA Res. Rep. No. 11, London.

Smith, A. G. (1981a). *Proc. Int. Palynol. Conf. 4th Lucknow (1976–1977)* **3**, 248–257.

Smith, A. G. (1981b). *In* "The Environment in British Prehistory" (I. G. Simmons and M. J. Tooley, eds.), pp. 125–290. Duckworth, London.

Smith, A. G. (1984). *New Phytol.* **98**, 35–55.

Smith, A. G., and Collins, A. E. P. (1971). *Ulster J. Archaeol.* **34**, 3–25.

Smith, A. G., and Pilcher, J. R. (1972). *Ulster J. Archaeol.* **35**, 17–21.

Smith, A. G., and Pilcher, J. R. (1973). *New Phytol.* **72**, 903–914.

Smith, A. G., and Willis, E. H. (1962). *Ulster J. Archaeol.* **24–25**, 16–24.

Tallis, J. H. (1964). *J. Ecol.* **52**, 333–344.

Ten Hove, H. A. (1968). *Palaeogeogr. Palaeoclimatol. Palaeoecol.* **5**, 359–369.

Teunissen, D., and Teunissen-Van Oorschot, H. G. C. M. (1980). *Acta Bot. Neerl.* **29**, 285–306.

Thompson, R., and Edwards, K. J. (1982). *Boreas* **11**, 335–349.

Thompson, R., Battarbee, R. W., O'Sullivan, P. E., and Oldfield, F. (1976). *Limnol. Oceanogr.* **20**, 687–698.

Tinsley, H. M., and Grigson, C. (1981). *In* "The Environment in British Prehistory" (I. G. Simmons and M. J. Tooley, eds.), pp. 210–249, Duckworth, London.

Tohall, P., De Vries, H. L., and Van Zeist, W. (1955). *J. R. Soc. Antiq. Irel.* **85**, 77–83.

Tooley, M. J. (1978). "Sea-level Changes in North-West England during the Flandrian Stage." Clarendon Press, Oxford.

Troels-Smith, J. (1960). *Dan. Geol. Unders. [Afh.] Raekke 4* **4**, 1–32.

Turner, J. (1981). *In* "The Environment in British Prehistory" (I. G. Simmons and M. J. Tooley, eds.), pp. 250–281. Duckworth, London.

Van Zeist, W. (1954). *Palaeohistoria* **3**, 220–224.

Watson, J. W. (1939). *Scott. Geogr. Mag.* **55**, 148–161.

Watts, W. A. (1961). *Proc. Linn. Soc. London* **172**, 33–38.

Watts, W. A. (1977). *Philos. Trans. R. Soc. London, Ser. B* **280**, 273–293.

Webb, D. A. (1943). "An Irish Flora." Dundalgan Press, Dundalk.

Weber, C. A. (1900). *Jahres-Bericht der Manner von Morgenstern* **3**, 3–23.

Whelan, C. B. (1933). *Ir. Nat. J.* **4**, 149.

Whelan, C. B. (1934). *Proc. R. Ir. Acad. Sect. C* **42**, 121–143.

Whelan, C. B. (1936). *Rep. XVI Int. Geol. Congr. Washington 1933,* 1209–1218.

Whelan, C. B. (1938). *Proc. R. Ir. Acad. Sect. C* **44**, 115–136.

Whittle, A. W. R. (1978). *Antiquity* **52**, 34–42.

van Wijngaarden-Bakker, L. H. (1974). *Proc. R. Ir. Acad., Sect. C* **74**, 313–383.

Woodman, P. C. (1977). *Proc. Prehist. Soc.* **43**, 155–199.

Woodman, P. C. (1978). "The Mesolithic in Ireland: Hunter–Gatherers in an Insular Environment." BAR Bri. Ser. No. 58, Oxford.

CHAPTER 10

QUATERNARY VERTEBRATES

Anthony J. Stuart
Department of Zoology
University of Cambridge
Cambridge, England

Louise H. van Wijngaarden-Bakker
Albert Egges van Giffen Instituut voor Prae- en Protohistorie
University of Amsterdam
Amsterdam, The Netherlands

INTRODUCTION

This chapter consists of two sections; the first dealing with the faunas of the Midlandian Stage, and the second covering those of the postglacial or Littletonian Stage. This arrangement is convenient since the nature of the available evidence, and the treatments of it, differs considerably for the two periods under discussion. The postglacial faunas are much better studied and dated than those of Midlandian age, as they have been the subject of much recent research in connection with archaeological excavations. In contrast, information on the earlier faunas is poor because most of the material is inadequately dated and little recent research work has been done. Nevertheless it is possible to extract a surprising amount of useful information from the Midlandian fossil record as it stands, and there is of course considerable potential for future work.

221

MIDLANDIAN FAUNAS*

Introduction

Fossil mammal remains dating from the Midlandian (last glacial) Stage occur fairly abundantly in Ireland and were recorded as long ago as the late sixteenth century AD. A skull of a giant Irish deer ("Irish elk") found in the Bog of Meath in 1588 and later sent to Hatfield House, Hertfordshire, England, is mentioned in the state papers of Elizabeth I (O'Riordan, 1980). Since that time very many more finds have come to light, but our current knowledge of the Quaternary vertebrates of Ireland is tantalizingly incomplete. Nearly all of the available fossils were obtained by chance finds, amateur digging (especially for the highly saleable giant deer "heads"), or cave excavations of a standard which would be unacceptable today. With the exception of numerous lateglacial records of giant deer and reindeer, which by good fortune can be assigned to deposits of which the age can be fixed within narrow limits, and a few specimens directly dated by radiocarbon, Irish fossil vertebrate finds are poorly stratified and dated. Moreover, there are no records whatsoever of finds from earlier than the middle of the Midlandian Stage.

It is clear, from both the above remarks and the details which follow, that there is much scope for future research, both in discovering new material and in the radiocarbon dating of finds. In particular, the peripheral and largely isolated situation of Ireland makes its Quaternary faunas of considerable interest from the viewpoint of palaeozoogeography. Recent reviews or general accounts of Quaternary vertebrates in Ireland include Savage (1966), Stuart (1977, 1982), Mitchell (1969, 1976), and O'Riordan (1980).

The localities mentioned in this section are shown in Fig. 1.

The Lack of Faunas Older Than the Middle Midlandian

Whereas Britain has yielded rich fossil vertebrate faunas covering the middle and upper Pleistocene (see Stuart, 1982), Ireland has no vertebrate records which can be shown to predate the middle of the Midlandian Stage, about 35,000 years bp. It is unlikely that no terrestrial vertebrates reached Ireland before the Midlandian. Within earlier cold stages, animals could well have crossed from Britain on floating ice or there may have been land connections with Britain at times of lowered sea level. During interglacial periods, the occasional large mammal may have succeeded in swimming to

*This section was written by Dr. Stuart.

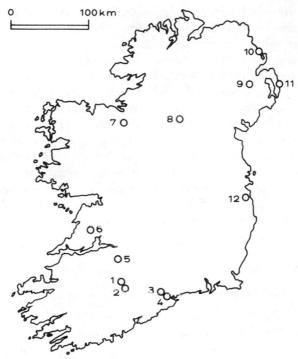

Fig. 1. Localities with fossil vertebrates of Midlandian age mentioned in the text: 1, Castlepook Cave, Co. Cork; 2, Castletownroche Caves, Co. Cork; 3, Shandon and Kilgreany Caves, Co. Waterford; 4 Ballynamintra Caves, Co. Waterford; 5, Red Cella Cave, Co. Limerick; 6, Edendale Caves, Co. Clare; 7, Keshcorran Caves, Co. Sligo; 8, Maghery, Co. Cavan; 9, Derriaghy, Co. Down; 10, Kilwaughter Quarry, Co. Antrim; 11, Roddans Port, Co. Down; 12, Ballybetagh, Co. Dublin.

Ireland, whereas small mammals and reptiles could have been transported across the Irish Sea as unwitting passengers on floating vegetation. Birds and bats would have encountered few obstacles to colonization, but they are usually poorly represented in fossil assemblages.

The lack of records of pre-Midlandian terrestrial vertebrates is no doubt mainly due to the fact that Ireland was extensively glaciated late in the Midlandian Stage. Significantly, few finds of pre-lateglacial age have been made from open sites in Britain within the area covered by late Devensian (cf. late Midlandian) ice. In both areas this phenomenon is probably the result of obliteration of earlier deposits by the ice sheets, and for similar reasons "pre-glacial" fossil vertebrates are extremely rare both in Scandinavia and in much of Canada.

The lack of vertebrate records from the organic muds of the Gortian

Interglacial are not likely to be of significance, as vertebrate fossils are relatively scarce in lacustrine sediments and soon disappear by solution in a noncalcareous environment.

Fossil vertebrates of Midlandian age are known from several caves in the Carboniferous limestone of Counties Clare, Cork, Limerick, Sligo, and Waterford, but no interglacial fauna has yet come to light; on the other hand, more than 14 bone caves in England and Wales have yielded "hippopotamus faunas" characteristic of the middle of the last (Ipwichian) interglacial (Sutcliffe, 1960; Stuart, 1976, 1982). Again, however, it needs to be stressed that the vast majority of British bone caves have produced only faunas of Devensian (cf. Midlandian) age. Since less than 20 Irish bone caves have been excavated, compared with probably several hundred in Britain (generally also excavated to a poor standard), the lack of an interglacial fauna from Ireland is probably more apparent than real. With luck, the discovery of an Irish last interglacial vertebrate fauna could reward the efforts of a future excavator.

Middle Midlandian Faunas

Without doubt, the most important bone cave in Ireland is Castlepook Cave, situated in a knoll of Carboniferous limestone about 4 km north of Doneraile, Co. Cork. The cave consists of a large and complex series of interconnecting galleries, none more than about 1.5 m wide or 2.7 m high, reaching to at least 120 m from the present entrance which originated by solution along a network of joints in the limestone (see Scharff *et al.*, 1918). The site was excavated by Ussher in 1904 (see Scharff *et al.*, 1918), who recorded a bed of sand, generally containing bones, throughout much of the cave system. From the sections given by Scharff *et al.* (1918), this sand appears to have been only about 0.6–0.9 m in thickness. Bones were found mainly on or near the surface of the sand, and, in places, stalagmite was recorded immediately overlying the sand, sometimes with bones adhering to the underside.

The area richest in bones, termed "Hyaena Land" by Ussher, lay farthest from the present entrance, suggesting the previous existence of an entrance on the far side of the knoll. Two radiocarbon dates on Castlepook material are available:

Material	Date (^{14}C years bp)	Lab number
Mammuthus primigenius, mammoth (molar)	33,500 ± 1200	D-122
Crocuta crocuta, spotted hyaena	34,300 ± 1800	I-13,265

Table I Mammal Fauna from Castlepook Cave

Carnivora
 Vulpes[a] or *Alopex,* red fox or arctic fox
 Canis lupus L.,[a] wolf
 Mustela erminea L.,[a] stoat
 Ursus arctos L.,[a] brown bear
 Crocuta crocuta (Erxleben), spotted hyaena

Lagomorpha
 Lepus timidus L.,[a] arctic or mountain hare

Rodentia
 Dicrostonyx torquatus (Pallas), arctic lemming
 Lemmus lemmus (L.), Norway lemming

Proboscidea
 Mammuthus primigenius (Blumenbach),[b] woolly mammoth

Artiodactyla
 Megaloceros giganteus (Blumenbach),[b] giant deer
 Rangifer tarandus (L.), reindeer

[a] Present in Ireland at the present day or within approximately the last 1,000 years.
[b] Extinct species.

Taken together these dates are of interest as they may indicate that the Castlepook Pleistocene fauna is of much the same age (i.e., about 34,000 bp), although clearly more ^{14}C dates would be welcome.

Some of the faunal remains comprised splintered, fragmentary material, much of it clearly chewed by hyaenas, whereas other remains, including many reindeer bones, were preserved intact. Unfortunately, in addition to the remains of exotic species, obviously of Pleistocene age, recent material including domestic animals, rabbit, and brown rat also occurred in the cave, so that, as Scharff *et al.* (1918) realized, it is unclear in certain cases whether a given species belongs to the Pleistocene fauna or not. Table I lists those taxa which are probably of Midlandian age. Scharff *et al.* (1918) also included wood mouse (*Apodemus sylvaticus*) in their list, although this is a woodland animal apparently quite out of place in a cold stage fauna. Their paper does, however, mention that the fossil mouse material was not actually found *in situ.* Sparse bird remains and bones of frog from the site could also be of Pleistocene age. Castlepook Cave is the only Irish site with spotted hyaena (*Crocuta crocuta*). With no proper descriptions of the material, or its stratigraphic provenance and spatial distribution at the site, it is not possible to say much about the taphonomy of the fossil assemblage. The fact that some material had clearly been chewed by spotted hyaena and the presence of hyaena coprolites strongly suggests that some of the bones were dragged in by this animal, either from its own kills or from scavenged carcasses. The numerous lemming remains were no doubt, as elsewhere, accumulated from the regurgitated pellets of roosting owls.

Other records are more tentatively assigned to a date prior to the main late Midlandian glaciation, solely on the presence of mammoth. Mammoth (*Mammuthus primigenius*) is generally absent from west European faunal assemblages after about 12,000 bp, and in Britain there is no strong evidence that it returned after the main late Devensian glaciation of about 18,000–15,000 bp (Stuart, 1982).

As long ago as 1715, mammoth remains were recorded from Co. Cavan by W. Molyneux (Charlesworth, 1963). Four molars, portions of the jaws, skull, and postcranial bones were found 1.2 m below the surface in river bank clay, near Maghery, 13 km from Belturbet. Savage (1966) discussed this and other finds of mammoth from Ireland. A humerus dredged from Galway Bay and described by Adams as mammoth could equally well be from a modern elephant. Savage also expressed doubts concerning several records from Co. Antrim. Part of a molar from a fissure in the chalk at Kilwaughter Quarry, near Larne, however, appears to be a reliable record, although only a cast of the original survives.

Mammoth is reliably recorded not only from Castlepook Cave but also from Shandon Cave, Co. Waterford (Brenan, 1860; Carte, 1860), and Foley Cave, Castletownroche, Co. Clare (Coleman, 1947). Shandon Cave also yielded bear, wolf, reindeer, and horse (apparently in association with the other fauna), while bear, wolf, Norway lemming, arctic lemming, giant deer, and reindeer make up the records from Foley Cave.

Other cave sites which have yielded a "cold" fauna could be either of late Midlandian age or date from earlier in the stage. For convenience, they will be discussed below with the lateglacial faunas.

Late Midlandian Faunas

Only two species, giant deer (*Megaloceros giganteus*) and reindeer (*Rangifer tarandus*), are definitely known to have been present in Ireland within late Midlandian (lateglacial) times, and their remains occur widely in lacustrine deposits of that age. A number of cave deposits with poorly provenanced faunal material probably also date from the lateglacial. The fossil occurrences of giant deer in Ireland are discussed in detail by Mitchell and Parkes (1949), while reindeer was dealt with by Mitchell (1941, 1976).

Reindeer is recorded from 13 caves and 13 open sites, whereas the giant deer has been found at over 12 cave sites and at more than 145 open sites (Mitchell, 1941, his Fig. 1; Mitchell and Parkes, 1949, their Fig. 1; Mitchell,

1976, p. 76). In almost every case where there is some documentary record of the stratigraphic provenance of the find, remains of giant deer and reindeer from open sites were found in calcareous lake marls, muds, or peats which can now be assigned to the Woodgrange (late Midlandian) Interstadial. These deposits are usually sealed by a layer of sandy clay dating from the succeeding Nahanagan Stadial (the "Younger *Dryas*") and, in turn, by postglacial peats. Controlled excavations of giant deer and reindeer remains *in situ* will be necessary before it is possible to establish the precise stratigraphic range of these animals in the late Midlandian. At present it is known that many fossils come from sediments broadly dated between 13,000 and 10,500 bp, and there are a few radiocarbon dates now available, obtained either directly from bone collagen or from associated plant material (Table II). These suggest that giant deer was present between about 11,380 and 10,900 bp and perhaps as far back as 12,200 bp, while dates of about 10,700 and 10,250 bp for reindeer suggest survival into Nahanagan Stadial times. Comparable late dates for reindeer have also been obtained from England (see summary in Stuart, 1982).

The remains of numerous individuals commonly occur at a single site. At Ballybetagh, near Dublin (Jessen and Farrington, 1938), sediments beneath the peats of three small bogs (the largest of which measures about 330 by 60 m) have yielded the remains of more than 60 giant deer plus several reindeer, and no doubt many more remain entombed in the deposits. The material from Ballybetagh and other such sites includes entire associated skeletons, skulls with antlers and associated mandibles, or portions of the skeleton (e.g., complete limbs). Male giant deer remains in museum collections far outnumber those of females, while fossils of giant deer are much better represented than those of reindeer.

As has been suggested previously (e.g., Coope, 1973; Mitchell, 1976), many giant deer may have died as a result of becoming mired in soft muds and clays at lake margins. Brecciation of sediments containing giant deer remains, thought to have been produced by the struggles of mired animals, were observed at Shortalstown, Co. Wexford, and Derriaghy, near Lisburn, Co. Down (Colhoun and Mitchell, 1971). Another possible natural trap mechanism could have been provided by thin ice on lakes, especially toward the end of winter. There are eyewitness accounts of modern North American moose (*Alces alces*) breaking through thin ice on lakes and rivers and subsequently drowning (see Peterson, 1955). Significantly, male moose are more at risk in this situation, being encumbered by their antlers which quickly become weighed down by ice from water freezing onto them. The fact that fossils of the male giant deer outnumber those of females in museum collections could be due in part to a collecting bias in favour of the

Table II Radiocarbon Dates Relevant to Irish Fossil Mammal Material

Species and sites	Material	Dates in years bp	Source
Mammuthus primigenius, mammoth			
Castlepook Cave, Co. Cork	Bone	33,500 ± 1200 (D-122)	Dresser and McAulay, 1974
Rangifer tarandus, reindeer			
Roddans Port, Co. Down	Antler[a]	10,250 ± 350 (LJ-658)	Hubbs et al., 1965
Kilgreany Cave, Co. Waterford	Bone	10,700 ± 100 (Pta-2378)	T. Molleson, personal communication
Megaloceros giganteus, giant deer			
Ballybetagh, Co. Dublin	Antler	15,170 ± 160[b] (BM-1794)	Burleigh et al., 1982
Locality unknown	Antler	11,380 ± 280 (BM-1904)	Burleigh et al., 1982
Locality unknown	Ribs	10,920 ± 250 (BM-1840)	Burleigh et al., 1982
Shortalstown, Co. Wexford	Plant macrofossils, broadly associated	12,160 ± 180 (I-4963)	Buckley and Willis, 1972
Crocuta crocuta, spotted hyaena			
Castlepook Cave, Co. Cork	Bone	34,300 ± 1800 (I-13,625)	G. F. Mitchell, personal communication

[a] Antler from pollen zone II horizon; date appears to be too young.
[b] Date thought too old due to incomplete removal of paraffin wax preservative.

more readily saleable antlered skulls, which are also easier to find and recognize in the field. Nevertheless, this preponderance of male remains may also be ascribed to a greater susceptibility to entrappment by miring or breaking through ice of the larger, heavier males with their enormous antlers. The relative scarcity of reindeer remains can be explained readily, since their much smaller body weight and proportionately large spreading feet would have usually kept them from danger. After death, corpses may have either stayed intact because of fairly rapid burial by sediment or floated and decayed, dropping off the skull and articulated portions of the skeleton onto the lake bottom, thus accounting for the variety of skeletal remains found.

Mitchell (1976) draws attention to the concentrations of giant deer finds in the lowlands, especially in Counties Limerick and Meath, and their absence in the upland areas of the west and southwest with their acid soils. He suggests that the distribution of giant deer in Woodgrange Interstadial times corresponded with areas of rich grazing. It should be stressed that the preservation of giant deer remains, on such a large scale, in the calcareous lacustrine deposits of the lowlands is most unusual. The absence of giant deer fossils elsewhere could be due largely to a lack of such conditions in upland acidic areas. Other aspects of the giant deer are discussed later in this section.

A few Irish bone caves have yielded "cold" faunas but not mammoth or hyaena. On comparison with radiocarbon-dated Devensian faunas from Britain (see Stuart, 1982), these faunas could either date from the lateglacial, or, if mammoth remains did not happen to reach the site, from earlier in the Midlandian Stage. None of these bone caves is satisfactory from the viewpoint of stratigraphy or excavation. Movius (1935) reexamined the sequence in Kilgreany Cave, near Dungarvan, Co. Waterford, and concluded that the stratigraphy of the loose deposits was hopelessly confused. In particular, he pointed out that earlier claims for the presence of man of Palaeolithic age at the site could not be demonstrated, a conclusion subsequently supported by radiocarbon dates of postglacial age on the human material (Oakley et al., 1971; but see Edwards and Warren, Chap. 1, this volume). The Kilgreany fauna, according to Movius (1935), includes giant deer, reindeer, wolf, bear, and arctic lemming (Dicrostonyx torquatus) together with many finds clearly of postglacial age. The record of lynx, if correct, is likely to be of postglacial (or interglacial) date, as the modern animal inhabits forested areas. A radiocarbon date of about 10,700 bp (Table II) on giant deer bone indicates that at least some of the fauna is of lateglacial age. A similar fauna is recorded from the Edenvale Caves, Co. Clare, and the Keshcorran Caves, Co. Sligo (Coleman, 1947), while arctic lemming is recorded from Ballynamintra Cave, Co. Waterford, and Red Cellar Cave, Co. Limerick (Sutcliffe and Kowalski, 1976).

Some Observations on the Giant Deer

The giant deer, or "Irish elk" (*Megaloceros giganteus* Blumenbach), has long attracted interest, principally because of its enormous antlers and large body size, and to many it symbolizes the Irish Quaternary. It seems appropriate, therefore, to discuss here some aspects of the anatomy and ecology of this fascinating beast, especially as a number of misconceptions occur in the literature.

As pointed out succinctly by Gould (1974) in his important biometrical study, the "Irish elk," like the Holy Roman Empire, is misnamed in all of its attributes, being neither an elk nor exclusively Irish. The living elk (as it is called in Europe), or moose (as it is called in North America), belongs to a different species and genus, namely, *Alces alces*, which has quite distinctive antler and body morphology. The giant deer, as it is called more appropriately, is assigned to the extinct genus *Megaloceros* (or *Megaceros*) and may be most closely related to the living fallow deer (*Dama dama*). Early species of *Megaloceros* are known from the European early middle Pleistocene (Azzaroli, 1953). Far from being restricted to Ireland, *M. giganteus* is known in England from the Hoxnian, Ipswichian, and Devensian Stages and from their equivalents on the Continent, extending as far as European Russia, while a related species was present in China. Gould's (1974) contention that *M. giganteus* appeared suddenly in the late Pleistocene fossil record due to "strong selective pressures exerted upon immigrants into recently glaciated territory" is therefore ill founded.

A further misconception about *M. giganteus* concerns its size. Although boasting the greatest antler span (up to 3.5 m or more) of any known deer, living or fossil, it is nevertheless quite untrue, in spite of statements to the contrary, that *M. giganteus* was also the largest in body size. The largest mounted skeleton of giant deer approaches 1.8 m in shoulder height. In body size, however, *M. giganteus* is exceeded by modern North American moose and even more so by an extinct elk (*Alces latifrons*) from the early Middle Pleistocene of Europe, which appears to have been the largest ever species of deer (Azzaroli, 1953).

Both Coope (1973) and Gould (1974) consider that the spectacular antlers carried by male *M. giganteus* were never put to the test in actual combat but were merely displayed in order to intimidate rivals for the affections of the females. The large antler size is therefore thought to have evolved by sexual selection for display. The view that the antlers were not used as weapons, however, is disputed by Clutton-Brock (1982). Both Coope and Gould also assert that the Irish animals were larger overall, and had larger antlers, than any *Megaloceros* known from elsewhere, because the absence of predators in Ireland allowed the deer to indulge the luxury of sexual selection to a point which would have prejudiced survival if predators had been present. Bearing in mind, however, the abundance of well-preserved complete fossils from Ireland and their scarcity elsewhere, there is insufficient evidence to suppose that the Irish giant deer were in fact larger than their foreign contemporaries. Moreover, Ipswichian (last interglacial) *M. giganteus* bones from England are as large as those from the Irish Midlandian (A. Lister, unpublished measurements, personal communication).

A final point concerns the widespread belief that the population density of giant deer was highest in late Midlandian Ireland; a belief based solely on the abundance of fossils in Ireland. The large numbers of giant deer finds probably reflect both the unusually widespread occurrence of shallow lakes which acted as traps for these animals and the disposition within the lakes of calcareous sediments which preserved the bones. Considerable numbers of fossils could have accumulated at any given site from the trapping of an occasional giant deer, or group of animals, over a period of, say, one or two thousand years. Nevertheless, because of the lack of ungulate competitors other than reindeer, it is quite possible that giant deer were in fact more abundant in Ireland than elsewhere. There is no way, however, that this can be demonstrated directly from the fossil record.

Faunal History of the Midlandian

The Midlandian age Irish faunas appear to be markedly impoverished in comparison with those of Britain, although the differences may be exaggerated by the small numbers of Irish fossil localities. Conspicuously absent from the Irish faunas, but present in those from Britain and Continental Europe, during the middle Midlandian are woolly rhino, an extinct bison (*Bison priscus*), red deer (common in British bone caves), and all species of vole. The absence of such animals as musk ox and lion, both rare in Britain, is, however, less remarkable.

Similarly, the Irish late Midlandian faunas lack the following: elk (*Alces alces*), recorded from late Devensian Interstadial sites in northern England (Hallam *et al.*, 1973); pika (*Ochotona pusilla*), common in English late Devensian faunas; and, again, all species of vole. Of particular importance also is the absence of any trace of man in Ireland until the postglacial (see Woodman, Chap. 11, this volume).

It appears unlikely that the absence of most of the above species in Ireland is due to their failure to reach southwest Scotland and from there cross a supposed land bridge to northeast Ireland, because woolly rhinoceros, elk, and several voles are recorded from Devensian deposits further north in Scotland. Clearly, some sort of filter was allowing only certain species to reach Ireland. There was probably no continuous land connection (although see Synge, Chap. 6, this volume), and the only successful immigrants were those able to swim a narrow strait or straits, or to cross the sea when frozen or on ice flows. Since the main late Midlandian ice covered most of Ireland, it is very probable that the earlier terrestrial fauna was entirely, or almost entirely, exterminated. The lateglacial fauna would then have reimmigrated via Britain.

The demise of the giant deer in Ireland may be ascribed to the climatic deterioration, and the attendant deterioration in the quality of grazing, at the onset of the Nahanagan Stadial (Mitchell, 1976). Reindeer, and perhaps arctic lemming and arctic fox, may have survived until the end of the stage, when ameliorating temperatures and the development of forests caused their extinction locally.

At several times in the past few hundred thousand years in Ireland, a rich vegetation cover probably supported large populations of terrestrial verte-brates. However, because of the tenuous nature and northerly position of past migration routes between Britain and Ireland, any major climatic changes might have left most animals no means of escape by migration elsewhere. In other words, Quaternary Ireland has been repeatedly a death trap for its terrestrial vertebrate faunas.

Acknowledgments

I am grateful to Dr. A. Lister for discussion with regard to giant deer, and to R. Burleigh, Ms. T. Molleson, and Professor G. F. Mitchell for supplying information on radiocarbon dates.

References

Azzaroli, A. (1953). *Bull. Br. Mus. (Nat. Hist.) Geol.* **2,** 3–96.
Brenan, E. (1860). *J. R. Dublin Soc.* **2,** 344–355.
Buckley, J., and Willis, E. H. (1972). *Radiocarbon* **14,** 120.
Burleigh, R., Ambers, J., and Matthews, K. (1982). *Radiocarbon* **24,** 263–264.
Carte, A. (1860). *J. R. Dublin Soc.* **2,** 351–357.
Charlesworth, J. K. (1963). *Proc. R. Ir. Acad., Sect. B* **62,** 295–322.
Clutton-Brock, T. H. (1982). *Behaviour* **79,** 108–125.
Coleman, J. C. (1947). *J. R. Soc. Antiq. Irel.* **77,** 63–80.
Colhoun, E. A., and Mitchell, G. F. (1971). *Proc. R. Ir. Acad., Sect. B* **234,** 211–245.
Coope, G. R. (1973). *Deer* **2,** 974–977.
Dresser, P. Q., and McAulay, I. R. (1974). *Radiocarbon* **16,** 8.
Gould, S. J. (1974). *Evolution* **28,** 191–220.
Hallam, J. S., Edwards, B. J. N., Barnes, B., and Stuart, A. J. (1973). *Proc. Prehist. Soc.* **39,** 100–128.
Hubbs, C. L., Bien, G. S., and Suess, H. E. (1965). *Radiocarbon* **7,** 103.
Jessen, K., and Farrington, A. (1938). *Proc. R. Ir. Acad., Sect. B* **44,** 205–260.
Mitchell, G. F. (1941). *Proc. R. Ir. Acad., Sect. B* **46,** 183–188.

Mitchell, G. F. (1969). *Bull. Mammal Soc.* **31,** 21–25.
Mitchell, G. F. (1976). "The Irish Landscape." Collins, London.
Mitchell, G. F., and Parkes, H. M. (1949). *Proc. R. Ir. Acad., Sect.* B **52,** 291–314.
Movius, H. L. (1935). *J. R. Soc. Antiq. Irel.* **65,** 254–296.
Oakley, K. P., Campbell, B. G., and Molleson, T. J. (eds.) (1971). "Catalogue of Fossil Hominids, Part II: Europe." British Museum (Natural History), London.
O'Riordan, C. E. (1980). "Extinct Terrestrial Mammals of Ireland in the National Museum." The Stationery Office, Dublin.
Peterson, R. L. (1955). "North American Moose." Univ. of Toronto Press, Toronto.
Savage, R. J. G. (1966). *Ir. Nat. J.* **15,** 117–130.
Scharff, R. F., Seymour, H. J., and Newton, E. T. (1918). *Proc. R. Ir. Acad., Sect.* B **34,** 33–72.
Stuart, A. J. (1976). *Philos. Trans. R. Soc. London, Ser.* B **276,** 221–250.
Stuart, A. J. (1977). *Philos. Trans. R. Soc. London, Ser.* B **280,** 295–312.
Stuart, A. J. (1982). "Pleistocene Vertebrates in the British Isles." Longman, London.
Sutcliffe, A. J. (1960). *Trans. Torquay Nat. Hist. Soc.* **13,** 3–28.
Sutcliffe, A. J., and Kowalski, K. (1976). *Bull. Br. Mus. (Nat. Hist.) Geol.* **27**(2), 33–147.

LITTLETONIAN FAUNAS*

Historical Background of Research

The restricted nature of Ireland's fauna has long attracted attention. The remarkable absence of snakes, for example, was noted by Solinus as early as the third century AD. The monk Augustine, writing in AD 655, is credited with the mention of the first list of Irish species: wolf, red deer, wild boar, badger, hare, and squirrel. The absence of snakes and frogs was remarked upon in the eighth century by Nennius and in the ninth century by the Irish monk Donatus (Praeger, 1950; Fairley, 1975). A vivid and much quoted description of the composition of the Irish fauna was given by Giraldus Cambrensis, who wrote his *Topographica Hibernica* in 1187. Although his descriptions contain many inaccuracies, there are also some remarkable observations, such as his comment on the generally smaller size of mammal populations on islands.

In the seventeenth and eighteenth centuries scientific interest in Ireland's natural history revived, and in the nineteenth century the first standard work on the Irish vertebrate fauna was written by William Thompson. Volumes I through III (on the birds) of the "Natural History of Ireland" were published between 1849 and 1851, while Volume IV (on the mammals) appeared posthumously in 1856. His work was continued by the zoologists Scharff and Barrett-Hamilton. The latter investigator published a detailed taxonomic study of the British mammals in which much space is devoted to Irish species (Barrett-Hamilton and Hinton, 1910–1921).

*This section was written by Dr. van Wijngaarden-Bakker.

Whereas the main interest of Thompson and Barrett-Hamilton lay in the present distribution and taxonomic status of the Irish vertebrates, Scharff's interests were in the former distribution of animal species. As keeper of the Natural History division of the National Museum in Dublin, he was also involved in the collection and identification of bones from Irish cave deposits and archaeological sites (Scharff, 1904, 1906, 1928; Scharff et al., 1906).

The postglacial (Littletonian Stage) history of the composition of the Irish fauna received attention in the review by Moffat (1928), who carefully listed the distribution and status of Irish mammals. A reinterpretation of the evidence was undertaken by Savage (1966) and Mitchell (1969). The first and only summary on the postglacial faunal history of Ireland in which archaeozoological evidence has been included appeared less than 10 years ago (van Wijngaarden-Bakker, 1974).

Data on the present land vertebrate fauna of Ireland have been assembled by Praeger (1950), O'Rourke (1970), and Fairley (1975). Data on Irish freshwater fish were published by Went (1946, 1950) and Wheeler (1977). Provisional distribution maps were published by Corbet (1961) and Herity (1975).

Composition of the Postglacial Vertebrate Fauna

Taphonomy

Taphonomy can be defined as the detailed study of the passage of organic remains from the biosphere into the lithosphere. The postmortem relations between organic material and its environment are subject to two kinds of processes: biostratinomic processes that take place between death and burial and diagenetic processes that take place between burial and excavation (Lawrence, 1968).

Biostratinomic processes that influence the recovery of prehistoric animal remains are butchering methods, human discard behaviour, carnivore gnawing, bone transport by carnivores and/or scavengers, and trampling. When viewed separately, these activities may leave recognizable patterns in a bone assemblage. In a prehistoric settlement, however, these activities will be superimposed; for example an animal may be butchered, its bones discarded, then gnawed and transported by dogs, and finally trampled by a passing herd of cattle.

The diagenetic processes that bones undergo are produced by physical and chemical agents of alteration. The visible effect of exposure of bone to the elements is called weathering. The degree to which bones remain preserved depends on their treatment by man (e.g., discarded in a fresh, boiled, or roasted state) and on local soil conditions (e.g., calcium content, water table level). Diagenetic processes also produce recognizable patterns of disintegration of bones (van Wijngaarden-Bakker, 1980). The reconstruction of animal populations from excavated bone assemblages can only be attempted with an insight into the taphonomic processes that have acted upon them.

Fishes

The native Irish fauna contains no primary freshwater fish. It is generally thought that during the last glaciation all freshwater fish species disappeared from Britain and Ireland and that recolonization was restricted to the English river catchments of the North Sea and the eastern English Channel rivers (Wheeler, 1977). Wheeler (1978) suggested that this recolonization may have taken place in early Boreal times, or around 7,000–6,000 bc. By that time the Irish Sea would have acted as a barrier to the further distribution of these fishes.

Two species that occur in a number of Irish lakes are generally considered to be glacial relics. Of these the pollan (*Coregonus autumnalis*) may have been the first colonizer, quickly followed by the charr (*Salvelinus alpinus*) (for present distributions, see diagrams in Wheeler, 1977). Both species are migratory—anadromous under arctic/subarctic conditions. They migrated into the lakes in early postglacial times and became isolated as a result of local changes in the configuration of effluent rivers. The twaite shad (*Alosa fallax killarniensis*) migrated into the Lakes of Killarney, but under more temperate conditions. Its isolation may have taken place somewhat later than that of the pollan and charr populations (Wheeler, 1977).

Three other migratory species, anadromous salmon (*Salmo salar*), anadromous trout (*Salmo trutta*), and catadromous eel (*Anguilla anguilla*), started entering Irish river systems in early postglacial times. Remains of these species have been found at two Mesolithic sites: Newferry, Co. Antrim, dated between 5,500 and 3,500 bc, and Mount Sandel upper, Co. Londonderry, dated around 6,600 bc (Woodman, 1978). At Newferry, Brinkhuizen (1977) found bones of eel and of a salmonid, while at Mount Sandel salmon, trout, and eel were identified (van Wijngaarden-Bakker, forthcoming). At both sites no primary freshwater fish remains were found.

All primary freshwater species that populate present-day Irish lakes and streams seem to have been distributed anthropogenically (Wheeler, 1977). Most species were introduced in historic times, although the date of introduction for four species (gudgeon, bream, rudd, and loach) is still unknown (Table III). Although a number of prehistoric and historic inland sites have produced a few fish remains, none of these has been identified to species level. Whether the observed scarcity of fish remains at crannog sites such as

Table III   Date of Introduction of Primary Freshwater Fish Species

Species	Date of introduction	Source
Esox lucius, pike	~Sixteenth century	Went, 1946
Cyprinus carpio, carp	1634–1640	Went, 1950
Gobio gobio, gudgeon	Unknown	Wheeler, 1977
Tinca tinca, tench	1634–1640	Went, 1950
Abramis brama, bream	Unknown	Wheeler, 1977
Phoxinus phoxinus, minnow	Nineteenth century?	Thompson, 1849
Scardinius erythrophthalmus, rudd	Unknown	Wheeler, 1977
Rutilus rutilus, roach	1889	Went, 1950
Leuciscus leuciscus, dace	1889	Went, 1950
Noemacheilus barbatulus, stone loach	Unknown	Wheeler, 1977
Perca fluviatilis, perch	Eighteenth century?	Went, 1946

Lagore, Co. Meath (Hencken, 1950), and Ballinderry 1 and 2, Co. Offaly (Roche and Stelfox, 1936; Stelfox, 1942), is due to taphonomic processes, excavation technique, or genuine absence of primary freshwater fish is difficult to say. Future excavation in conjunction with a large-scale sieving programme might provide an answer to this problem.

Amphibians and Reptiles

There are but three species of Amphibia in Ireland: *Triturus vulgaris* (common newt), *Bufo calamita* (natterjack toad), and *Rana temporaria* (common frog). Of these, the common (or smooth) newt is generally considered to be native, although fossil evidence is lacking. It is a typical lowland species that does not range further north than mid-Scandinavia (Arnold and Burton, 1978). This would not make it a likely candidate for an early postglacial entry.

The natterjack toad occurs only in Co. Kerry, where it was first reported in 1805 (Gresson and O'Dubhda, 1974). On the basis of its west European distribution, Praeger (1950) had no doubt that the natterjack was an indigenous relict species and part of the so-called Lusitanian element in the Irish fauna. Corbet (1961), however, has argued that this "Lusitanian" element originated entirely through human introduction. Its limited distribution can also be taken as an argument in favour of introduction through anthropogenic factors. It has recently been suggested that the natterjack toad was introduced in ballast sand (Frazer, 1983). The report of natterjack bones in grave 27 at Carrowmore, Co. Sligo (Ove and Persson, 1980a), needs further investigation.

Frog bones mixed with those of late Pleistocene mammals have frequently been found in caves, but on careful examination all these bones could be designated as recent intrusions (Praeger, 1950; O'Rourke, 1970). Frog bones have also been found at Newgrange, Co. Meath, both within the tomb chambers and in the Beaker settlement. In both cases there was evidence that the remains are from recent specimens that died during hibernation (van Wijngaarden-Bakker, 1974, 1982a). Praeger (1939) assumed the frog to be an "ancient native," but recent opinion favours a human introduction either in Norman times or in the seventeenth century (O'Rourke, 1970). Spawn was introduced in AD 1697 into the ditch of College Park, Dublin, and from then on frogs spread quickly over the whole country.

Until about 25 years ago the common lizard (*Lacerta vivipara*) was the only reptile found in Ireland, and it is widespread over the whole country (Herity, 1975; Arnold and Burton, 1978). The presence of lizards was specifically mentioned by Giraldus Cambrensis. Although fossil evidence is still lacking, there seems to be no reason not to accept the common lizard as a native species which entered the country in early postglacial times. The unsuccessful introduction of the green lizard (*Lacerta viridis* L.) between 1958 and 1962 was described by Cabot (1965), while McCarthy (1977, and personal communication) has reported the thus far successful introduction of the slow worm (*Anguis fragilis* L.) in the Burren, Co. Clare.

Birds

Because birds are easily dispersed, the Irish avifauna is far less restricted in composition than the remaining vertebrate fauna. As soon as the climate ameliorated at the end of the lateglacial period, birds could migrate into Ireland. The richness of the bird fauna can be demonstrated by a taubulation of all the species that have been recorded in archaeological excavations (Fig. 2; archaeological sites with bird remains mentioned in the text are shown in Fig. 3). The bird remains that are recovered, however, represent a biased sample. Man chooses which species to hunt and subsequently to bring back to his settlements. Only a small number of species (among them the barn owl and some members of the Corvidae family) might be interpreted as remains of the natural thanatocoenosis. Further selection is imposed by the taphonomic factors that have acted on fragile bird bones. Thus it should be borne in mind that Fig. 3 presents only part of the former avifauna.

The scarcity of bird remains from Bronze and Iron Age sites is striking. The reason for this is the small number of settlements dating to these periods that have been excavated. Regrettably, at the important Iron Age sites at Carrowmore, Co. Sligo, the bird remains have not yet been identified (Ove and Persson, 1980a,b). Another noticeable feature of Fig. 2 is the absence of goose and swan bones from Mesolithic sites. This could be due

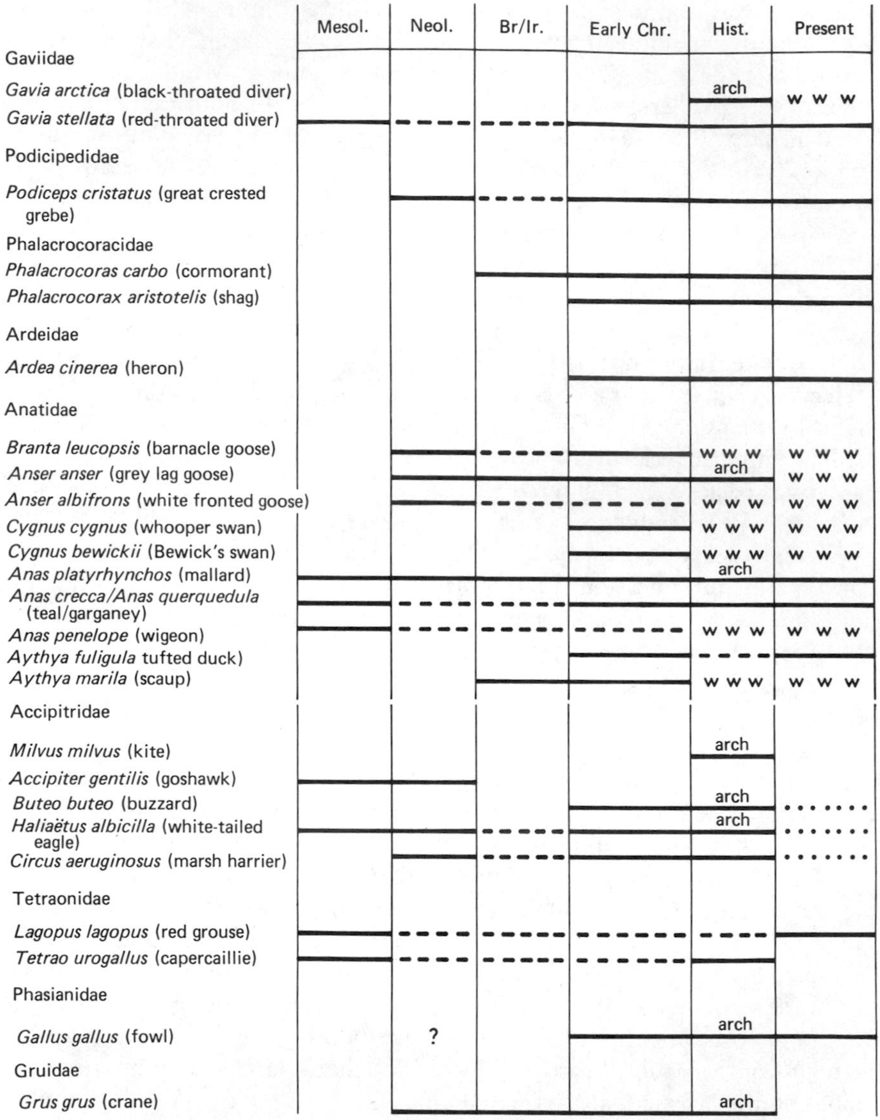

	Mesol.	Neol.	Br/Ir.	Early Chr.	Hist.	Present
Gaviidae						
Gavia arctica (black-throated diver)					arch	w w w
Gavia stellata (red-throated diver)						
Podicipedidae						
Podiceps cristatus (great crested grebe)						
Phalacrocoracidae						
Phalacrocoras carbo (cormorant)						
Phalacrocorax aristotelis (shag)						
Ardeidae						
Ardea cinerea (heron)						
Anatidae						
Branta leucopsis (barnacle goose)					w w w	w w w
Anser anser (grey lag goose)					arch	w w w
Anser albifrons (white fronted goose)					w w w	w w w
Cygnus cygnus (whooper swan)					w w w	w w w
Cygnus bewickii (Bewick's swan)					w w w	w w w
Anas platyrhynchos (mallard)					arch	
Anas crecca/Anas querquedula (teal/garganey)						
Anas penelope (wigeon)					w w w	w w w
Aythya fuligula tufted duck)					– – – –	
Aythya marila (scaup)					w w w	w w w
Accipitridae						
Milvus milvus (kite)					arch	
Accipiter gentilis (goshawk)						
Buteo buteo (buzzard)					arch	• • • • • •
Haliaëtus albicilla (white-tailed eagle)					arch	• • • • • •
Circus aeruginosus (marsh harrier)						• • • • • •
Tetraonidae						
Lagopus lagopus (red grouse)						
Tetrao urogallus (capercaillie)						
Phasianidae						
Gallus gallus (fowl)		?			arch	
Gruidae						
Grus grus (crane)					arch	

Fig. 2. Presence of bird species in archaeological sites of different periods: (—), present c.q. breeding; (– – –), supposedly present; (**ww**), winter migrant; (•••), rare vagrant; (**arch**), historic and excavated evidence for presence.

	Mesol.	Neol.	Br/Ir.	Early Chr.	Hist.	Present
Rallidae						
Rallus aquaticus (water rail)						
Crex crex (corncrake)						
Gallinula chloropus (waterhen)						
Fulica stra (coot)						
Scolopacidae						
Gallinago gallinago/Scolopax rusticola (snipe/woodcock)						
Laridae						
Larus ridibundus (black headed gull)						
Larus canus (common gull)						
Larus argentatus (herring gull)						
Alcidae						
Alca impennis (great auk)						
Alca torda/Uria calge (razorbill/guillemot)					w	w
Fratercula arctica (puffin)						
Columbidae					arch	
Columba palumbus (wood pigeon)						
Columba livia (rock dove)						
Tytonidae						
Tyto alba (barn owl)						
Turdidae						
Turdus merula (blackbird)						
Turdus philomelos (song thrush)						
Sturnidae						
Sturnus vulgaris (starling)						
Corvidae						
Garrulus glandarius (jay)						
Pyrrhocorax pyrrhocorax (chough)						
Corvus monedula (jackdaw)						
Corvus frugilegus (rook)						
Corvus corone (hooded crow)						
Corvus corax (raven)						

Fig. 2. (*Continued*)

to sampling error, as only two Mesolithic sites have produced bird remains in appreciable quantities [Dalkey Island, Co. Dublin (Hatting, 1968), and Mount Sandel upper (van Wijngaarden-Bakker, forthcoming)]. At Mount Sandel there was evidence for a winter occupation, so that if geese and swans had been present and had been hunted, their bones might be expected

Fig. 3. Bird remains: archaeological sites mentioned in the text 1, Dalkey Island, Co. Dublin; 2, Mount Sandel, Co. Londonderry; 3, Lough Gur, Co. Limerick; 4, Newgrange, Co. Meath; 5, Tramore Bay, Co. Waterford; 6, Whitepark Bay, Co. Antrim; 7, Carrowmore, Co. Sligo. ▲, Iron age; ○, Neolithic; ●, Mesolithic.

to turn up amongst the faunal remains. Alternative explanations might be differences in migration patterns of the species in question, absence of a suitable habitat in the neighbourhood of the settlement, or differences in the dietary habits of the Mesolithic hunter–gatherers. These hypotheses remain to be tested.

Figure 2 also provides archaeological evidence for the historically documented extinction or near extinction of a number of bird species in Ireland. Remains of five species of birds of prey have been found at archaeological sites: kite, goshawk, buzzard, white-tailed eagle, and marsh harrier. With the exception of the goshawk, these species were widespread as breeding species until the last century (Ruttledge, 1966). They subsequently became either extinct (kite), rare vagrants (white-tailed eagle, marsh harrier), or rare breeders (buzzards) owing to intensive and direct persecution by man (Table IV). The fact that their remains are found at several archaeological excavations is evidence that their persecution started in prehistoric times.

Table IV Present and Former Status of Extinct or Near-Extinct Bird
Species of which Remains Have Been Identified
at Archaeological Sites[a]

Species	Present Status	Date of extinction
Milvus milvus, kite	Extinct in Ireland	~1840
Accipiter gentilis, goshawk	Very rare vagrant	?
Buteo buteo, buzzard	Rare	Still breeding
Haliaetus albicilla, white tailed eagle	Very rare vagrant	~1850
Circus aeruginosus, marsh harrier	Rare vagrant	~1840
Tetrao urogallus, capercaillie	Extinct in Ireland	~1790
Grus grus, crane	Extinct in Ireland	Middle ages?
Alca impennis, great auk	Extinct	~1840

[a] Data from Kennedy (1961) and Ruttledge (1966).

Bones of the goshawk have been found at the Mesolithic sites of Dalkey
Island (Hatting, 1968) and Mount Sandel upper (van Wijngaarden-Bakker,
forthcoming) and at the Beaker site of Newgrange (van Wijngaarden-Bakker, 1974). The goshawk, a bird of prey inhabiting deciduous woodlands,
occurs in very low densities [about one breeding pair per 5000 hectares
(ha)]. As a result of the large-scale deforestation of the Dowris phase of the
Bronze Age (Mitchell, 1976), the goshawk may well have ceased to breed at
that time in Ireland. The closely allied sparrow hawk (*Accipiter nisus*), which
is better adapted to different environments and occurs at higher densities
(about one pair per 500–700 ha), is still breeding in every county (Kennedy,
1961; Ruttledge, 1966).

Both the capercaillie (*Tetrao urogallus*), a typical woodland species, and the
crane (*Grus grus*), an inhabitant of open wetlands, were formerly abundant.
Bones of the capercaillie were identified at Mount Sandel (van Wijngaarden-
Bakker, forthcoming), and it has been suggested (C. H. Maliepaard, personal communication) that the fowl bones which were recovered at the Neolithic
sites at Lough Gur, Co. Limerick (Ó Ríordáin, 1951, 1954), might well
belong to either red grouse or capercaillie. The capercaillie and the crane both
became (near) extinct due to the rapid expansion of cultivated land in the
historical period.

The last species, the great auk (*Alca impennis*), presumably bred on the
Irish coast, as is suggested by the well-known catching of a juvenile spec-
imen in Waterford Harbour in 1834 and by the bones that have been found
in kitchen middens at Tramore Bay, Co. Waterford, and Whitepark Bay,
Co. Antrim (Newton, 1891). The species became totally extinct around
1844 as a result of relentless direct persecution by man.

The only domestic bird species of which remains have been found at Irish

Table V   Species of the Irish Terrestrial
Mammal Fauna which Extend
to Subarctic Regions of Europe

Species	Present status
Sorex minutus, pygmy shrew	Widespread
Canis lupus, wolf	Extinct
Vulpes vulpes, fox	Widespread
Ursus arctos, brown bear	Extinct
Martes martes, pine marten	Widespread
Mustela erminea, stoat	Widespread
Lutra lutra, otter	Widespread
Lepus timidus, mountain hare	Widespread

archaeological sites is the chicken. The species was domesticated in India in the first millennium bc, was first brought to Europe in the La Tène period, and was distributed over Europe by the Romans (Bökönyi, 1974). The occurrence of fowl bones in a Neolithic context at Lough Gur can presumably be ascribed either to an intrusion of recent bones or to misidentification.

Mammals

Ireland's native terrestrial mammal fauna is severely restricted in its composition. It is made up of three constituents: species extending to subarctic regions of Europe, a more temperate woodland element, and a large anthropogenic element. The eight species of the first group are shown in Table V.

Bones of most of the species in Table V have been found frequently in Irish caves (Savage, 1966; Mitchell, 1969; Stuart, this chapter, p. 225), but the species presumably disappeared by the beginning of the Younger *Dryas* period (cf. the Nahanagan Stadial). This period was characterized as a short (about 500 years) period of intense cold with a mean annual temperature of −5°C (Mitchell, 1976). With the single exception of reindeer, no vertebrate remains have so far been dated to this period (Stuart, this chapter, p. 227). At the end of the Younger *Dryas,* the temperature rose quickly, and, as soon as the vegetation reverted to a tundra/park–tundra type, those mammals that were well adapted to cold conditions migrated into Ireland. The species mentioned in Table V all show a wide adaptive range insofar as their present distribution is from the most northern part of Europe south to the Mediterranean.

Presumably owing to its small size, bones of the pygmy shrew have

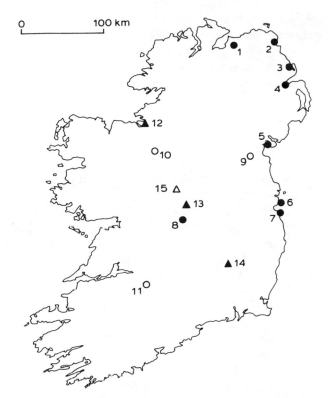

Fig. 4. Mammal remains: archaeological sites mentioned in the text. 1, Mount Sandel, Co. Londonderry; 2, Cushendun, Co. Antrim; 3, Curran Point, Larne, Co. Antrim; 4, Belfast Lough; 5, Rockmarshall, Co. Louth; 6, Sutton, Co. Dublin; 7, Dalkey Island, Co. Dublin; 8, Lough Boora, Co. Offaly; 9, Newgrange, Co. Meath; 10, Carrowkeel, Co. Sligo; 11, Lough Gur, Co. Limerick; 12, Carrowmore, Co. Sligo; 13, Ballinderry 2, Co. Offaly; 14, Freestone Hill, Co. Kilkenny; 15, Uisneach, Co. Westmeath. △, Early Christian; ▲, Bronze/Iron Age; ○, Neolithic; ●, Mesolithic.

never been recorded from any Irish archaeological site. The species has been included in Table V on the basis of its present distribution. There is some evidence for the likely presence of the wolf at the Mesolithic sites of the Curran at Larne, Co. Antrim, and at Sutton, Co. Dublin (Hatting, 1968) (Fig. 4). The species was definitely identified at Lough Gur, Knockadoon stone circle K (van Wijngaarden-Bakker, 1974), but not at any later pre-historic site. The presence of wolf in historic times is attested by a great number of written sources, place names, and folk tales (Moffat, 1938; Savage, 1966; O'Rourke, 1970; Fairley, 1975). As a result of human pressures, wolves almost became extinct in the eighteenth century; the last one may have been killed as late as 1810 (McMillan, 1971).

Bones of the fox have been found in small numbers at archaeological sites from Neolithic times onward. Apart from the sites mentioned by van Wijngaarden-Bakker (1974) the species has been reported from Iron Age sites at Carrowmore, Co. Sligo (Ove and Persson, 1980a).

Remains of the brown bear have been recorded from the Mesolithic layers at Dalkey Island, Co. Dublin (Hatting, 1968), and from the Neolithic sites of Newgrange, Co. Meath (van Wijngaarden-Bakker, 1980), Lough Gur, Co. Limerick (Ó Ríordáin, 1954; van Wijngaarden-Bakker, 1974), and Carrowkeel, Co. Sligo (MacAlister et al., 1912). The species may have become extinct before AD 850 (Moffat, 1938).

Perhaps due to their relatively small size, bones of stoat and pine marten have not been reported from prehistoric sites, but they were found at some early Christian sites. The absence of otter bones at Neolithic sites such as Newgrange and Lough Gur is surprising. The earliest record of the species is from the Bronze Age layer at Ballinderry 2 (Stelfox, 1942). In early Christian and historic sites it is fairly common.

Hare bones have been found at sites dating to all periods. The species was present at the early Mesolithic site of Mount Sandel upper.

The second element of the Irish mammal fauna consists of a small number of woodland species that generally are less well adapted to cold conditions (Table VI). Their present distribution does not range farther north than southern or mid-Scandinavia. It is in this group specifically that the restricted nature of the Irish mammal fauna becomes apparent. When compared to the indigenous British mammal fauna there is a conspicuous absence, in the past and present fauna of Ireland, of large herbivores such as elk (*Alces alces*) (present in Britain in the early postglacial), aurochs (*Bos primigenius*), and roe deer (*Capreolus capreolus*), as well as of a whole group of small woodland animals such as mole (*Talpa europaea*), common shrew (*Sorex araneus*), weasel (*Mustela nivalis*), polecat (*M. putorius*), brown hare (*Lepus capensis*), beaver (*Castor fiber*), bank vole (*Clethrionomys glareolus*), water vole (*Arvicola terrestris*), and field vole (*Microtus agrestis*).

Table VI   The Temperate Woodland Element of the Irish Terrestrial Mammal Fauna

Species	Present status
Meles meles, badger	Widespread
Felis silvestris, wild cat	Extinct
Sus scrofa, wild boar	Extinct
Cervus elaphus, red deer	Restricted
Sciurus vulgaris, red squirrel	Fluctuating
Apodemus sylvaticus, wood mouse	Widespread

Yalden (1981) argued that the pygmy shrew may have entered Ireland by a partially flooded "filter" land bridge in the period between about 6,600 and 5,500 bc. Although it is conceivable that such a land bridge would present a barrier for the common shrew (*Sorex araneus*), it is difficult to understand why large herbivores such as elk, aurochs, and roe deer did not cross the bridge.

Corbet (1961, 1969), who several times has drawn attention to the fact that voles are absent from Ireland, suggests that the last land connection was severed at a very early climatic stage (see Synge, Chap. 6, this volume). The large herbivores mentioned above were well established in England in the Preboreal (Grigson, 1981). The evidence thus suggests that the normal postglacial colonization sequence in Ireland stopped at the transition of the Younger *Dryas* and the Preboreal periods and that the typical woodland animals of the Irish fauna arrived as a result of accidental dispersal in the postglacial period.

The earliest evidence for the presence of the badger in Ireland comes from the Neolithic wedge grave at Lough Gur, Co. Limerick (Ó Ríordáin and Óh-Iccadha, 1955), and from the Bronze Age level at Ballinderry 2, Co. Offaly (Stelfox, 1942) (Table VII).

As the domestic cat is known to have been brought to England and subsequently to Ireland in the first centuries AD (Zeuner, 1963; van Wijn-

Table VII Wild Mammal Species Recorded from Irish Archaeological Sites

Species	Mesolithic	Neolithic	Bronze/Iron Age	Early Christian	Present
Canidae					
Canis lupus, wolf	(X)[a]	X		(X)	[b]
Vulpes vulpes, fox		X	X	X	X
Ursus arctos, brown bear	X	X			[b]
Martes martes, pine marten				X	X
Mustela erminea, stoat				X	
Meles meles, badger		X	X	X	X
Lutra lutra, otter			X	X	X
Felis silvestris, wild cat		X	(X)	(X)	[b]
Suidae					
Sus scrofa, wild boar	X				[b]
Cervidae					
Cervus elaphus, red deer	X	X	X	X	X
Leporidae					
Lepus timidus, mountain hare	X	X	X	X	X

[a] X, Present.
[b] Extinct.

gaarden-Bakker, 1974), any bones found in earlier, uncontaminated levels must belong to the wild cat (*Felis silvestris*). Wild cat was present at the late Neolithic sites of Newgrange (van Wijngaarden-Bakker, 1974) and Lough Gur (Ó Ríordáin, 1954; Ó Ríordáin, and Óh-Iceadha, 1955; van Wijngaarden-Bakker, 1974). There is, furthermore, some doubtful evidence from the Bronze Age layer at Ballinderry 2 (Stelfox, 1942) and from the Early Christian site of Uisneach, Co. Westmeath (Scharff, 1928) (Table VII).

Bones of the wild boar have been identified at two early Mesolithic sites both dating to the seventh millennium bc: Mount Sandel, Co. Londonderry (van Wijngaarden-Bakker, 1978, forthcoming), and Lough Boora, Co. Offaly (Ryan, 1980). Its presence was also reported in a Boreal layer at Belfast Lough (Praeger, 1893; Savage, 1964), at the late Mesolithic sites of Dalkey Island and Sutton, both in Co. Dublin (Hatting, 1968), and at Cushendun, Co. Antrim (Movius, 1940). Although evidence is lacking, the wild boar may have become extinct in the late prehistoric period as a result of the vast expansion of cultivated land. Literary sources for wild boar probably all refer to feral "greyhound" pigs. The early postglacial presence of wild boar may be explained by the fact that the animal is an excellent swimmer (IJsseling and Scheygrond, 1962).

Remains of red deer have been found at archaeological sites from all periods. The absence of red deer bones in early Mesolithic sites, together with a very low proportion of scrapers and burins at these sites, was noted by Woodman (1974), and the recently completed analysis of the Mount Sandel bones by the present author confirmed this absence. The available evidence suggests that the red deer arrived no earlier than the Atlantic period in Ireland. In Early Christian times deer were widely distributed, but hunting pressure may have been severe (van Wijngaarden-Bakker, 1982b). At present, their distribution is limited to three "wild" herds in Counties Donegal, Kerry, and Wicklow.

Presumably due to their small size, bones of the squirrel have not been found at archaeological sites, but the presence of woodmouse was reported from the Iron Age hillfort at Freestone Hill, Co. Kilkenny (Raftery, 1969). Both Corbet (1961) and Savage (1966) accept the presence of the woodmouse as due to accidental human introduction. There is evidence from epigenetic studies that *Apodemus sylvaticus* in Ireland originated with Norwegian stock (Berry, 1969; Yalden, 1982).

The third component of the Irish mammal fauna consists of a group of animals that have arrived through deliberate or accidental human introduction. Data on wild and domestic species have been assembled in Table VIII. Several of the species merit a short commentary. The hedgehog is mentioned by Giraldus Cambrensis as absent from Ireland. The species may

Table VIII The Anthropogenic Element of the Irish Terrestrial Mammal Fauna[a]

Species	Type of introduction	Date of introduction
Erinaceus europaeus, hedgehog	Accidental?/deliberate?	Norman times?
Canis familiaris, dog	Deliberate (domestic)	~6500 bc?
Mustela vison, mink	Deliberate	1950 onward
Felis catus, cat	Deliberate (domestic)	First centuries AD
Cervus nippon, sika deer	Deliberate	1884
Dama dama, fallow deer	Deliberate	Norman times?
Sus domesticus, pig	Deliberate (domestic)	~3500 bc
Bos taurus, cattle	Deliberate (domestic)	~3500 bc
Ovis aries, sheep	Deliberate (domestic)	~3500 bc
Capra hircus, goat	Deliberate (domestic)	~3500 bc
Equus caballus, horse	Deliberate (domestic)	~2000 bc
Equus asinus, donkey	Deliberate (domestic)	Eighteenth century AD
Lepus capensis, brown hare	Deliberate	1850 onward
Oryctolagus cuniculus, rabbit	Deliberate	Thirteenth century AD
Sciurus carolinensis, grey squirrel	Deliberate	1890, 1911
Glis glis, edible dormouse	Deliberate	1885 (unsuccessful)
Mus musculus, house mouse	Accidental	Roman times or earlier
Rattus rattus, black rat	Accidental	Before AD 1187
Rattus norvegicus, brown rat	Accidental	1722 onward
Clethrionomys glareolus, bank vole	Accidental	1964
Ondatra zibethica, musk rat	Deliberate	1927 (extinct 1934)

[a] Data from Moffat (1938), Claassens and O'Gorman (1965), Savage (1966), Deane and O'Gorman (1969), and van Wijngaarden-Bakker (1974).

have been imported for food (Savage, 1966; Fairley, 1975). Remains of the dog have been identified at a number of late Mesolithic sites such as Rock-marshall, Co. Louth, and Dalkey Island, Co. Dublin (Hatting, 1968). A radius fragment of a canid from Mount Sandel, Co. Londonderry, could not be identified to species level (van Wijngaarden-Bakker, forthcoming). Among the introduced wild species only the hedgehog, rabbit, house-mouse, and brown rat are widespread. The remaining wild species of Table VIII all present a restricted distribution.

Conclusions

The Irish postglacial vertebrate fauna is characterized first by its restricted nature (with the exception of the avifauna) and second by the anthropogenic factors that have affected it. A great number of archaeological excavations allow the documentation of the effect of these factors on the native fauna. A number of bird and mammal species have become extinct or near extinct, either through direct persecution by man or through manipulation of their biotopes from prehistoric to recent times. On the other hand, from early times man has also been an agent in the deliberate or accidental introduction of a number of species of fish, amphibia, birds, and mammals. Due to its island situation the profound effect of man on his natural environment can be detected and followed more easily in Ireland than in possibly any other part of Europe.

References

Arnold, E. N., and Burton, J. A. (1978). "A Field Guide to the Reptiles and Amphibians of Britain and Europe." Collins, London.
Barrett-Hamilton, G. E. H., and Hinton, M. A. C. (1910–1921). "A History of British Mammals." Gurney and Jackson, London.
Berry, R. J. (1969). *J. Zool.* **159**, 311–328.
Brinkhuizen, D. C. (1977). *Proc. Prehist. Soc.* **43**, 197.
Bökönyi, S. (1974). "History of Domestic Mammals in Central and Eastern Europe." Akademiai Kaido, Budapest.
Cabot, D. (1965). *Ir. Nat. J.* **15**, 111.
Claassens, A. J. M., and O'Gorman, F. (1965). *Nature (London)* **205**, 923–924.
Corbet, G. B. (1961). *Nature (London)* **91**, 1037–1040.
Corbet, G. B. (1969). *Mammal Soc. Bull.* **31**, 14–16.
Deane, C. D., and O'Gorman, F. (1969). *Ir. Nat. J.* **16**, 198–202.
Fairley, J. S. (1975). "An Irish Beast Book." Blackstaff, Belfast.
Frazer, D. (1983). "The British Amphibians and Reptiles." Collins, London.
Gresson, R. A. R., and O'Dubhda, S. O. (1974). *Ir. Nat. J.* **18**, 97–103.
Grigson, C. (1981). In "The Environment in British Prehistory" (I. G. Simmons and M. J. Tooley, eds.), pp. 110–124. Duckworth, London.
Hatting, T. (1968). *Proc. R. Ir. Acad., Sect. C* **66**, 172–174.
Hencken, H. O'N. (1950). *Proc. R. Ir. Acad., Sect. C* **53**, 1–247.
Herity, P. (ed.) (1975). "Provisional Distribution Maps of Amphibians, Reptiles and Mammals in Ireland." Folens, Dublin.
IJsseling, M. A., and Scheygrond, A. (1962). "Onze Zoogdieren." Thieme, Zutphen.
Kennedy, P. G. (1961). "A List of the Birds of Ireland." Stationery Office, Dublin.
Lawrence, D. L. (1968). *Geol. Soc. Amer. Bull.* **79**, 1315–1330.
MacAlister, R. A. S., Armstrong, E. C. R., and Praeger, R. Ll. (1912). *Proc. R. Ir. Acad., Sect. C* **29**, 311–347.
McCarthy, T. K. (1977). *Ir. Nat. J.* **19**, 49.
McMillan, N. F. (1971). *Ir. Nat. J.* **17**, 103.
Mitchell, G. F. (1969). *Mammal Soc. Bull.* **31**, 21–25.
Mitchell, G. F. (1976). "The Irish Landscape." Collins, London.
Moffat, C. B. (1938). *Proc. R. Ir. Acad., Sect. B* **44**, 61–128.
Movius, H. L. (1940). *Proc. R. Ir. Acad., Sect. C* **46**, 1–48.
Newton, E. T. (1891). *Proc. R. Ir. Acad. Ser. III* **1**, 624–625.
Ó Ríordáin, S. P. (1951). *Proc. R. Ir. Acad., Sect. C* **54**, 37–74.
Ó Ríordáin, S. P. (1954). *Proc. R. Ir. Acad., Sect. C* **56**, 297–459.
Ó Ríordáin, S. P., and Óh-Iceadha, G. (1955). *J. R. Soc. Antiq. Irel.* **85**, 34–50.
O'Rourke, F. J. (1970). "The Fauna of Ireland: an Introduction to the Land Vertebrates." Mercier Press, Cork.

Ove, P., and Persson, E. (1980a). *Theses Pap. North-European Archaeol.* **9,** 117–129.
Ove, P., and Persson, E. (1980b). *Stockholm Archaeol. Rep.* **7,** 140–148.
Praeger, R. Ll. (1893). *Proc. R. Ir. Acad.* **2,** 212–289.
Praeger, R. Ll. (1939). "The Way that I Went." Figgis, Dublin.
Praeger, R. Ll. (1950). "Natural History of Ireland: a Sketch of Its Flora and Fauna." Collins, London.
Raftery, B. (1969). *Proc. R. Ir. Acad., Sect. C* **68,** 1–108.
Roche, G., and Stelfox, A. W. (1936). *Proc. R. Ir. Acad., Sect. C* **43,** 230–235.
Ruttledge, R. F. (1966). "Ireland's Birds." Witherby, London.
Ryan, M. (1980). *Antiquity* **54,** 46–47.
Savage, R. J. G. (1964). *Ir. Nat. J.* **14,** 303–304.
Scharff, R. F. (1904). *Proc. R. Ir. Acad., Sect. C.* **25,** 16–19.
Scharff, R. F. (1906). *Proc. R. Ir. Acad., Sect. B* **26,** 1–12.
Scharff, R. F. (1928). *Proc. R. Ir. Acad., Sect. C* **38,** 122–124.
Scharff, R. F., Ussher, R. J., Cole, A. J., Newton, E. T., Dixon, A. F., and Westropp, T. J. (1906). *Trans. R. Ir. Acad.* **33B,** 1–76.
Stelfox, A. W. (1942). *Proc. R. Ir. Acad., Sect. C* **47,** 20–21 and 67–74.
Thompson, W. (1849–1856). "Natural History of Ireland," 4 volumes. Bohn, London.
Went, A. E. J. (1946). *Salmon Trout Mag.* **118,** 248–256.
Went, A. E. J. (1950). *J. Dept. Agric. Dublin* **47,** 119–124.
Wheeler, A. (1977). *J. Biogeogr.* **4,** 1–24.
Wheeler, A. (1978). *J. Archaeol. Sci.* **5,** 85–89.
van Wijngaarden-Bakker, L. H. (1974). *Proc. R. Ir. Acad., Sect. C* **74,** 313–383.
van Wijngaarden-Bakker, L. H. (1980). "An Archaeozoological Study of the Beaker Settlement at Newgrange, Ireland." Unpublished Ph.D. thesis, Univ. of Amsterdam, Amsterdam.
van Wijngaarden-Bakker, L. H. (1982a). In "Newgrange. Archaeology, Art, and Legend" (M. J. O'Kelly, ed.), pp. 214–218. Thames and Hudson, London.
van Wijngaarden-Bakker, L. H. (1982b). In "Irish Antiquity" (D. Ó. Corráin, ed.), pp. 78–81. Tower Books, Cork.
van Wijngaarden-Bakker, L. H. (forthcoming). In "Mount Sandel" (P. C. Woodman, ed.). H. M. Stationery Office, Belfast.
Woodman, P. C. (1974). *Ulster J. Archaeol.* **36/37,** 1–16.
Woodman, P. C. (1978). In "The Early Postglacial Settlement of Northern Europe" (P. Mellars, ed.), pp. 333–369. Duckworth, London.
Yalden, D. W. (1981). *J. Zool. London* **195,** 147–156.
Yalden, D. W. (1982). *Mamm. Rev.* **12,** 1–57.
Zeuner, F. E. (1963). "A History of Domesticated Animals." Harper and Row, New York.

CHAPTER 11

PREHISTORIC SETTLEMENT AND ENVIRONMENT

Peter C. Woodman

Department of Archaeology
University College
Cork, Ireland

INTRODUCTION

A study of prehistoric settlement in Ireland has to rely on rather scanty evidence which, considering the richness of the archaeological material, is at first sight rather paradoxical. It may be, however, because of the richness of Irish archaeology that the problems of settlement were considered to be secondary. The main topics of discussion will be as follows: (1) How did any period–economy–technology begin? (2) What was the main basis of the economy? (3) Did the changing postglacial environment have any impact on economic strategies, and how was the environment used? (4) Were there any problems associated with the procurement of raw materials for the manufacture of tools? These topics are discussed in relation to the problem of assessing the validity of interpretations based on slight evidence and the problem of the representativeness of our data.

Irish archaeology only began to organize itself on a professional and state-supported basis at the end of the nineteenth century, which coincided with the end of one of the most productive periods of antiquarianism. Land enclosure, population increase, and land tillage during the later eighteenth and nineteenth centuries at the same time produced a wealth of archaeological material and destroyed a significant selection of field monuments. Therefore it is hardly surprising that one of the major preoccupations of Irish archaeology during the last two generations could be described as

251

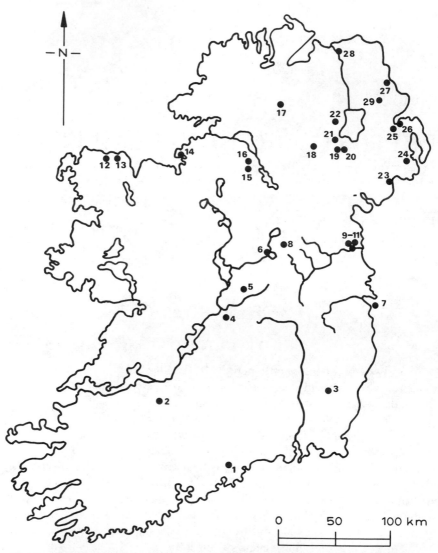

Fig. 1. Location map of selected sites referred to in the text. 1, Kilgreaney Cave, Co. Waterford; 2, Lough Gur, Co. Limerick, 3, Rathgall, Co. Wicklow; 4, Lough Boora, Co. Offally; 5, Ballinderry 2, Co. Offally; 6, Lough Derravarragh, Co. Westmeath; 7, Dalkey Island, Co. Dublin; 8, Loughcrew, Co. Meath; 9, Newgrange, Co. Meath; 10, Monknewton, Co. Meath; 11, Townleyhall, Co. Louth; 12, Behy/Belderg, Co. Mayo; 13, Ballyglass, Co. Mayo; 14, Carrowmore complex, Co. Sligo; 15, Poll-na-Gollum Cave, Co. Fermanagh; 16, Belmore Mountain, Co. Fermanagh; 17, Island MacHugh, Co. Tyrone; 18, Clogher, Co. Tyrone; 19, Navan Fort, Co. Armagh; 20, Armagh City, Co. Armagh; 21, Lough Eskragh, Co. Tyrone; 22, Ballynagilly, Co. Tyrone; 23, Dundrum, Co. Down; 24, Downpatrick, Co. Down; 25, Giant's Ring, Co. Down; 26, Glenulra, Co. Down; 27, Carnlough, Co. Antrim; 28, Mount Sandel, Co. Derry; 29, Donegore, Co. Antrim.

stock taking. Emphasis has been placed on the completion and publication of surveys of monuments, for example, the Megalithic Survey (de Valera and Ó Nualláin, 1961), and on listing the artifacts in museum collections, which can entail the searching of many museums outside of Ireland. Here the Bronze Age with its relatively finite numbers of objects has particularly benefited (see, for instance, Harbison 1969a,b, Kavanagh, 1973, Eogan, 1983, and Raftery, 1984). Since the early 1930s, the desire to understand the chronology of the field monuments and their context has necessitated excavation, which received an initial impetus through the work of E. Estyn Evans and Oliver Davies in the north and of Seán Ó Ríordáin at Lough Gur, Co. Limerick in particular (Fig. 1). The work of the Harvard Archaeological Mission in Ireland under the direction of Hencken gave a useful boost to the study of several periods, for example, Movius (1942) on the Irish Mesolithic, while those connected with the archaeological survey of Co. Down (Jope, 1966) carried out a much more structured research and excavation programme to elucidate the problems of many monument types.

With the introduction of radiocarbon dating as a normal aid on excavations, providing help in solving problems of chronology, there should have been within the last decade a switch in the emphasis of excavation policy toward excavations geared to help our understanding of the economic and social bases of prehistoric Ireland. While certain excavations, as will be shown below, have provided tremendous insight into these aspects of Irish prehistory, the trend in excavation policy and comparison with other areas can only lead to a pessimistic conclusion that, unless there is a major change, comparatively little fresh information will be produced before the end of the century.

There are numerous contributory factors which have produced the present trend, such as the growing awareness of an ecological perspective and the whole process of excavation, which has become more careful. Thus instead of five sites per year, the archaeologist may investigate only one site per 5 years (within the last decade, ten sites in Ireland have required three or more seasons of excavation). It is of course only correct that excavation should take as long as is needed, but it does mean that we are looking at a diminishing number of sites from which our generalizations are made. A second equally valid factor is the more equitable distribution of resources between prehistoric and later archaeology.

The emergence of rescue and salvage archaeology has resulted in a reduction in the number of research excavations and, owing to the importance of the cult of rescue, in a diminution of the control over what sites are excavated. One major problem with small numbers of anything is that they rarely result in normal distributions; the range of prehistoric excavations during the 1970s (Fig. 2) illustrates this point very clearly. The numbers are

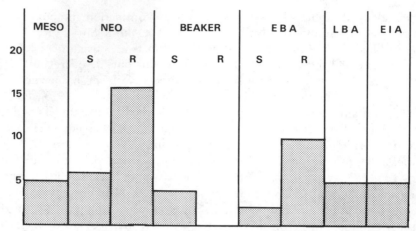

Fig. 2. Histogram of excavations of prehistoric sites carried out during the 1970s. The **S** column represents settlement while the **R** column represents ritual sites.

too low to ensure an even distribution across all periods, and the monument orientation in recent years shows in the number of burial sites which have been excavated. If rescue excavation of individual cist graves had been included, then the numbers of burial sites excavated would have been even greater. These were excluded as they have usually been very quick, often 1 day, recovery tasks which cannot, due to the extenuating circumstances, be considered as normal excavations.

In summary, at a time when there should be a shift to problem-orientated excavation there has been a significant reduction in our capacity to carry out research programmes. Recent improvements in excavation techniques in terms of sampling and environmental studies have been applied to a restricted number of sites which may not be typical, so there is a great danger that we may overgeneralize from the particular.

A simple chronological column for the main phases of Irish prehistory may be found in Tables I and II of Edwards and Warren (Chap. 1, this volume). A study of this nature cannot hope to examine all aspects of prehistoric settlement, therefore problems of chronology have only been discussed in any detail where they have an important bearing on other aspects of Ireland's prehistory.

THE MESOLITHIC

While there is no evidence as yet for a Palaeolithic in Ireland, the fact that animals such as reindeer and giant Irish deer apparently existed in abundance during lateglacial times shows that certain mammals were able to colonize Ireland successfully, and so we should not assume that man did not also arrive here before the end of the Ice Age. Green (1981) found early stone tools and the remains of hominids in Wales, which again reminds us that man could have been in Ireland during the Palaeolithic.

At the moment, the earliest evidence for man in Ireland comes from Mount Sandel, Co. Derry (Woodman, 1981a), where a series of ^{14}C dates for the main settlement range from 7,000 to 6,400 bc but with a significant number lying between 7,000 and 6,800 bc (dates given as bc are radiocarbon dates; those given as BC are calibrated dates). It is possible that there could have been earlier settlement, traces of which have been lost through rising sea levels (see Synge, Chap. 6, this volume), growth of peat, and the silting up of river valleys. However, unless human occupation was shown to go back into the Palaeolithic (Edwards and Warren, Chap. 1, and Stuart, Chap. 10, this volume), earlier finds would not significantly alter the present picture.

The characteristic implements found at Mount Sandel are microlithic, small, elongated, scalene triangles and rods and some local implements which imply at least a short phase of earlier human settlement. Some of this material can be paralleled in northern England on sites of roughly similar date (e.g., Filpoke Beacon, Co. Durham). On the other hand, it cannot be paralleled in western Scotland (Mercer, 1974), where the narrow blade scalene triangle industry only appears later in the sequence after 6,000 bc. Combining this with the growing distribution of similar material through-out a large part of Ireland, it could be argued that these people came to Ireland across the Isle of Man basin, at a time when the sea level was considerably lower than today (Woodman, 1981b). As there was perhaps no postglacial land bridge (see Synge, Chap. 6, this volume), these island colonists not only would have faced economic problems in adapting to a new environment but would have encountered the social problems associated with perpetuating a population from an initially small group. In spite of these difficulties, a strong case can be made for a successful colonization resulting in a continuous occupation of Ireland throughout the Mesolithic. The traditional view of the Irish Mesolithic was of a group of impoverished "strandloopers" confined to the flint-rich areas of north eastern Ireland (Mitchell, 1970). Recent research and excavation, in particular the seventh millenium bc site at Lough Boora, Co. Offaly (Ryan, 1980), has produced evidence that even in the earliest stages of the Mesolithic man had left traces in many parts of Ireland (Fig. 3).

On typological grounds after 6,000 bc there would appear to have been a major change in the flint artifacts used. This change is from composite artifacts with microlith insets to a large heavy-bladed industry which lacks

Fig. 3. Distribution map of findspots of Mesolithic artifacts.

microliths and where many large blades were probably used without further modification. Occasional traces of human activity have been found to date to between these two main phases of the Mesolithic so that there would appear to be a record of continuous human occupation, and certainly the late Irish Mesolithic is so different that no outside progenitor can be found for it. While we can observe this change in artifacts at the moment, no explanation can be offered as to why such a radical alteration took place within a relatively short time. Woodman (1981b) has argued that whatever the actual reason for these changes, it was only after a viable population level was achieved in Ireland and the Irish Mesolithic became socially self-sufficient that these changes could have taken place.

The problem of how Mesolithic communities coped with the rather restricted range of resources is one of the most fascinating aspects of Irish prehistory. As has been noted by Fairley (1975) and van Wijngaarden-Bakker (1980, Chap. 10, this volume), Ireland had a very restricted mammalian fauna and may also have had a restricted range of freshwater fish (Woodman, 1978). In particular, wild cattle, roe deer, and elk could not have been hunted in Ireland, while pike (an important fish in some other Mesolithic economies) was also absent. In the case of the mammalian fauna, many of the ungulates would have specific environments, and so a reduction in the range of species would not result in complete compensation through increased numbers of those species living here. The major food gathering problem for hunter–gatherers would probably have been the restricted range of species, resulting in a restricted choice.

Unfortunately, faunal remains are rare on many prehistoric sites in Ireland. However, at Mount Sandel it would appear that extensive use was made of the migratory fish (salmon and eels) and that pig was extensively hunted. At this site, which appears to have been occupied throughout the winter, storage may have been used to supplement the scarce winter food sources. Two aspects of the Mount Sandel site appear to be typical of the Irish Mesolithic. First, red deer may have played a rather limited role in the economy, and, second, Mount Sandel, like nearly all Irish Mesolithic sites, is low lying and was positioned carefully to rely extensively on fishing. This contrasts strongly with parts of northern England, where an extensive series of sites are found above 300 m OD in the Pennines (Jacobi et al., 1976). At the moment the faunal remains from Mount Sandel loom very large in any examination of the economy, but it should not be assumed just yet that red deer was not hunted more extensively in other parts of Ireland (see discussion of cooking places in the section on the Bronze Age for possible bias in site data). Similarly, the presence of faunal remains on coastal sites such as estuaries and islands has shown us how these groups could survive on the coast. There are also an increasing number of inland sites, not only on major rivers such as the Bann but often on small lakes, and at the moment it is impossible to assess the seasonal extent and significance of inland settlement. We do not know if resident fish populations were established in these lakes. The report on the early Mesolithic site of Lough Boora is awaited with interest.

The early postglacial environment was not a static phenomenon. In particular the vegetation cover would have changed considerably. At the time of presumed colonization, a relatively boreal-type environment would have still existed. The forest cover would have consisted of pine, hazel, and birch and as such would have been ideally suited to species such as wild cattle and elk, animals not native to Ireland. This would suggest that human population levels would have been relatively low. The gradual change to the Atlantic deciduous forest is often presumed to have had a paradoxical effect on the economic potential of late Mesolithic environments. If held open, it might have been considerably richer than the earlier Boreal environment (Mellars, 1976), but if the forest had closed over, only burning and deliberate forest clearance would release its potential. Many have argued that there is some evidence for forest clearance in the Mesolithic of Britain (Smith, 1970; Jacobi et al., 1976), but the evidence, as for Ireland, is still equivocal (but see Smith, 1981, and Edwards, Chap. 9, this volume). Any real absence of vegetational impact may, of course, reflect a possible extensive reliance on fishing in the Irish Mesolithic.

On the other hand, the extensive growth of peat bogs during the Atlantic period (Mitchell, 1976) may have reduced the number of open water areas available for fishing in parts of the Bann Valley and in particular the Midlands. This could have had a significant effect in forcing settlement onto major rivers, large lakes, and the sea. While relative sea level was changing rapidly in the early Holocene, the effect of its change may be more apparent than real, that is, caused by the loss of many early shoreline sites which existed at a period when the sea level was lower than today. Unlike the case for the North Sea basin, the transgressions in the Irish Sea did not inundate a massive area, and, as it is possible that areas of newly transgressed coastline were exceptionally rich in nutrients, the transgression may have increased the economic potential of the coastline around the isostatically active northern part of Ireland.

Two other problems require considerably more research. These are the extent to which Ireland was occupied throughout the Mesolithic and the type of social organization which existed. The first problem is really a question of whether there was a limit to the potential for increasing population within Ireland (Fig. 2). Recent fieldwork in the south and west has produced sites in the Cork Blackwater Valley where microliths have been found—presumably indicating occupation before 6,000 bc. Probable Late Mesolithic sites have been found in south Cork and at the western end of the Dingle Peninsula, Co. Kerry (Woodman, 1984).

The second problem is whether there is any evidence for long-term near-sedentary settlement of large groups of hunter–gatherers during the Mesolithic. Ironically, with the exception of Mount Sandel, it is difficult to find sites which could be described as extensive base camps, and it may be that only a very small group lived at Mount Sandel (Woodman, forthcoming). Many other Mesolithic sites appear to be rather specialized components of a complex cycle of human settlement, such as the industrial sites at Bay Farm I, Carnlough, Co. Antrim, Lough Derravarragh, Co. Westmeath (Mitchell, 1972), or the short-term coastal sites of Dalkey Island, Co. Dublin (Liversage, 1968), and Rough Island, Co. Down (Movius, 1942).

The presence of sites without industrial waste in the Bann Valley and sites on the Antrim Coast, in particular, Bay Farm I, where few retouched tools are found on an industrial site, is highly suggestive of a population able to transport raw material in the form of readily produced blades to areas of extensive settlement (Woodman, 1980a). Attempts are being made to determine whether the flint in use in the Bann Valley comes from the Antrim coast or other inland sources. This divergence of industrial areas from those associated with food procurements illustrates the weakness in believing that procurement of raw material would be a major restriction on the distribution of Mesolithic settlement. Instead, the movement of flint and other raw materials such as chert and baked mudstone shows a remarkable ability on the part of these communities to exploit a total landscape.

THE NEOLITHIC

The traditional concept of the Irish Neolithic economy is of arable farming in the form of cereal crops and stock rearing of animals such as pigs, cattle, and sheep. The material culture associated with this life style included pottery, a new range of artifacts including piercing arrowheads, and the building of communal burial monuments.

Waddell (1978) noted that the introduction of this life style through population movement was such a truism that it did not require discussion. This view was supported by Woodman (1980b) who, in examining this problem from a Mesolithic perspective, noted that (1) the Mesolithic communities with their lowlying water-side distribution were not living in areas where agriculture could be easily practiced; (2) they had achieved a viable life style already; (3) the Irish Mesolithic appears to be isolationist in character and not easily susceptible to change; and (4) there is more evidence that the Neolithic way of life influenced surviving Mesolithic communities rather than vice versa. Burenhult (1980a,b), on the basis of his work on the Carrowmore area, Co. Sligo, has challenged these traditional concepts, and these points of difference provide a useful basis for the discussion of the settlement and economy of the Irish Neolithic. Burenhult has claimed that the megalithic tombs at Carrowmore may be the monuments of a Mesolithic community and that our concept of a Neolithic life style has been oversimplified. He considers that stock rearing was used extensively only at the end of the Neolithic, while arable farming did not play a prominent role until the Iron Age.

The idea of the native Mesolithic communities being receptive to outside influences as a result of contacts through exchange systems and thus importing a Neolithic life style to Ireland has many attractions in that it avoids the problems of establishing initial populations and the need for the archaeologist to identify these. Instead, reliance can be placed on the concept of gradual change in which few mixed points need to be identified. Burenhult's argument for Mesolithic communities building the Carrowmore tombs is based on a series of absences, namely the tombs have produced little distinctively Neolithic material, therefore they must have Mesolithic origins. This is of course a dangerous argument as these tombs have not produced anything Mesolithic. Many English long barrows are also lacking in grave goods as are many French megaliths, but this does not make the sites Mesolithic. The presence of shells also suggested a Mesolithic life style. Considering that shellfish are used throughout Irish prehistory, then neither is this a valid argument. The early ^{14}C dates at Carrowmore can be matched by others from the Neolithic site at Ballynagilly, Co. Tyrone (ApSimon, 1976). Similarly, the interesting argument for a continuation of a Mesolithic life style is based on the absence of domesticated animal bones from the lower part of the Culeenamore shell midden. Again, this could be used to argue that the middens were more or less the product of a specialized component of a broader spectrum economy which could contain agriculture. Thus, the evidence which has been used to support the strong Mesolithic component in this area is also open to an entirely different interpretation. It does, however, make us reexamine the evidence on which our concepts of the Neolithic is based.

Unfortunately, aside from the palynological evidence (Edwards, 1979, Chap. 9, this volume), there is surprisingly little evidence for the economic basis of the Irish Neolithic. It is difficult to quantify the importance of agriculture from the evidence of pollen diagrams, and there is surprisingly little material equipment which can be confidently claimed to be used in cereal production and preparation. At the moment we cannot identify the equivalent of the sickle blades which are found in the English Neolithic, and querns have yet to be found on an Irish Neolithic site [saddle querns abound on late Bronze Age sites, e.g., at Lough Eskragh (Williams, 1978)]. Also, there are only five known grain impressions on Irish Neolithic pottery, and few archaeological sites have produced cereal grains. There are of course explanations for each of these gaps. The absence of the sickles could be due to hand plucking of grain, perhaps with the aid of the hollow scrapers which are peculiar to Ireland (Flanagan, 1965); the cereals could be used in a porridge mixture without grinding; and the absence of cereal impressions could simply be due to the fact that recently no one has looked. The relative lack of prehistoric sites under excavation since flotation became common practice has resulted in a virtual absence of seeds from any prehistoric sites.

In spite of Burenhult's (1980b) suggestions, there is clear evidence for the presence of domesticated cattle, sheep, and pigs in Ireland throughout the Neolithic. While most of this comes from either Mesolithic survival shell middens or from the late Neolithic/Beaker settlement at Newgrange, Co. Meath, occasional bones of domesticated animals have been found in Neolithic contexts in megalithic tombs (van Wijngaarden-Bakker, 1980). While the domesticated animals are usually seen as of secondary importance in the Neolithic, it must be remembered that the introduction of cattle and sheep doubled the number of large mammals in Ireland, and so they must be expected to have made a major impact.

Unfortunately, due to the fact that many Neolithic sites are found on acid soils, no faunal remains have survived at sites such as Ballynagilly, Co. Tyrone (ApSimon, 1976), while the faunal remains from the Neolithic house at Lough Gur, Co. Limerick (Ó Ríordáin, 1954), were treated as a unit and because later material was also present, it cannot be assumed that all the material is Neolithic. However, even if the bones were mostly Neolithic (all 28 sacks), the overwhelming presence of cattle could be exaggerated by the methods of excavation, namely, the largest bones are those which are usually kept where little concern is felt for faunal remains.

From these very sparse facts, a working hypothesis can be established for the early Neolithic economy of Ireland. A mixed economy would appear to have been introduced, and, from the distribution of sites, there appears to have been a preference for upland light soils, where cultivation and forest clearance would have been slightly easier. It could be argued that sites such as megalithic tombs which lay on the heavier lowland clays could have been destroyed by recent agriculture. A more reliable indicator could be the major concentration of flint artifacts. These show concentrations on the edge of the higher ground and back up Watson's suggestion, made as far back as 1945, that on the basis of the distribution of tombs in Co. Antrim, primary settlement was on the morainic sands and gravels which often lay at about 90–150 m above sea level. This of course shows a considerable shift in areas of occupation from the preceeding Mesolithic (Figs. 3 and 4). An extreme example of Neolithic cultivation is that suggested by Kirk (1974) for Slieve Croob, Co. Down. Here the presence of over 5% of cereal-size Gramineae pollen may be an indication of arable farming at an extremely high altitude, although the pollen grains may have been transported by upwelling currents of air from lower ground (see Edwards, 1979).

Three relatively substantial houses at Ballynagilly (ApSimon, 1976),

Lough Gur (Ó Ríordáin, 1954), and Ballyglass, Co. Mayo (O Nualláin, 1972), have been excavated. These imply communities which were sufficiently sedentary to have a permanent base rather than to rely on a pastoralist nomadic life style.

One almost embarrassing problem is the extent to which early Neolithic communities occupied Ireland. Until recently there was almost a consensus that the earliest Neolithic settlements were in the northern part of the island and in particular in the east (Waddell, 1978). Now there is not only the early evidence from Carrowmore but pollen evidence for early cereal cultivation in Co. Cork (Lynch, 1981; and see Groenman-van Waateringe, 1983, and Edwards and Hirons, 1984). In fact, without Ó Ríordáin's work in the Lough Gur area of County Limerick, it might have been assumed that much of the extreme south of Ireland was not occupied by man until the final stages of the Neolithic. There is no doubt that the tendency to derive data from the exacavation of burial monuments such as megaliths, where probable early forms are found only in the northern two-thirds of the island, creates a negative bias in areas where what were considered to be classic early burial monuments were presumed to be absent. Yet we must note O'Kelly's discovery of a Neolithic burial under Moneen Cairn, Co. Cork (O'Kelly, 1952). Could the Neolithic burial rites of parts of Munster be much more nonmegalithic? In the extreme southwest, a fragment of a possible planoconvex knife has been found at Ballyferriter in Corca Dhuibhne (Vernon, 1976). This site is at present under investigation (Woodman, 1984).

Although early Neolithic sites are not usually associated with further signs of settlement, the idea of the lone pioneer log cabin is probably also inappropriate. Small family units could almost certainly not survive on their own for either social or economic reasons. In England, Mercer (1980) has found a large settlement enclosure close to the Hambledon Hill causewayed camp complex. This and other sites are also producing evidence that warfare was endemic in the English early Neolithic, which may suggest relatively high populations, and we must ask whether such posited high population levels were achieved in Ireland. Recent excavations at Donegore Hill, Co. Antrim (Mallory and Hartwell, 1984), and outside the Tomb 1 at Knowth, Co. Meath (Eogan, 1984), have produced evidence that large enclosures were also built in Ireland during the Neolithic period.

One major point of debate has been the question of social and economic changes during the latter part of the British Neolithic. Whittle (1978) and Bradley (1978) argued for the collapse of arable farming, although Edwards (1979, Chap. 9, this volume) questioned the way in which the palynological evidence had been interpreted. It is initially tempting to apply the Whittle and Bradley model to Ireland, but a more complex one may be necessary. Herity (1974) noted that there is a significant number of low-lying riverside and coastal settlements which have produced decorated Neolithic pottery.

Three main types of sites can be identified: (1) transitory sites such as Island MacHugh, Co. Tyrone (Davies, 1950), with its scatter of material on the island shore, or Dundrum sand dunes (Collins, 1952); (2) ditched sites such as Townleyhall I, Co. Meath (Liversage, 1960), or Armagh (Scotch Street site D) (C. Lynn, personal communication); and (3) field systems such as those being investigated by Herity and Caulfield in Co. Mayo (Caulfield, 1978). The first two categories with their rather faint traces of structures have coloured our concept of the late Neolithic and created an impression of a very transitory community, although, as McInnes (1971) has noted, substantial houses can be built without leaving convenient signs for the archaeologists. As a contrast, the work in north Mayo suggests a rather different life style in which sedentary occupation had developed to the extent that permanent land division was taking place. At Glenulra, large open fields on a ladder system were probably used for a paddock system of pasture. Caulfield (1978) has noted that with the longer growing season of the west of Ireland, perhaps combined with a marginally better climate, animals could have been grazed virtually all year round. At Belderg a mixture of Neolithic and Bronze Age settlement has been found during the excavation of smaller tillage plots. Here both traces of ploughing and ridge and furrow were found. The excavator has inferred that at this site cultivation and land division began in the late Neolithic. The implications of this land division are that occupation was relatively sedentary and high levels of population existed. The Mayo evidence does suggest that a concept of highly mobile pastoralists in the late Neolithic may be an oversimplification. It is tempting to believe that these field systems might have existed in the east of Ireland as well.

Evidence for a rather broader spectrum economy can be seen from the location of sites in more marginal areas. Burenhult (1980b) has found small camp sites with numerous hollow scrapers on Knocknarea Mountain in Co. Sligo. The shell midden at Culeenamore could be part of the same system. At Carnlough, where at higher altitudes there is a relatively narrow band of sands and gravels with very thin poor soils and heavy boulder clay below, there is also evidence in certain areas for a range of specialist sites. This is mostly based on surface collection and occasional excavation in the Glencloy area, where a possible base settlement has been found at Galdanagh, lying some distance from the coast (Fig. 4). In contrast, small scatters of flintwork with numbers of hollow scrapers have been found on the edge of the Garron Plateau, while a rather specialized site was found at the edge of the Glencloy River and chipping floors found on the coast. This illustrates that Neolithic communities must not be seen as static groups exploiting only the immediate environs of their farms but, particularly if domesticated animals were important, would have used the total landscape.

The ability to concentrate large groups of people in the later Neolithic is

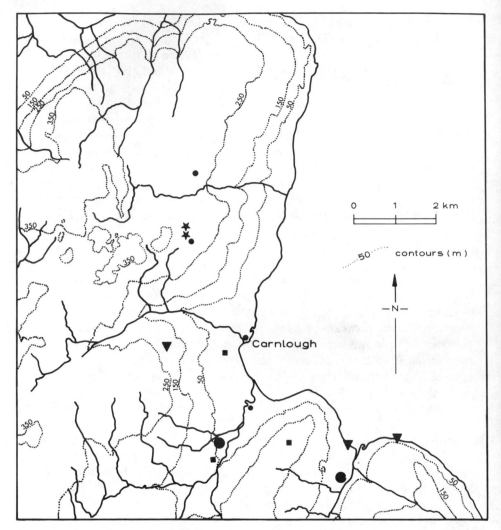

Fig. 4. Map of the Carnlough, Co. Antrim, area showing distribution of known Neolithic groups of material: ●, possible main settlements; ★, hollow scraper dominated sites; ■, end scraper dominated sites; ▼, industrial sites; •, miscellaneous.

shown by the complex of monuments in the Bend of the Boyne cemetery. Other concentrations of tombs in cemeteries are known in Ireland, and it is tempting to see the massive earthwork at the Giant's Ring, Co. Down, as part of a similar complex in which several small passage graves, which existed until the early nineteenth century, played an important role. These ritual complexes could be compared with such areas as Avebury in Wessex.

Herity (1974) suggested that Newgrange would require 5,000 people for its construction, although it was likely that much smaller numbers would gather seasonally to build such tombs over several generations. The building of the tomb itself may have been the main reason for annual gatherings of a dispersed population.

If the few pieces of evidence about the late Neolithic economy of eastern Ireland are pointing in the right direction, the move to a broader economy may have also resulted in more population mobility. Amongst the grave goods are items which could not be procured locally such as chalk balls, whose value most probably lay in the difficulty of procuring the chalk, since none have been found in the few northeastern tombs with easy access to the raw materials. These could imply a well-established exchange system in nonessentials. Similarly, the presence of a few exotic shells such as *Pecten* at Belmore Mountain may be a product of an exchange system, but the large quantities of shells found at Loughcrew, Co. Meath (Cairn K), could represent a population movement from the coast. Certainly if there is greater continuity of population with the Beaker period than has been supposed, the evidence from Newgrange (van Wijngaarden-Bakker 1980; see below) does point to a society in which transhumance played a leading role. Therefore these concentrations of tombs may have been the focal centre of seasonal or annual gatherings in which the building of the tomb may have been the reason for gathering. It is possible that one of the reasons for the absence of a passage grave cemetery in Co. Mayo is that the Neolithic people of that area continued to be rather more sedentary than elsewhere. Contrasting the economic strategies of north Mayo with the neighbouring Carrowmore area must lead one to believe that variety of economic strategy is the main characteristic of the late Neolithic in Ireland.

Exchange systems would have existed not only for the high-status or ritual objects associated with the passage tombs but would have been necessary to procure many of the stone artifacts. The existence of Tievebulliagh and Brockley, Co. Antrim, as quarry and factory sites for the production of porcellanite axes, has long been known (Jope, 1952). Other factory sites must exist in Ireland, and it is a matter of regret that no careful work has attempted to identify them. Knowles (1901) noted one possible axe factory on the coast of Co. Clare, while many of the axes from the Lough Gur area are made of the same distinctive greenstone (L. N. W. Flanagan, personal communication). Due to the hardness of Irish chalk, mining for flint is a much more restricted phenomenon. Open cast mining is known at Ballygalley Head, Co. Antrim (Collins, 1978), another site was investigated at Black Mountain, Co. Antrim, near Belfast, and a third possible site has been identified on Slieve Gullion, Co. Derry. Again, as in the Mesolithic, these sites need not necessarily be in areas of intensive Neolithic settlement.

THE PROBLEM OF THE SO-CALLED
BEAKER FOLK

The traditional concept of the so-called Beaker Folk stems from the belief that there was a very cohesive group of material which was associated with Beaker pottery. This belief was so strong that the Beaker folk were heralded as both the first Celts (Harbison, 1975) and the introducers of metallurgy (de Valera and Ó Nualláin, 1961), and the material culture was undoubtedly assumed to be related to an actual invasion of Ireland. Many would now believe that there are very significant differences within the range of Beaker material found throughout Europe, so much so that it may not have a common origin (Harrison, 1974).

The original concept was that Beaker pottery was associated with stone wrist guards, barbed and tanged arrowheads, copper flat axes, tanged daggers, and perhaps the wedge-shaped gallery graves. In fact de Valera and Ó Nualláin (1961), in suggesting that wedge-shaped gallery graves were introduced to Ireland from Brittany by the Beaker Folk, felt confident enough to suggest that the first representatives landed in Co. Kerry, where they were interested in exploiting Irish copper. Only Flanagan's coining of the phrase "Beaker compatible" seemed to show any caution (see Harbison, 1976).

Careful examination shows that there is little evidence this material group belonged together and that it is impossible to find a common point of origin for it. Sheridan (1979), in a reappraisal of the origins of Irish metallurgy, suggests that the role of exchange systems has been underestimated and that there is no reason why the earliest copper axes should not have arrived in Ireland during the late Neolithic, and, as has already been shown, exchange systems for objects which were not entirely functional could have already been established. At the other extreme, Harbison (1976) has suggested that most of the wrist guards could be rather later than Beaker pottery.

This leaves the apparent "association" of wedge-shaped gallery graves and beaker pottery. Even this now appears unlikely because, as Waddell (1978) has noted, the best parallels for tombs and pottery lie in different areas, thus weakening the case for a Breton origin of the Beaker culture in Ireland. The case for a Kerry landfall was already weakened by the fact that the tombs in that area have more differences from the Breton tombs than is usual even for Irish tombs in general.

More recent work on the material from Newgrange, in particular by Cleary (1980), shows a continuum in the pottery from the late Neolithic and through the Beakers. This includes methods of manufacture and sources of clay. Similarly, in spite of the presence of occasional barbed and tanged flint arrowheads, the flint work of many of the Beaker sites looks remarkably like a continuation from the late Neolithic as typified by Newgrange

(O'Kelly, 1973; Lehane, 1980) and Monknewton (Sweetman, 1976). The very fact that substantial quantities of Beaker pottery are found in the area of the Bend of the Boyne passage grave cemetery and at Monknewton reinforces the continuity of occupation associated with Beaker material. Therefore, as in many parts of Europe, it might be possible to argue that the appearance of this group of material is not necessarily to be associated with large-scale movements of people and certainly cannot be used as a convenient archaeological peg on which other technological or linguistic changes can conveniently be hung.

In a study of Newgrange material, van Wijngaarden–Bakker (1980) has argued for an economy in which transhumance played a major role. Cereal production would appear to have been of relatively minor importance in the economy, while beef and pork played a major part in the diet of the Beaker community; indeed, there is evidence for the slaughtering of swine in this extremely rich site. There is evidence from elsewhere for a greater reliance on pork. Doughty (personal communication), for example, has found that pig bones were, after those of cattle, by far the most common at the cave of Poll-na-gollum, Co. Fermanagh. This appears to contain a natural deposit where animals had fallen into the cave. However, most of the species recovered are domesticated animals, and the deposit appears to have formed during the late Neolithic and early Bronze Age. One interesting addition to the fauna is the domesticated horse which appeared, as in many parts of western Europe, at the same time as Beaker pottery. One horse bone was found in Audleystown court cairn, Co. Down (Collins, 1954), but as sherds of a food vessel were found in the same chamber, it would be dangerous to assume that horse was found in Ireland earlier than the Beaker period.

As in many parts of Europe, there is a virtual lack of structural remains from this period apart from scatters of stake holes at Ballynagilly (Ap-Simon, 1976), a small hut 3 m across at Newgrange (O'Kelly, 1973), and a dwelling hollow at Monknewton (Sweetman, 1976). Slight structural remains have often caused archaeologists to make the presumptuous extrapolation that these sites must have been used by pastoralists who presumably could not stay long enough to dig convenient holes for posterity. Ironically, several Beaker houses were found at Lough Gur.

The idea that the users of Beaker pottery were pastoralists was of course favoured by those who saw the spread of Beaker pottery as a sign of rapid population movement. In many areas, however, the trend toward pastoralism may have already been established. De Valera and Ó Nualláin (1961) have argued for a Beaker pastoralist economy on the basis of their association of beakers with wedge graves. As numerous examples of these occur on the Burren in Co. Clare, they argued that the thin soils which exist at the moment could only have supported pastoralists. Drew (1983), however, has shown that a well-developed soil may have covered the Burren in prehistoric times. Thus, while there is some evidence for transhumance, the case for a totally pastoralist economy is not proved.

THE BRONZE AGE

Considering the numerous typologies of pottery and bronzes and the discussion of their relative merits, a cynic might be inclined to wonder whether the major aim of future research in this period was to establish whether human life was present at all! On the other hand, the early Bronze Age is the first period for which there is clear evidence of the occupation of the whole of Ireland, and there is less need to worry about potential bias in the data recovered.

Little can be said about the settlement and economy of this time. It is usually assumed that this is a period of continuity in which there was relatively little change. Only one adequately excavated settlement exists, at Downpatrick, Co. Down, and here the site may have been mostly obliterated by later building development. It should be pointed out that these are not Beaker houses as claimed by many (e.g., Megaw and Simpson, 1979). Some houses at Lough Gur could date to the Bronze Age, but there is the usual difficulty of associating material with structures. On the other hand, Caulfield (1978) has Bronze Age ^{14}C dates from a large round house at Belderg, Co. Mayo. The relative lack of known settlement sites in the early Bronze Age makes it difficult to assess the economy. Certainly the numerous discoveries of Bronze Age cist burials in certain lowland areas, for instance, around Ballymena, Co. Antrim, would suggest a continued expansion of the area utilized. T. Reeves-Smyth (personal communication) has suggested, on the basis of an examination of distribution sites in Counties Derry and Tyrone, that there is a continued occupation of the upland areas with a certain amount of expansion outward from initial areas of Neolithic utilization. This he would attribute to a growing population which he feels is gradually placing more emphasis on arable farming. O'Kelly (1952) found numerous quern stones in the early Bronze Age Cairn of Moneen, Co. Cork. These must be amongst the earliest known in Ireland and could reflect a greater reliance on cereal production, as could numerous examples of saddle querns from the round house at Belderg.

One major problem in Ireland is the extent to which upland forest clearance and farming may have influenced the development of blanket peat. Smith (1975) has noted that most upland blanket peat in the northeast of Ireland would seem to begin during the Bronze Age. This he feels is at least in part due to anthropogenic factors, although climatic factors could also have helped cause its inception. This topic is explored more fully by Edwards (Chaps. 9 and 12, this volume).

The continued use of a broad spectrum economy is best illustrated by the cooking places, which are often in historical sources called "fulachta fiadh." These are known to have existed in the Bronze Age. O'Kelly (1954) has noted several which are associated with early Bronze Age axes, and some have been radiocarbon dated to the Bronze Age. These sites were usually placed in a damp area where an artificial trough would fill with water. The water was heated by dropping in stones heated in a nearby hearth. This use of heated stones usually resulted in the large mound of burnt stones by which these sites are usually identified. Small huts are often associated with these cooking places. O'Kelly (1954), using historical sources, has suggested that these sites are similar to those used by the Fianna Eireann (warriors or young men of Eireann) when they were deer hunting. This type of cooking is often associated with aceramic cultures.

Such sites are one of the few indications that hunting played an important part in Irish prehistory. Deer bones are virtually absent on archaeological sites, but Stelfox noted that numerous deer had died naturally in the lake at Ballinderry 2 (Hencken, 1942). The recent excavations of a cooking place at Carran, Co. Clare, has produced both cattle bones and deer antlers (Ó Drisceoil, personal communication). Monk (personal communication) has suggested that these sites were associated with hunting and the boiling of beef from free-range cattle. If animals such as deer were butchered and consumed off site, the data derived from principal settlements will be biased, and if the historical data are correct, it is possible to have a mobile component within a society rather than to have the whole society move. Although these sites may not have existed in the Neolithic, they imply that our usual models for population mobility are oversimplified. The exploitation of a broad range of environments can also be seen in the continued use of marine resources, since shell middens in the west of Ireland have also produced Bronze Age materials.

At the moment the major problems of this period are those connected with finding the source of certain raw materials, namely, copper and gold (Ryan, 1979). Hartman (1979) maintains that the trace element analysis of gold points to a foreign origin, while many local archaeologists feel that insufficient samples of local gold were analyzed for comparison (Taylor, 1979; see note no. 6 in Scott, 1979). One worrying aspect of this discussion is that no consideration has been given to social factors. Ireland was a major producer of bronzes during the early Bronze Age (Fig. 5), and it would appear that its products were either widely exported or copied; therefore, it could have been a rich society, well able to afford to import status symbols. As Randsborg (1979) has pointed out, numerous copper artifacts are found

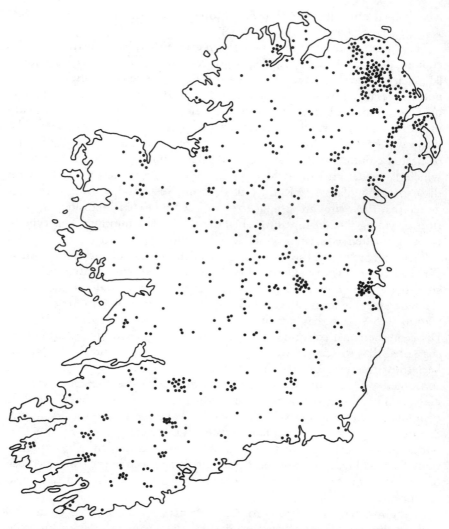

Fig. 5. Distribution map of findspots of early Bronze Age axes (after Flanagan, 1979).

in Denmark in the early Neolithic although there is no copper ore in Denmark.

While there is some disagreement about the extent of climatic deterioration in the late Neolithic and early Bronze Age, there would seem to be a consensus of opinion (Mitchell, 1976; Lynch, 1981; Simmons and Tooley, 1981) that the climate was wetter and cooler during at least part of the first millennium BC. At numerous sites such as Cashelkeelty, Co. Kerry

(Lynch, 1981), and Sluggan, Co. Antrim (Goddard, 1971), there is evidence of prolonged extensive clearance with the woodlands never completely regenerating. There is at this period a significant concentration of sites around the edges of lakes, for example, at Ballinderry (Hencken, 1942) and Lough Eskragh (Williams, 1978). Could it be that the initial expansion during the early Bronze Age, onto more marginal land and brittle environments, was brought to an end by climatic deterioration and that this in turn led to both the development of blanket peat and the move into lower lying areas?

There is also some evidence for continued enclosure of land. At Belderg, Co. Mayo, a wooden fence has been dated to the transition from the early to late Bronze Age, and Lynch (1981) would argue that enclosures existed at Cashelkeelty, Co. Kerry, during the late Bronze Age. A number of these later sites have produced saddle querns in substantial numbers, and at one site, Bay Farm I (Upper), Carnlough, Co. Antrim, J. Mallory (personal communication) has recovered barley. This evidence and presence of bronze sickles suggest a greater reliance on cereal production. Unfortunately, these sites have not produced significant quantities of faunal remains, but it is possible to argue that cattle rearing played a major role. At Ballinderry 2, Co. Offaly (Hencken, 1942), a brief report notes that cattle made up 80% of the fauna while pigs were of secondary importance. Sheep again are of lesser importance. Some confirmation of this range of animals can be found in the small fauna sample found at the Kings Stable, Co. Armagh (Lynn, 1977), where, in spite of the small sample size, the dominance of cattle appears to be fairly certain.

Unlike the case for the early Bronze Age, there is no problem identifying dwellings and settlements of this period. Larger substantial circular dwellings have been found at sites such as Navan, Co. Armagh (Waterman and Selkirk, 1970), and Rathgall, Co. Wicklow (Raftery, 1976a). These are houses up to 10 m across, and, in the case of Navan, the house was rebuilt on numerous occasions on the same spot. These houses sometimes have compounds associated with them, and sometimes they were placed in ditched enclosures (this may be the source of a typological problem discussed below).

There is also growing evidence that hillforts may have had their origin in the late Bronze Age. At Downpatrick, Co. Down, Warner (1980) has suggested that the first phase of fortification was begun in the late Bronze Age. Several other sites where hillforts were built have traces of extensive late Bronze Age occupation, for example, Navan Fort, Co. Armagh, Rathgall, Co. Wicklow, and Clogher, Co. Tyrone (Warner, 1974). It is tempting to see these substantial sites and hillforts as part of an increased, more organized, population. Unfortunately, the virtual absence of identifiable burials for the late Bronze Age makes it difficult to identify evidence of social stratification (see Raftery, 1982).

At the end of the Bronze Age there is a problem deciding when iron metallurgy was introduced and, second, when it made a significant contribution to equipment in everyday use in the early Iron Age. A continuum of settlement can be identified at Navan and to a lesser extent at Rathgall, but it is impossible to know how late the floruit of late Bronze Age metalwork really came (see Scott, 1974; Warner, 1974; Raftery, 1976b). It has even been suggested that there may have been a period at the end of the Bronze Age when the use of bronze dropped away.

THE EARLY IRON AGE

A problem of the early Iron Age is the provision of an agreed set of chronological indicators. Thus there is a dichotomy between those who would start the Iron Age in Ireland with an early date in the Halstatt Iron Age (e.g., Rynne, 1961) and those who believe that native iron products were in general use only much later (Warner, forthcoming). Due to a lack of objects in context and to a scarcity of objects which can be easily paralleled outside Ireland, much of this period has to suffice with relative typological frameworks for a range of objects which often cannot be securely related one to another. This does not imply that there is a scarcity of objects— rather a relative lack of contexts.

Aside from the problem of the status of the Halstatt objects in Ireland, there is a group of objects which relate much more clearly to the La Tène culture of Europe, for example, scabbards and swords. The point of origin for even these is disputed; Jope (1954) would suggest Europe while Rynne (1961) would suggest that some at least derive from Yorkshire. Beside these disagreements over origins, there would seem to be no consensus as to how this range of material was probably introduced—by the small intrusion of a new aristocracy or group of craftsmen into Ireland or by migration (see Raftery, 1982, and Caulfield, 1982).

One characteristic traditionally claimed of this period has been the hill-fort. As has been noted earlier, these may have their origin in the late Bronze Age, but it might be expected that they could continue in use into the early Iron Age. Large enclosures of this type are relatively rare in Ireland and still remain largely unexcavated. In the east is a series of mostly univallate hillforts, while in the west there is a group of hillforts with widely spaced ramparts (Raftery, 1972).

The chronological problems of these sites are proving to be remarkably intractable. This is partially due to a lack of excavation; in fact, none of Raftery's Class II western sites have been examined in recent years. Even in the east, no certain early Iron Age occupation can be identified at Clogher, Co. Tyrone, or Rathgall, Co. Wicklow. The presence of ritual sites has been confirmed at Dún Áilinne, Co. Kildare (Wailes, 1970). This is one of the few sites to have produced evidence of an early Iron Age phase of occupation. There is now growing evidence that some of the large linear earthworks could belong to this period. In particular, the construction of the Dorsey enclosure has been dated to about 200 bc. This and the ritual phase at Navan Fort have been shown to be contemporary through dendro-chronology (Lynn, 1982). In Ireland, occupation of hillforts can be shown to exist contemporaneously with the Roman period. In particular, Roman material has been found at the Rath of the Synods at Tara, Co. Meath, while sub-Roman pottery has been found in the inner citadel at Clogher, Co. Tyrone, although in this case the hillfort may have been long abandoned.

It is only when one moves away from the large field monuments and ritual sites that the problems of settlement emerge. This is exacerbated by the tendency to look for the origins of the field monuments of the following Early Christian period (around AD 400 and later) in the early Iron Age. These include ringforts (or raths) and crannogs. In the case of ringforts, sites such as Carrigillihy, Co. Cork (O'Kelly, 1951), Cush, Co. Limerick (Ó Ríordáin, 1940), and Feerwore, Co. Clare (Raftery, 1944), have been claimed to prove that ringforts go back to before the Early Christian period. The rather substantial ringfort at Cush, Co. Limerick, was initially dated to the late Bronze Age following Ó Ríordáin's observation that urn burials were found in a late stratigraphic context. However, as these are now presumed to be early Bronze Age, many would doubt Ó Ríordáin's in-ference (Proudfoot, 1961; Herity and Eogan, 1977). At Feerwore the pre-sumed early material was found under the bank of a ringfort and so cannot be used to date the ditch within which it lay. Carrigillihy, on the other hand, is probably prehistoric, although Kelly (1978) has suggested, on the basis of a reexamination of the pottery, that a late Bronze Age date is more likely rather than the Neolithic date that was originally proposed. Howev-er, as this is a very slight structure, it is worth noting Warner's (1980) comments that ringforts are rather more substantial and may, in fact, mea-sure 4 m from the base of the ditch to the lip of the bank.

There have also been suggestions that some of the sites referred to above show that ringforts in the south occur earlier and that the idea spread from the south. Yet, the presence of a ditch around a house at Navan Fort (Waterman and Selkirk, 1970) and the presence of a circular stone wall on the hilltop at Clogher, Co. Tyrone (R. B. Warner, personal communica-tion) show that simple enclosures also existed in the north during the late Bronze Age.

There is also extensive evidence of lake shore settlements in the later part of Ireland's prehistory. Some of these, such as Knocknalappa (Raftery,

1942), could go back to the late Bronze Age. Similarly, at Rathtinaun (Raftery, 1976a) a late Bronze Age/early Iron Age date has been suggested for another shoreline site which may have been slightly raised artificially. It is questionable whether these sites should be regarded as direct ancestors of the large, artificial, defended islands often known as crannogs. In fact, artificial platforms can be shown to have existed as far back as the Mesolithic, while Mitchell (1976) has also noted that in some instances the heightening of these islands could be natural.

Evidence of farming itself is rather contradictory. Rotary querns, highly suggestive of arable farming, appear for the first time. Some, as has been pointed out by Caulfield (1972), are decorated in a fashion reminiscent of La Tène art. These are usually beehive querns, yet none have been found in an early Iron Age context. On the other hand, Mitchell (1976) has claimed that this same period is characterized by a decrease in farming, as shown in numerous pollen diagrams from across Ireland (but see Edwards, Chap. 9, this volume).

The problems of this period are finally compounded by recent suggestions that the so-called early sagas, such as the Ulster Cycle, do not reflect accurately the way of life in the early Iron Age but may be rather anachronistic in character (Mallory, 1982). The presence of numerous rotary querns also highlights a paradox between the archaeological material and the literary record. The literary record suggests that cattle are the all-important source of wealth, yet it is apparent from the querns that arable farming also played an important role.

Unlike the case for Britain and many parts of Europe, it is impossible to document the end of the Irish early Iron Age. The Romans do not provide a convenient chronological horizon, therefore the introduction of Christianity into Ireland—perhaps imported with wine from the Mediterranean—has been used to provide a break between prehistory and the early historic periods. The notional end of Ireland's prehistory is therefore usually taken to be sometime around AD 400. However, the classic ringfort-using, monastic society may not have emerged until some time after this date. The end of the early Iron Age may instead be seen in the collapse of highly organized federations, such as that controlled from Navan, followed by a virtual *Volkwanderungzeit* which can be seen in the earliest stratum of heroic and analystic literature and could even have been the result of a major intrusion of peoples from late Roman Britain.

CONCLUSIONS

Beside the problems associated with the differential biases of material remains from different periods, the overall impression of the economic patterns in Ireland's prehistory is one in which changes were more gradual than has often been presumed in the past. If the evidence for prehistoric economies is examined *in toto,* there may have been a tendency to over-emphasize different aspects of the economies of different periods.

The major problems of Ireland's prehistory may be suggested to be (1) the relative absence of excavated settlement sites in many periods; (2) the lack of environmental backgrounds from locations adjacent to areas of intensive settlements (there is a need for more coring of lakes rather than of peat bogs); and (3) the creation of a series of priorities in an archaeological research policy rather than a reliance on the windfall benefits of rescue archaeology.

Acknowledgments

A paper of this type cannot be written without discussion with other colleagues. I would like to thank in particular L. N. W. Flanagan, who provided Fig. 5, R. B. Warner, J. Mallory, C. Lynn, and E. Grogan for information and helpful criticism; Miss G. Sands for providing the drawings; and Miss P. Auterson who typed and coped patiently with the numerous versions of the text.

References

ApSimon, A. (1976). *In* "Acculturation and Continuity in Atlantic Europe. Proc. of the 4th Atlantic Colloquium" (S. J. De Laet, ed.), pp. 15–30. Acta Archaeologica Gandensis, Ghent.

Bradley, R. (1978). "The Prehistoric Settlement of Britain." Routledge & Keegan Paul, London.

Burenhult, G. (ed.) (1980). "The Archaeological Excavation at Carrowmore, Co. Sligo, Ireland: Excavation Seasons 1977–1979." Theses and Papers in North European Archaeology, No. 9. Inst. of Archaeol., Univ. of Stockholm, Stockholm.

Burenhult, G. (1984). "The Archaeology of Carrowmore: Environmental Archaeology and the Megalithic Tradition at Carrowmore, County Sligo, Ireland." Theses and Papers in North European Archaeology, No. 14. Inst. of Archaeol., Univ. of Stockholm, Stockholm.

Caulfield, S. (1972). *J. R. Soc. Antiq. Irel.* **107,** 104–138.

Caulfield, S. (1978). *In* "Early Land Allotment" (H. C. Bowen and P. J. Fowler eds.), pp. 137–143. BAR Brit. Ser. No. 48, Oxford.

Caulfield, S. (1982). *In* "Irish Antiquity" (D. Ó Corráin, ed.), pp. 205–215. Tower Books, Cork.

Cleary, R. (1980). "The Late Neolithic/Beaker Period Ceramic Assemblage from Newgrange, Ireland." Unpublished M.A. thesis, University College Cork (N.U.I.), Cork.

Collins, A. E. P. (1952). *Ulster J. Archaeol.* **15,** 2–30.

Collins, A. E. P. (1954). *Ulster J. Archaeol.* **17,** 7–56.

Collins, A. E. P. (1978). *Ulster J. Archaeol.* **41,** 15–32.

Davies, O. (1950). "Excavations at Island McHugh." Suppl. Proc. Rep. Belf. Nat. Hist. Philos. Soc., Belfast.

de Valera, R., and Ó Nualláin, S. (1961). "Survey of the Megalithic Tombs in Ireland." Stationery Office, Dublin.

Drew, D. P. (1983). *J. Hydrol.* **61,** 113–124.

Edwards, K. J. (1979). *J. Archaeol. Sci.* **6,** 255–270.

Edwards, K. J., and Hirons, K. R. (1984). *J. Archaeol. Sci.* **11,** 71–80.

Eogan, G. (1983). "Hoards of the Irish Later Bronze Age." University College, Dublin.

Eogan, G. (1984). "Excavations at Knowth (1)." Monographs in Archaeology. Royal Irish Academy, Dublin.

Fairley, J. S. (1975). "An Irish Beast Book." Blackstaff Press, Belfast.

Flanagan, L. N. W. (1965). *In* "Atti del VI Congresso Internazionale Delle Scienze Pre-istoriche et Protostoriche" (G. C. Sansoni, ed.), pp. 323–328. Rome.

Flanagan, L. N. W. (1979). *Ir. Archaeol. Res. Forum* **6.**

Goddard, I. (1971). "The Palaeoecological Investigations of Some Sites in N. E. Ireland." Unpublished M.Sc. thesis, The Queen's Univ. of Belfast, Belfast.

Green, S. (1981). *Antiquity* **55,** 184–196.

Groenman-van Waateringe, W. (1983). *In* "Landscape Archaeology in Ireland" (T. Reeves-Smyth and F. Hammond, eds.), pp. 217–232. BAR Brit. Ser. No. 116, Oxford.

Harbison, P. (1969a). "The Daggers and Halberds of the Early Bronze Age in Ireland. Prahistorische Bronzefunde." Abteilung VI, I. C. H. Beck'sche Verlagsbuchandlung, Munich.

Harbison, P. (1969b). "The Axes of the Early Bronze Age in Ireland. Prahistorische Bronze-funde." Abteilung IX, I. C. H. Beck'sche Verlagsbuchandlung, Munich.

Harbison, P. (1975). *J. Indo-Eur. Stud.* **3,** 101–119.

Harbison, P. (1976). *Archaeologica Atlantica Res. Rep.* No. 1. Swapach, Hamburg.

Harrison, R. J. (1974). *Antiquity* **48,** 99–109.

Hartman, A. (1979). *In* "The Origins of Metallurgy in Atlantic Europe. Proc. of the 5th Atlantic Colloquium" (M. Ryan, ed.), pp. 215–228. Stationery Office, Dublin.

Hencken, H. O. (1942). *Proc. R. Ir. Acad., Sect. C* **47,** 1–76.

Herity, M. (1974). "Irish Passage Graves." Irish Univ. Press, Dublin.

Herity, M., and Eogan, G. (1977). "Ireland in Prehistory." Routledge & Keegan Paul, London.

Jacobi, R. M., Tallis, J., and Mellars, P. A. (1976). *J. Archaeol. Sci.* **3,** 307–320.

Jope, E. M. (1952). *Ulster J. Archaeol.* **15,** 31–55.

Jope, E. M. (1954). *Ulster J. Archaeol.* **17,** 31–91.

Jope, E. M. (1966). "An Archaeological Survey of County Down." HM Stationery Office, Belfast.

Kavanagh, R. (1973). *Proc. R. Ir. Acad., Sect. C* **73,** 507–617.

Kelly, J. (1978). *Ir. Archaeol. Res. Forum* **5,** 23–27.

Kirk, S. (1974). *Ulster J. Archaeol.* **36–37,** 99–100.

Knowles, W. J. (1901). *Proc. R. Ir. Acad.* **22,** 331–389.

Lehane, D. (1980). "An Analysis of the Flintwork from the Late Neolithic/Beaker Period at Newgrange, Co. Meath." Unpublished M.A. thesis, Univ. College Cork (N.U.I.), Cork.

Liversage, D. (1960). *J. R. Soc. Antiq. Irel.* **90,** 49–59.

Liversage, D. (1968). *Proc. R. Ir. Acad., Sect. C* **66,** 53–233.

Lynch, A. (1981). "Man and Environment in S. W. Ireland 4000 B.C.–A.D. 800, A Study of Man's Impact on the Development of Soil and Vegetation." BAR Brit. Ser. No. 85, Oxford.

Lynn, C. J. (1977). *Ulster J. Archaeol.* **40,** 42–62.

Lynn, C. J. (1982). "Studies on Early Ireland" (B. G. Scott, ed.), pp. 121–128. Assoc. Young Ir. Archaeologists, Belfast.

Mallory, J. (1982). "Studies on Early Ireland" (B. G. Scott, ed.), pp. 99–114. Assoc. of Young Ir. Archaeologists, Belfast.

Mallory, J., and Hartwell, B. (1984). *Curr. Archaeol.* **92,** 271–274.

McInnes, I. J. (1971). *In* "Economy and Settlement in Neolithic and Early Bronze Age Britain and Europe" (D. D. A. Simpson, ed.), pp. 113–130. Leicester Univ. Press, Leicester.

Megaw, V., and Simpson, D. D. A. (eds.) (1979). "An Introduction to British Prehistory." Leicester Univ. Press, Leicester.

Mercer, J. (1974). *Antiquity* **48**, 65–66.

Mercer, R. (1980). "Hambledon Hill." Edinburgh Univ. Press, Edinburgh.

Mitchell, G. F. (1970). *Ulster J. Archaeol.* **33**, 3–14.

Mitchell, F. (1972). *J. R. Soc. Antiq. Irel.* **102**, 160–173.

Mitchell, F. (1976). "The Irish Landscape." Collins, London.

Movius, H. L. (1942). "The Irish Stone Age, Its Chronology, Development and Relationships." Harvard Univ. Press, Cambridge, Massachusetts.

O'Kelly, M. J. (1951). *J. Cork Hist. Archaeol. Soc.* **56**, 69–86.

O'Kelly, M. J. (1952). *Proc. R. Ir. Acad., Sect. C* **54**, 121–159.

O'Kelly, M. J. (1954). *J. R. Soc. Antiq. Irel.* **84**, 105–155.

O'Kelly, M. J. (1973). "Megalithic Graves and Ritual" (G. Daniel and P. Kjaerum, eds.), pp. 137–146. Jutland Archaeol. Soc. Publics., Aarhus.

Ó Nualláin, S. (1972). *J. R. Soc. Antiq. Irel.* **102**, 49–57.

Ó Ríordáin, S. P. (1954). *Proc. R. Ir. Acad., Sect. C* **56**, 297–459.

Proudfoot, V. B. (1961). *Medieval Archaeol.* **5**, 94–122.

Raftery, J. (1944). *J. R. Soc. Antiq. Irel.* **74**, 23–52.

Raftery, J. (1942). *North Munster Archaeol. J.* **3**, 53–72.

Raftery, B. (1972). *In* "The Iron Age in the Irish Sea Province" (C. Thomas, ed.), pp. 37–58. CBA Res. Rep. No. 9, London.

Raftery, B. (1976a). *In* "Later Prehistoric Earthworks in Britain and Ireland" (D. W. Harding, ed.), pp. 339–357. Academic Press, London.

Raftery, B. (1976b). *In* "Acculturation and Continuity in Atlantic Europe. Proc. of the 4th Atlantic Colloquium" (S. J. De Laet, ed.), pp. 189–197. Acta Archaeologica Gandensis, Ghent.

Raftery, B. (1982). *In* "Irish Antiquity" (D. Ó Corráin, ed.), pp. 173–204. Tower Books, Cork.

Raftery, B. (1984). "La Tene in Ireland: Problems Origin and of Chronology." Veroffentlichung des Vorgeschichtlichen Seminars Marburg, Marburg.

Randsborg, K. (1979). *In* "The Origins of Metallurgy in Atlantic Europe. Proc. of the 5th Atlantic Colloquium" (M. Ryan, ed.), pp. 303–318. Stationery Office, Dublin.

Ryan, M. (ed.) (1979). "The Origins of Metallurgy in Atlantic Europe. Proc. of the 5th Atlantic Colloquium." Stationery Office, Dublin.

Ryan, M. (1980). *Antiquity* **54**, 46–47.

Rynne, E. (1961). *In* "Bericht uber den V Internationalen Kongress fur Vor und Fruhgeschichte, Hamburg, 1958" (G. Bersu, ed.), pp. 705–709. Berlin.

Scott, B. G. (1974). *Ir. Archaeol. Res. Forum* **1**, 9–24.

Scott, B. G. (1979). *In* "The Origins of Metallurgy in Atlantic Europe. Proc. of the 5th Atlantic Colloquium" (M. Ryan, ed.), pp. 189–204. Stationery Office, Dublin.

Sheridan, A. (1979). *Ir. Archaeol. Res. Forum* **6**.

Simmons, I. G., and Tooley, M. J. (eds.) (1981). "The Environment in British Prehistory." Duckworth, London.

Smith, A. G. (1970). *In* "Studies in the Vegetational History of the British Isles: Essays in Honour of Harry Godwin" (D. Walker and R. G. West, eds.), pp. 81–96. Cambridge Univ. Press, London.

Smith, A. G. (1981). *Proc. 4th Int. Palynol. Conf. Lucknow (1976–1977)* **3**, 248–257.

Sweetman, P. D. (1976). *Proc. R. Ir. Acad., Sect. C* **76**, 25–72.

Taylor, J. (1979). *In* "The Origins of Metallurgy in Atlantic Europe. Proc. of the Fifth Atlantic Colloquium" (M. Ryan, ed.), pp. 229–250. Stationery Office, Dublin.

Vernon, P. V. (1976). *J. Cork Hist. Archaeol. Soc.* **81,** 118–119.

Waddell, J. (1978). *Antiquity* **52,** 121–128.

Wailes, B. (1970). *J. R. Soc. Antiq. Irel.* **100,** 79–90.

Warner, R. (1974). *Ir. Archaeol. Res. Forum* **1,** 45–47.

Warner, R. (1980). *Bull. Ulster Place-Name Soc.* **3,** 45–53.

Warner, R. (forthcoming). *In* "The Archaeology of Ulster" (C. Lynn, L. N. W. Flanagan, and T. McNeill, eds.), Ulster Archaeol. Soc., Belfast.

Waterman, D., and Selkirk, A. (1970). *Curr. Archaeol.* **22,** 304–308.

Whittle, A. W. R. (1978). *Antiquity* **52,** 34–42.

van Wijngaarden-Bakker, (1980). "An Archaeological Study of the Beaker Settlement at Newgrange, Ireland." Unpublished Ph.D. thesis, Univ. of Amsterdam, Amsterdam.

Williams, B. B. (1978). *Ulster J. Archaeol.* **41,** 37–48.

Williams, P. C. (1978). "The Mesolithic in Ireland" BAR Brit. Ser. No. 58, Oxford.

Woodman, P. C. (1980a). *In* "Proc. 3rd International Flint Conference" (G. Engelen, ed.), pp. 113–115. Nederlandse Geol. Vereniging, Maastricht.

Woodman, P. C. (1980b). *In* "Proceedings of the 2nd Mesolithic in Europe Congress" (B. Gramsch, ed.), pp. 201–210. Veroffentlichungen des Museums fur ur und fruhgeschichte, Potsdam.

Woodman, P. C. (1981a). *Sci. Am.* **245,** 120–132.

Woodman, P. C. (1981b). *In* "Irish Antiquity" (D. Ó Corráin, ed.), pp. 93–110. Tower Books, Cork.

Woodman, P. C. (1984). *J. Cork Hist. Archaeol. Soc.* **89,** 1–11.

Woodman, P. C. (forthcoming). "Excavations at Mt. Sandel." HM Stationery Office, Belfast.

CHAPTER 12

CHRONOLOGY

Kevin J. Edwards
Department of Geography
University of Birmingham
Birmingham, England

Michael G. L. Baillie,* Jonathan R. Pilcher,† and
Kenneth R. Hirons‡
Departments of Archaeology,* Botany,† and Geography‡
and Palaeoecology Centre
The Queen's University of Belfast
Belfast, Northern Ireland

Roy Thompson
Department of Geophysics
University of Edinburgh
Edinburgh, Scotland

INTRODUCTION

The placing of Quaternary events within a time frame has long exercised the minds of researchers. The traditional stratigraphic approach involves the dating of strata "relative" to each other, but current work frequently permits the use of so-called absolute techniques where a metric in years may be established for a site and its contained deposits (Bowen, 1978). Although numerous nonrelative dating techniques are available to the Quaternary scientist, the three discussed below have had the widest application within Ireland.

THE QUATERNARY HISTORY OF IRELAND
279

RADIOCARBON DATING*

Introduction

Ireland was able to benefit from the radiocarbon dating revolution from a very early stage. The first date list of Willard Libby's Chicago laboratory, published in *Science* (Arnold and Libby, 1951), reported a ^{14}C date of 11,310 ± 720 radiocarbon years bp (laboratory code C-355) for a zone II lake mud sample from Knocknacran, Co. Monaghan, and a date of 5824 ±300 bp (C-358) for a peat deposit from Clonsast, Co. Offaly. Both palaeoecological samples had been submitted by G. F. Mitchell. The second and fifth Chicago date lists also reported results from Irish materials (Libby, 1951, 1954). The fifth list presented data for oak wood from "ancient Irish cooking places" at Killeens, Co. Cork. The dates of 3,506 ± 230 bp (C-877) and 3,713 ± 270 bp (C-878) (Libby, 1954) were the first from archaeological contexts in Ireland, obtained from materials submitted by M. J. O'Kelly and E. de Valera. Dates on Irish samples from the Yale, Groningen, and Cambridge laboratories appeared in the late 1950s (Barendsen *et al.*, 1957; de Vries *et al.*, 1958; Godwin and Willis, 1959).

From early 1958 until early 1960, a radiocarbon system using methyl alcohol scintillation techniques was operated in the Physical Laboratory of Trinity College, University of Dublin. The dating system employed was described in Delaney and McAulay (1959), where results from an initial palaeoecological programme were presented. A modified system based on benzene scintillation techniques was also in use in Dublin in 1974 (Dresser and McAulay, 1974). Only three Dublin date lists appeared (McAulay and Watts, 1961; Dresser and McAulay, 1974; Dresser, 1980), with final determinations on some 87 Irish samples.

Compensation for the demise of the Dublin laboratory was signalled by the first date list from Belfast which appeared in 1970 (Smith *et al.*, 1970). The radiocarbon dating unit of the Palaeoecology Laboratory (now Centre) of The Queen's University of Belfast employs both gas proportional and benzene scintillation techniques [the latter developed especially for high precision counting (Pearson, 1979)]. The Belfast laboratory has specialized in dates for palynological, archaeological, and dendrochronological purposes. In this last sphere it has made major contributions to the European tree-ring chronologies (e.g., Pearson *et al.*, 1977, 1983; Baillie *et al.*, 1983; Pearson and Baillie, 1983). The Belfast unit was also one of the 13 laboratories selected for the international collaborative effort to calibrate the oxalic acid reference standard (Mann, 1983).

*This section was written by Dr. Edwards.

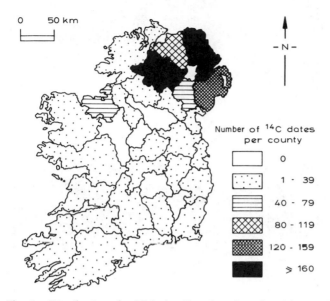

0   50 km

– N –

Number of ¹⁴C dates
per county

☐	0
⬚	1 - 39
▤	40 - 79
▨	80 - 119
▦	120 - 159
■	≥ 160

Fig. 1.   Distribution of published radiocarbon dates for Irish counties.

At least 18 radiocarbon dating laboratories have provided ¹⁴C determinations on Irish materials. In excess of 960 dates had been published in the pages of the journal *Radiocarbon* by early 1984 or were available from other sources (e.g., *Science, Ulster Journal of Archaeology, Journal of the Royal Society of Antiquaries of Ireland;* Woodman, 1978, Caulfield, 1978, 1983).

The data available to the author are summarized in Table I and Figs. 1 and 2. The dates are based on the 5568-year half-life with AD 1950 as the reference year. "Non-date" and non-Quaternary context results on recent biospheric samples measured by the Glasgow laboratory (Baxter and Walton, 1970) are excluded, as, for example, are the non-date results on carbonate materials (Dresser, 1980). Where subsamples from a single sample have been dated in different laboratories (e.g., the Clonsast pine root—D-29, Gro/GrN-271, Y-94), each result has been included as a separate entry. Where one laboratory has reported multiple ¹⁴C determinations on a single sample, yet provided a single laboratory sample number (e.g., the measurements on Fallahogy peat—Q-558), then the result is entered as a single averaged determination. Where multiple entries on different fractions of a single sample have been published, only that in the conventional heavy type used in the journal *Radiocarbon* has been recorded (e.g., UB-434 E but not humic acid date UB-434 C), unless stated otherwise in the text. The dates are placed in categories according to their research purpose (e.g. palaeoecological, archaeological, geological) (Table I). Such classification is a

Table I Summary of Published Radiocarbon Dates for Ireland

Laboratory	Laboratory code	Number of dates					
		Palaeoecological	Archaeological	Timber	Geological	Geochemical	Totals
Belfast University	UB	353	208	83	9	19	672
Dublin University	D	38	41	3	5	—	87
Lund University	Lu	11	34	—	—	—	45
Cambridge University	Q	21	4	—	2	—	27
Smithsonian Institution	SI	—	24	—	—	—	24
Groningen University	Gro/GrN	13	7	—	—	—	20
Teledyne Isotopes	I	5	4	—	5	—	14
Scottish Universities Research & Reactor Centre	SRR	—	—	—	13	—	13
British Museum	BM	—	12	—	—	—	12
Harwell	HAR	—	12	—	—	—	12
Birmingham University	Birm	2	1	—	8	—	11
Institute of Geological Sciences/Stockholm	IGS-C14/St	—	3	—	5	—	8
La Jolla, University of California	LJ	—	—	—	6	—	6
Chicago University	C	3	2	—	—	—	5
Pennsylvania University	P	—	4	—	—	—	4
Yale University	Y	2	1	—	—	—	3
Lamont, Columbia University	L	—	2	—	—	—	2
Totals		448	359	86	53	19	965

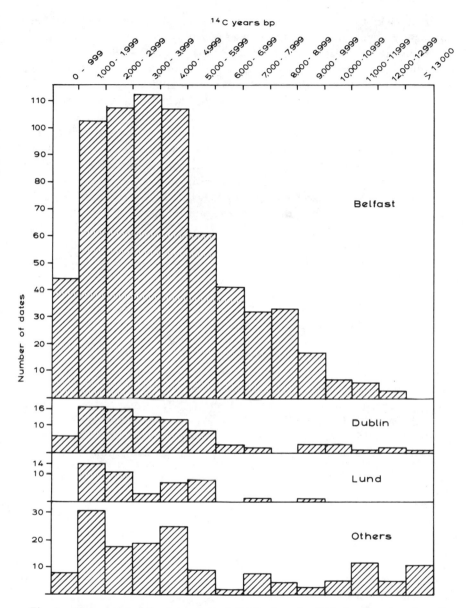

Fig. 2. Histograms of the chronological distribution of published radiocarbon dates for Ireland.

difficult task, for example, where a palaeoecological sample is obtained from an archaeological context. In cases of ambiguity, the purpose stated by the submitter of a date is used.

Figure 1 indicates that the geographical location of ^{14}C-dated sites within Ireland is unbalanced. Almost three-quarters of all dates (697) are from sites in the six counties which constitute Northern Ireland; 54.8% of all Irish dates are from the three counties of Antrim, Tyrone, and Down. This concentration in the north, explicable particularly by the research focus of many researchers based in Belfast, also extends to numbers of sites and not just location within county boundaries. The rest of Ireland is not so well served apart from the large number of dates from Co. Sligo, most of which are associated with the archaeological project in the Carrowmore area (Burenhult, 1980). Eighteen of the southern counties have less than 20 reported dates each, while five counties (Carlow, Cavan, Laois, Longford, and Roscommon) appear to have none.

Figure 2 shows the distribution of ^{14}C dates in histogram form. The dates are grouped at intervals of 1,000 ^{14}C years bp, and a separation has been made of the determinations from the Belfast, Dublin, and Lund Laboratories, which collectively have produced 83.3% of published dates. It is evident that most dates cover the postglacial (Littletonian stage) with relatively few for the lateglacial (late Midlandian) and even fewer for older time periods. The large number of determinations for the period subsequent to 7,000 bp reflects the great interest shown in man-based studies—indeed, the Belfast laboratory was set up to investigate such topics. This orientation is further underscored by the data in Table I, which show that dating for archaeological and palaeoecological purposes has been of major concern. Over 36% of dates are represented by the Belfast palaeoecological contribution, much of which stems from anthropogenic pollen studies.

Table II indicates that woody materials (wood and charcoal) have been most favoured for ^{14}C dating, with a combined figure of 45.0% of total determinations. This is almost certainly an underestimate of the true situation, since many samples measured for dendrochronological purposes in the Belfast laboratory have yet to be published. Peat samples represent 34.2% of the total, with organic muds constituting another 13.3%. Most of the peat and mud samples were measured in connection with palynological studies, and they represent the bulk of the materials dated for palaeoecological purposes. At the opposite end of the spectrum, shells and bones represent the least favoured dating materials (0.4 and 0.9% of the total, respectively), and this reflects not only their relative scarcity in a Quaternary Period context but also the related physical and contextual problems of interpreting dates from such materials.

Table II Summary of Materials Used for
Radiocarbon Dating in Ireland

Material	Number of dates	Percentage of dates
Peat	330	34.2
Charcoal	267	27.7
Wood	167	17.3
Mud	128	13.3
Soil	34	3.5
Other plant	26	2.7
Bones	9	0.9
Shells	4	0.4
Totals	965	100.0

The large number of ^{14}C dates from Irish contexts and the difficulties for one person in becoming familiar with every research problem for which the date was sought make invidious any discussion of the significance of selected dates. Consequently, the comments which follow might justifiably be considered primarily to reflect the writer's interests.

Early Dates

Notwithstanding the debate concerning the Gortian Interglacial and its European affinities (Warren, 1979; Warren, Chap. 3, and Watts, Chap. 8, this volume), there are several "infinite" dates from deposits ascribed to this phase from both the north and the south of Ireland. At Newtown, Co. Waterford, wood pieces separated from peat in silt beneath Ballyvoyle Till (Watts, 1959, 1967; Mitchell, 1970) were dated to >38,000 bp (Birm-89). This date would appear to preclude a Littletonian interpretation for the silt (Orme, 1966). At Ballymakegogue, near Tralee, Co. Kerry, a laminated peat of proposed late Gortian/early Munsterian age (Mitchell, 1970) produced a date of >42,500 bp (Birm-165). In the north, lignite deposits underlying boulder clay were exposed in the bank of a narrow river to the southwest of Lough Neagh and were referred to the Gortian Interglacial on palynological criteria (Harkness and Wilson, 1979). These deposits were dated at >46,600 bp (SRR-716) and >46,450 bp (SRR-717). At Hollymount, Co. Fermanagh, vegetable debris from silt beneath a drumlin was dated to >41,500 (Birm-309). An interstadial origin has been suggested for the deposit (McCabe et al., 1978). Shells from a till in Ballykelly, Co. Londonderry, and derived from marine sediments of Lough Foyle or the North Channel, were dated at >40,000 bp and 28,720 ± $^{910}_{820}$ bp (IGS-C14/St-101, inner and outer fractions, respectively). In a comment associated with the dates, R. A. B. Bazley suggested that the till may be of Munsterian age (Wellin *et al.*, 1974).

Dating assignations for materials of middle and late Midlandian age are also available. At Castlepook Cave, Co. Cork, part of the femur of a mammoth (*Mammuthus primigenius*) was dated to 33,500 ± 1200 bp (D-122). Plant debris sandwiched between till deposits at Derryvree, Co. Fermanagh, produced a date of 30,500 ± $^{1170}_{1030}$ bp (Birm-166) (Colhoun *et al.*, 1972). Marine shells in a till from Glastry claypit on the Ards Peninsula, Co. Down, were dated at 24,050 ± 650 bp (I-3268) (Hill and Prior, 1968). An antler of the giant Irish deer (*Megaloceros giganteus*) from Ballybetagh, Co. Dublin, gave a date of 15,170 ± 160 bp (BM-1794).

The Lateglacial–Postglacial Transition

The lateglacial–postglacial boundary had been formally equated with the 10,000-bp chronozone in northwest Europe (Mitchell *et al.*, 1973; Mangerud *et al.*, 1974; Mitchell, 1982), a view which is not accepted by all (Watson and Wright, 1980; Watts, Chap. 8, this volume). Spatial and probably chronological variations in vegetation type, site lithological differences, and a confusing history of pollen zone conventions (Watts, 1977) have led to an environmental picture which is far from clear. This situation has not been helped by the difficulties over the radiocarbon dating of deposits. There are, for example, striking inconsistencies in dating for the Roddans Port, Co. Down, profile (Godwin and Willis, 1964; Morrison and Stephens, 1965; Dresser, 1970). A humic acid date for the base of pollen zone III–IV [10,430 ± 150 bp (UB-399 C); see comments by Dresser in Smith *et al.* (1971)] may prove the most reliable date for the horizon at this site, but it should be noted that it is correlated with the pollen zones of Morrison and Stephens (1965) on lithostratigraphic criteria while Mitchell (1982) has selected a different date from the site [10,130 ± 170 bp (Q-371)] for the base of his Littletonian stage. Rootlet penetration at some of the bog monolith sites may well have led to contamination, giving falsely young dates [e.g., at Slieve Gallion (UB-298 D) and Altnahinch (UB-411)]. On the other hand, falsely old dates may well have arisen due to the "hardwater" effect (Olsson, 1979; Sutherland, 1980), especially in areas with calcareous substrates, such as at Killymaddy Lough and Weir's Lough (UB-2471, UB-2483) in Co. Tyrone (Hirons, 1984). A range of radiocarbon dates from pollen sites displaying lateglacial–postglacial transitional phenomena are displayed in Fig. 3. The diagram includes a date estimated from the dating sequence of the Co. Tyrone site of Meenadoan (Pilcher and Larmour, 1982). Following a discussion of the radiocarbon chronology of the lateglacial in Ireland, which included consideration of six of the nine sites of Fig. 3, Watts (1977, p. 291)

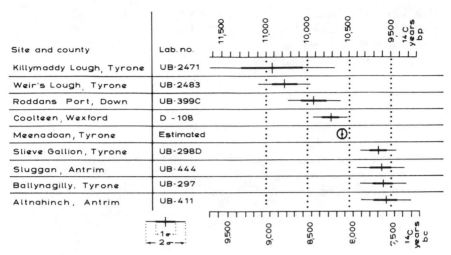

Fig. 3. Selected radiocarbon dates from pollen sites at the lateglacial–postglacial transition.

concluded that a uniform biostratigraphically based chronology was "not to be expected." The data from the additional sites included in Fig. 3 do not force a change in this view.

Early Agriculture

The earliest evidence for man in Ireland comes from Mount Sandel in Co. Londonderry (Woodman, 1978, Chap. 11, this volume). A series of ^{14}C dates for this Mesolithic hunter–gatherer settlement range from 8,960 ± 70 bp (UB-925) to 8,400 ± 65 bp (UB-2008). However, evidence for cereal crop agriculture in Ireland, as elsewhere in northwest Europe, is usually associated with the elm (*Ulmus*) decline and Neolithic agriculturalists (Edwards, Chap. 9, this volume). Four Irish pollen sites for which ^{14}C dates are available have produced cereal pollen grains before their elm decline levels are reached. This raises the possibility of either pioneer Neolithic agriculture or arable activity by indigenous Mesolithic communities which had obtained knowledge of farming (Edwards and Hirons, 1984). Thus, Cashelkeelty I in Co. Kerry has an early cereal pollen date of 5,845 ± 100 bp (UB-2413) and an elm decline date of 4,915 ± 95 (UB-2367). At Ballynagilly, Co. Tyrone, the estimated first cereal date is *5750 bp (Pilcher and Smith, 1979) (* indicates a date estimated from a ^{14}C depth–time sequence). At Weir's Lough, also in Co. Tyrone, pre–elm decline cereal pollen was detected at *5,620 bp, with the *Ulmus* fall at 5,295 ± 85 bp (UB-2488) (Edwards and Hirons, 1984; Hirons, 1984). Analysis at a Co. Antrim site, Newferry, revealed cereal pollen prior to the elm decline which itself was dated to 5,320 ± 170 bp (D-36) (Smith and Collins, 1971). Figure 4 in Edwards (Chap. 9, this volume) shows the chronological relationships between these sites and selected dated archaeological contexts from Ireland.

The Elm Decline

The elm decline is not only the most dated pollen horizon in the postglacial of Ireland and Britain but also, within the limits of the radiocarbon method, the most synchronous (Smith and Pilcher, 1973). In Fig. 4 the elm decline dates for 20 sites are presented. Where a number of dates from across the horizon are available, only that from the start is used (see legend). The

Site and county	Lab. no.	
Weir's Lough, Tyrone	UB - 2488	S
Beaghmore, Tyrone	UB - 99	D
Lough Catherine, Tyrone	UB - 2266	D
Fallahogy, Londonderry	Q - 555	S
Ballynagilly, Tyrone	Estimated	S
Lough Catherine VI Tyrone	UB - 2389	D
Red Bog, Louth	D - 4	S
Lomcloon, Sligo	D - 12	D
Ballyscullion, Antrim	UB - 115	S
Gortcorbies, Londonderry	UB - 232	S
Lackan, Down	UB - 801	S
Killymaddy Lough, Tyrone	UB - 2475	S
Treanscrabbagh, Sligo	D - 13	D
Sluggan, Antrim	UB - 441	D
Cashelkeelty I, Kerry	UB - 2367	E
Slieve Gallion, Tyrone	UB - 275	S
Altnahinch, Antrim	UB - 423	E
Derryandoran, Tyrone	UB - 2234	D
Meenadoan Tyrone	UB - 2110	D
Garry Bog, Antrim	UB - 2191	D

Fig. 4. Radiocarbon dates for the elm decline. Sample position in relation to the elm decline indicated by: **S**, start; **D**, during; **E**, end.

date of 5,228 ± 120 bp used for Fallahogy is a mean of two determinations (Q-555). The ^{14}C date of 5,830 ± 65 bp (Lu-1961) for the elm decline at Carrowkeel, Co. Sligo, has been excluded because of the probable "hard-water" error applicable to the site (Göransson, 1981). Dates of 4,750 ± 85 (UB-870) and 4,685 ± 85 (UB-833) from the Co. Down sites of Carriv-moragh and Slieve Croob do not appear to date any well-defined classic elm decline (Holland, 1975; Pearson and Pilcher, 1975) and are also excluded. The dates for the elm decline range from 5,295 ± 85 bp (UB-2488) at Weir's Lough to 4,590 ± 85 bp (UB-2191) at Garry Bog, both in the north of Ireland. This narrow band (with a mean date of 5,055 bp for all dates and 5,132 bp for the dates which appear to measure the start of the elm decline) is from materials which may in some cases be too old (hardwater errors or inwashed soil at the lake sites) or too young (rootlet penetration at the peat sites). The determinations are from sites in many parts of Ireland, they point to a high degree of synchroneity, and there would appear to be no drastic "wiggles" in dendrochronological calibration curves for the period (Klein *et al.*, 1982). This may provide support for a catastrophic explanation for events at the elm decline, such as an elm disease (Watts, 1961, Chap. 8, this volume; Groenman-van Waateringe, 1983). Man, a most destructive force, was active at this time, however (Edwards, Chap. 9, and Woodman, Chap. 11, this volume), and the arguments over the elm decline are as controversial as they are long (Ten Hove, 1968).

Blanket Peat Initiation

The factors responsible for blanket peat formation have been the subject of much discussion (Goddard, 1971; Mitchell, 1972, 1976; Dimbleby, 1975; Moore, 1975; Smith, 1975; Lynch, 1981). Whether climatic, pedological, topographic, or anthropogenic factors, individually or collectively, are responsible is still undetermined. Sampling problems associated with basal blanket peat dates, for example, variable thicknesses of dated samples, intrusive "younger" roots, *in situ* "older" tree stumps growing in subpeat soils, and the lack of subpeat surveys, hinder many considerations of the problem (Edwards and Hirons, 1982). In Fig. 5, an attempt has been made to assemble all ^{14}C dates for the basal layers of Irish blanket peat. Where ascertainable, the dates selected do not include determinations from contexts which might be most likely to give rise to anomalous dates: for example, where blanket peat was growing above other peat types as at Slieve Gallion (UB-274), Slieve Croob (UB-828), and Breen Bog (UB-367 F); where the blanket peat covers a ditch deposit, as at Pubble (UB-193); or where the ^{14}C date does not appear to be from the basal sample, as at Behy (UB-153 F, UB-158 F) and Ballypatrick (UB-265, UB-403). Dates on

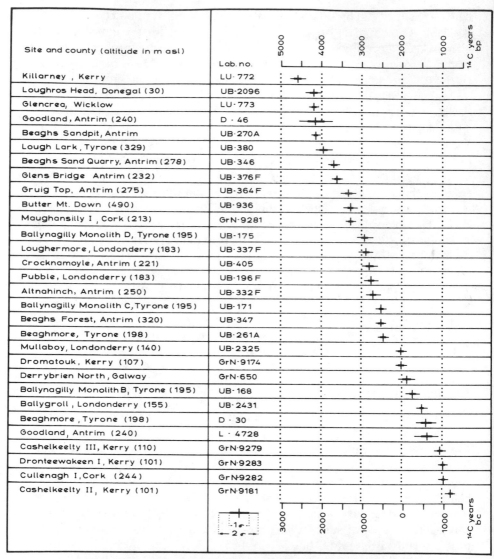

Fig. 5.   Radiocarbon dates for blanket peat initiation.

"wood from pine stump layer in contact zone between peat and underlying mineral soil" (Håkansson, 1974, notes to Lu-772 and Lu-773) were included, perhaps wrongly, because of insufficient evidence to demonstrate whether they were *in situ* tree stumps. Excluded from the dates which appear in Fig. 5 are some which have appeared in other peat initiation lists (Smith, 1975; Lynch, 1981).

The 30 dates for peat initiation of all ages cover many parts of the island.

Most dates come from the Bronze Age, but they span the whole temporal period from the middle Neolithic through to the late Bronze Age. There is no concentration of dates, such as to prove the sudden onset of a climatic deterioration, but past initiation processes may well involve the crossing of complex pedogenic thresholds where delays or accelerations are common (Ball, 1975), but where an initial shift towards wetter climatic conditions is a necessary prerequisite (Mitchell, 1972; Smith, 1975). Twenty of the dates come from the northern counties of Tyrone, Londonderry, Antrim, and Down, yet they embrace many different altitudes and many different times. Dates from southwest Ireland represent both the oldest and youngest determinations. Dates from blanket peat overlying the same apparent soil surface can even suggest a time difference of half a millenium, as at Ballygroll and Mullaboy (Edwards, 1983) or even almost 3 millenia as at Goodland (Case, et al., 1969). Only a most thorough analysis of the many possible variables and their interaction will illuminate the pattern displayed in Fig. 5. The unfortunate fact is that the available dates were obtained by many different research projects, making it impossible to hold constant the many possible factors involved in blanket peat initiation (Edwards and Hirons, 1982).

Conclusions

Although Ireland has long benefited from the provision of radiocarbon dates, some disquiet must be voiced over the geographical coverage of the technique. Of the 965 determinations considered above, over 72% relate to materials from the north. In the main, this reflects the energies of past and present workers in the Palaeoecology Centre of The Queen's University of Belfast. In the Republic of Ireland, admittedly a larger area with a less healthy funding system but with many more workers, there are far fewer dates while five counties appear to have none. Since 1974, Belfast has provided the only ^{14}C dating facility in Ireland. Radiocarbon dating is an expensive procedure. The Queen's University laboratory is funded within the framework of the United Kingdom, it is not a commercial laboratory, and it would be unreasonable to expect its workers to divert their attention to geographically new areas unless resulting research represented a necessary complement to ongoing projects. Bearing such strictures in mind, an improved ^{14}C coverage would seem to lie with four partly overlapping possibilities: (1) collaborative research with other workers who may have access to dating facilities; (2) the search for improved funding for dates to be obtained commercially; (3) the block purchase of dates from (or a subsidy to increase the capacity of) an existing laboratory; or (4) the establishment of a new radiocarbon dating laboratory in the Republic of Ireland. The realization of any of these possibilities rests largely with the initiatives of workers in the south.

Acknowledgments

Helpful comments on the text were received from Drs. Michael Baillie, Ken Hirons, and Jonathan Pilcher of the Palaeoecology Centre of The Queen's University of Belfast. Dr. Gordon Pearson and associates, also of Queen's University, are thanked for the provision of radiocarbon dates in connection with the author's joint research with Robert Larmour.

References

Arnold, J. R., and Libby, W. F. (1951). *Science* **113,** 111–120.
Baillie, M. J. L., Pilcher, J. R., and Pearson, G. W. (1983). *Radiocarbon* **25,** 171–178.
Ball, D. F. (1975). *In* "The Effect of Man on the Landscape: The Highland Zone" (J. G. Evans, S. Limbrey, and H. Cleere, eds.), pp. 20–27. CBA Res. Rep. No. 11, London.
Barendsen, G. W., Deevey, E. S., and Gralenski, J. L. (1957). *Science* **126,** 908–919.
Baxter, M. S., and Walton, A. (1970). *Radiocarbon* **12,** 496–502.
Bowen, D. Q. (1978). "Quaternary Geology." Pergamon, Oxford.
Burenhult, G. (1980). "The Archaeological Excavation at Carrowmore, Co. Sligo, Ireland: Excavation Seasons 1977–1979." Theses and Papers in North-European Archaeology, No. 9. Inst. of Archaeology, Univ. of Stockholm, Stockholm.
Case, H. J., Dimbleby, G. W., Mitchell, G. F., Morrison, M. E. S., and Proudfoot, V. B. (1969). *J. R. Soc. Antiq. Irel.* **99,** 39–53.
Caulfield, S. (1978). *In* "Early Land Allotment" (H. C. Bowen and P. J. Fowler, eds.), pp. 137–143. BAR Brit. Ser. No. 48, Oxford.
Caulfield, S. (1983). *In* "Landscape Archaeology in Ireland" (T. Reeves-Smyth and F. Hammond, eds.), pp. 195–215. BAR Brit. Ser. No. 116, Oxford.
Colhoun, E. A., Dickson, J. H., McCabe, A. M., and Shotton, F. W. (1972). *Proc. R. Ir. Acad., Sect. B* **180,** 273–292.
Delaney, C. F. G., and McAulay, I. R. (1959). *Sci. Proc. R. Dublin Soc. Ser. A* **1,** 1–20.
de Vries, H., Barendsen, G. W., and Waterbolk, H. T. (1958). *Science* **127,** 129–137.
Dimbleby, G. W. (1975). *Nature (London)* **256,** 265–267.
Dresser, P. Q. (1970). "A Study of Sampling and Pretreatments of Materials for Radiocarbon Dating." Unpublished Ph.D. thesis, The Queen's University of Belfast, Belfast.
Dresser, P. Q. (1980). *Radiocarbon* **22,** 1028–1030.
Dresser, P. Q., and McAulay, I. R. (1974). *Radiocarbon* **16,** 6–9.
Edwards, K. J. (1983). *Ulster J. Archaeol.* **44–45** (1981–1982), 43–46.
Edwards, K. J., and Hirons, K. R. (1982). *Quat. Newsl.* **36,** 32–37.
Edwards, K. J., and Hirons, K. R. (1984). *J. Arcaheol. Sci.* **11,** 71–80.
Goddard, A. (1971). "Studies of the Vegetational Changes Associated with Initiation of Blanket Peat Accumulation in North-east Ireland." Unpublished Ph.D. thesis, The Queen's Univ. of Belfast, Belfast.
Godwin, H., and Willis, E. H. (1959). *Radiocarbon* **1,** 63–75.
Godwin, H., and Willis, E. H. (1964). *Radiocarbon* **6,** 116–137.
Göransson, H. (1981). *Stockholm Archaeol. Rep.* **8,** 180–195.
Groenman-van Waateringe, W. (1983). *In* "Landscape Archaeology in Ireland" (T. Reeves-Smyth and F. Hammond, eds.), pp. 217–232. BAR Brit. Ser. No. 116, Oxford.
Harkness, D. D., and Wilson, H. W. (1979). *Radiocarbon* **21,** 203–256.
Hill, A. R., and Prior, D. B. (1968). *Proc. R. Ir. Acad., Sect. B* **66,** 71–84.
Hirons, K. R. (1984). "Palaeoenvironmental Investigations in East Co. Tyrone, Northern Ireland." Unpublished Ph.D. thesis, The Queen's Univ. of Belfast, Belfast.
Holland, S. M. (1975). "Pollen Analytical Study Concerning Settlements and Early Ecology in

Co. Down, Northern Ireland." Unpublished Ph.D. thesis, The Queen's Univ. of Belfast, Belfast.
Klein, J., Lerman, J. C., Damon, P. E., and Ralph, E. K. (1982). *Radiocarbon* **24,** 103–150.
Libby, W. F. (1951). *Science* **114,** 291–296.
Libby, W. F. (1954). *Science* **120,** 733–742.
Lynch, A. (1981). "Man and Environment in South-west Ireland." BAR Brit. Ser. No. 85, Oxford.
Mangerud, J., Andersen, S. T., Berglund, B. E., and Donner, J. J. (1974). *Boreas* **3,** 109–128.
Mann, W. B. (1983). *Radiocarbon* **25,** 519–527.
McAulay, I. R., and Watts, W. A. (1961). *Radiocarbon* **3,** 26–38.
McCabe, A. M., Mitchell, G. F., and Shotton, F. W. (1978). *Proc. R. Ir. Acad., Sect. B* **78,** 77–89.
Mitchell, G. F. (1970). *Proc. R. Ir. Acad., Sect. B* **70,** 141–162.
Mitchell, G. F. (1972). *24th. Int. Geol. Congr. Symp.* **1,** 59–68.
Mitchell, G. F. (1976). "The Irish Landscape." Collins, London.
Mitchell, G. F. (1982). *J. Earth Sci. (Dublin)* **4,** 97–100.
Mitchell, G. F., Penny, L. F., Shotton, F. W., and West, R. G. (1973). *Spec. Publ. Geol. Soc. London* **4,** 1–99.
Moore, P. D. (1975). *Nature (London)* **250,** 439–441.
Morrison, M. E. S., and Stephens, N. (1965). *Philos. Trans. R. Soc. London, Ser. B.* **249,** 221–255.
Olsson, I. U. (1979). *In* "Palaeohydrological Changes in the Temperate Zone in the Last 15,000 Years" (B. E. Berglund, ed.), pp. 1–38. IGCP, Lund.
Orme, A. R. (1966). *Trans. Inst. Br. Geogr.* **39,** 127–140.
Pearson, G. W. (1979). *Radiocarbon* **21,** 1–21.
Pearson, G. W., and Baillie, M. J. L. (1983). *Radiocarbon* **25,** 187–196.
Pearson, G. W., and Pilcher, J. R. (1975). *Radiocarbon* **17,** 226–238.
Pearson, G. W., Pilcher, J. R., Baillie, M. J. L., and Hillam, J. (1977). *Nature (London)* **270,** 25–28.
Pearson, G. W., Pilcher, J. R., and Baillie, M. J. L. (1983). *Radiocarbon* **25,** 179–186.
Pilcher, J. R., and Larmour, R. (1982). *Proc. R. Ir. Acad., Sect. B* **82,** 277–295.
Pilcher, J. R., and Smith, A. G. (1979). *Philos. Trans. R. Soc. London, Ser. B* **286,** 345–369.
Smith, A. G. (1975). *In* "The Effect of Man on the Landscape: the Highland Zone" (J. G. Evans, S. Limbrey, and H. Cleere, eds.), pp. 64–74. CBA Res. Rep. No. 11, London.
Smith, A. G., and Collins, A. E. P. (1971). *Ulster J. Archaeol.* **34,** 3–25.
Smith, A. G., and Pilcher, J. R. (1973). *New Phytol.* **72,** 903–914.
Smith, A. G., Pearson, G. W., and Pilcher, J. R. (1970). *Radiocarbon* **12,** 285–290.
Smith, A. G., Pearson, G. W., and Pilcher, J. R. (1971). *Radiocarbon* **13,** 450–467.
Sutherland, D. G. (1980). *In* "Studies in the Lateglacial of North-west Europe" (J. J. Lowe, J. M. Gray, and J. E. Robinson, eds.), pp. 139–149. Pergamon, Oxford.
Ten Hove, H. A. (1968). *Palaeogeogr. Palaeoclimatol. Palaeoecol.* **5,** 359–369.
Warren, W. P. (1979). *Geol. Surv. Irel. Bull.* **2,** 315–332.
Watson, R. A., and Wright, H. E., Jr. (1980). *Boreas,* **9,** 153–163.
Watts, W. A. (1959). *Proc. R. Ir. Acad., Sect. B* **60,** 79–134.
Watts, W. A. (1961). *Proc. Linn. Soc. London* **172,** 33–38.
Watts, W. A. (1967). *Proc. R. Ir. Acad., Sect. B* **65,** 339–348.
Watts, W. A. (1977). *Philos. Trans. R. Soc. London, Ser. B* **280,** 273–293.
Wellin, E., Engstrand, L., and Vaczy, S. (1974). *Radiocarbon* **16,** 95–104.
Woodman, P. C. (1978). "The Mesolithic in Ireland: Hunter–Gatherers in an Insular Environment." BAR Brit. Ser. No. 58, Oxford.

DENDROCHRONOLOGY*

Introduction

In the United States, the presence of two long-lived tree species, sequoia and bristlecone pine, enabled dendrochronologists to construct very long sequences of ring widths. In round figures, these ring width chronologies spanned 3 and 8 millennia, respectively. Because of dry conditions at the high-altitude bristlecone pine sites, stumps and remnants of dead trees have persisted for thousands of years. This ancient timber enabled dendrochronologists to extend a chronology far beyond the span of the 4,000-year-old living trees. At present, the maximum extent of this chronology is 8,681 years (Ferguson and Graybill, 1983).

In Europe, two sources of ancient timbers have rivalled bristlecone pine. In both Germany and Ireland, large numbers of oak trunks and stumps have been preserved from virtually every period of the last 8,000 years. In Germany the oaks occur in Holocene river gravels (Becker and Schirmer, 1977), while in Ireland preservation has been in the waterlogged conditions of peat bogs (Pilcher et al., 1977). As a result of work carried out during the 1970s, ring-width chronologies for oak have been outlined for almost the whole of the last 8 millennia in both areas.

The research aims behind the construction of the German and Irish chronologies were originally rather different. The river-gravel oaks were regarded as a source of direct dating information which would allow the history of the development of the major German river valleys to be traced. Since most of the river-gravel oaks occurred in the silted meanders of active river systems, the dating of oaks supplied information on river course changes with time. In Ireland, where the equivalent information relating to peat bogs was of less significance, the original aim behind the chronology construction was the calibration of the radiocarbon time scale. Thus the work at Belfast on chronology construction was paralleled by an intensive investigation into high-precision radiocarbon measurement with the aim of producing a definitive calibration (Pearson et al., 1977, 1983; Pearson and Baillie, 1983). Inevitably, as the German chronologies developed, they too formed an increasingly important source of wood samples, of precisely known relative dates, for calibration purposes (De Jong et al., 1979; Becker, 1983).

In considering the construction of very long tree-ring chronologies in Europe, there are several points which require clarification. First, given that the aim was to produce 6,000 to 8,000-year chronologies, oak was the only

*This section was written by Drs. Baillie and Pilcher.

timber which could be used. In most of northern Europe, and in Britain and Ireland in particular, oak has been used consistently as a building timber from earliest times. Since oak also occurs widely as subfossil "bog oak," it is inevitable that any long chronology building effort had to be based on this species. The only other timber preserved in large quantities in bogs is Scots pine (*Pinus sylvestris*), but as this species virtually died out in Ireland around 2,000 ^{14}C years bc (Smith and Pilcher, 1973), a pine chronology continuous to the present day is not possible. Moreover, even if a continuous spectrum of pines had existed, the building of a pine chronology would have involved the added difficulties of working with a species which exhibits missing and double rings.

Returning to oak, the second major consideration in the Irish work was the division of the chronology building into two clear spheres of interest. To construct a chronology of the order of 6,000 years in length, it was necessary to think in terms of building two overlapping chronologies. The first, running from the present day to around 500 calendar years BC, would exploit mainly historical and archaeological sources of timber with the use of relatively little subfossil material. Of necessity, this chronology would be aimed at the resolution of specific archaeological questions for a period where the coincidence of tree-ring and historical dates could be of greatest interest. The second chronology, running far back into the BC era, would make use almost exclusively of the subfossil timbers from a wide range of contexts scattered throughout the 7,500 km^2 area centered on Lough Neagh. As a tool for the direct resolution of archaeological dating problems, this prehistoric chronology is limited by the paucity of archaeological sites yielding oak timbers.

Chronology Building

Modern Oaks

While there are two species of oak indigenous to Britain and Ireland, *Quercus robur* and *Q. petraea,* in practice it is impossible to conduct species determinations of most historical, archaeological, and subfossil timbers in Ireland. Thus, in building a long chronology it is important that the matching of the ring patterns should not be dependent on the species in any limiting sense. In general, therefore, no distinction has been made between the two species, and it has been assumed that cross-dating can be obtained both within and between species. The results of an extensive survey of living trees in Britain and Ireland has so far failed to reveal any significant difference between the two species from a tree-ring viewpoint (Baillie, 1982).

In setting out to construct a tree-ring chronology it is necessary to define a basic area within which cross-dating can be expected. One way to do this is to ascertain if the ring patterns of modern trees can be cross-dated both visually and statistically within the chosen area. The basic premise of the method is that, in general, any two trees of the same species, growing over the same period of years, within a uniform climatic area should exhibit recognizably similar patterns of wide and narrow rings. Had it been impossible to cross-date modern ring patterns it would have been a hopeless task to attempt the construction of a long chronology. Thus, two studies were carried out: one to show that cross-dating could be found within the Lough Neagh basin area (Baillie, 1973), and the second, using modern site chronologies scattered around the whole of Ireland (Pilcher and Baillie, 1980), to ascertain the regional limits of cross-dating (Baillie, 1981, 1982). Both of these studies showed that good cross-dating could be obtained between the ring patterns of Irish oaks and strongly suggested that all of the trees were responding to the same basic climatic signal. If this is so, then the differences between site chronologies must be largely due to site factors rather than to spatial variations in the overall climatic signal to which the trees are responding. This is probably reasonable as, in an area as small as Ireland, temperature and even rainfall are relatively homogeneous.

One significant finding in our study of modern trees is the youthfulness of most of the oaks in Ireland. Almost all of the present-day oaks were planted in the eighteenth and nineteenth centuries. Only one site, Shane's Castle, Co. Antrim, produced trees which started life in the seventeenth century. It would appear that, to a large extent, the forests existing at the time of the Plantation around AD 1600 had been fully utilized by shortly after 1700 (Baillie, 1977a; McCracken, 1971). Given this relatively short age span of Irish living trees, compared to 600 years for oaks from the Spessart forests in Germany, post medieval timbers had to be sought which would overlap with the modern ring patterns.

Chronology Extension

Using timbers from a variety of seventeenth and early eighteenth century buildings, it was possible to overlap and to extend the sequence back to the fourteenth century (Baillie, 1974). Considerable difficulty was experienced in extending the chronology back across the fourteenth century, and this is clearly reflected in the fact that a parallel Dublin chronology consists only of the sections AD 855–1306 and AD 1357–1556 (Baillie, 1977b). Although it was possible in the north of Ireland to extend back to AD 919 using sub-fossil oaks from Toomebridge and the River Blackwater, it is interesting that in at least three areas in Britain, where chronology building has been attempted similar difficulties have been found. At the present time, a depletion/regeneration phase for oaks centred on 1350 is suggested. The coincidence of this date with the onset of the Black Death has been noted elsewhere (Baillie, 1979). One implication of this regeneration in the late fourteenth century is that the forests present in 1600 were not primeval as once thought.

While the chronologies of the last millennium were under construction, additional work had been continuing on the construction of a chronology for the first millennium AD. An initial 490-year chronology derived from timbers from an early Christian crannog at Teeshan, Co. Antrim, had been extended to a length of 795 years by 1975 (Baillie, 1975). Radiocarbon evidence suggested that this chronology covered the first 8 centuries AD. By 1979, a parallel chronology for the southern half of Ireland, derived exclusively from horizontal mill timbers, spanned the third to the ninth centuries AD and matched with the northern chronology (Baillie, 1979, 1981). Although the Dublin chronology mentioned above ran back to AD 855, it was not possible to find any definitive match between either of the dark age chronologies and the Dublin chronology. In fact, the dating of the whole dark age complex was resolved only with the aid of a chronology from Tudor St. London supplied by Jennifer Hillam (Baillie, 1979, 1980). Some justification for using English material in the completion of an Irish chronology is provided by extensions of the studies on modern trees alluded to above (Baillie, 1982, 1983a). The Tudor St. link allowed the end of the extended northern dark age chronology to be specified as AD 894 and meant that a continuous chronology existed from the present back to 13 BC.

While the AD chronologies were taking shape, similar progress was being made with the prehistoric chronologies. Almost all prehistoric oaks in Ireland are naturally occurring subfossil or "bog" oaks. At the start of the Belfast work on these subfossil timbers, it was feared that they might all fall into one or two narrow time ranges, related perhaps to the Boreal/Atlantic or Sub-Boreal/Sub-Atlantic transitions. In the first few years of fieldwork and sampling extensive use was made of radiocarbon dating, and, on the basis of some 51 dates available by 1972, it became clear that the oaks were in fact distributed through time and not clustered [see Fig. 1 in Smith *et al.* (1971)].

The approach to building a long subfossil chronology followed a rather different course from that adopted in the AD time period. Groups of material, usually of unknown age but from the same site, were processed into site chronologies. Thus, although the individual timbers might have only 150–300 rings it was possible to construct units of chronology of 600–1000 years. These site units were then placed approximately in time using routine radiocarbon determinations. Eventually, as more and more site units were constructed, matching between sites was achieved and the resulting chronologies replicated. Thus, by 1980 three major chronology units, namely, GBX (1,550 years in length covering the approximate period 4000–5500 BC), the Long Chronology [2,990 years spanning approximately 1000–4000 BC (Pilcher et al., 1977)] and GB2 (719 years spanning approximately 200–900 BC) had been constructed.

In the 10 years between 1970 and 1980 progress had been rapid; however, from 1980 on, the problems of chronology construction changed significantly. Random sampling had left gaps in the overall chronology, and hence timbers of *specific date* had to be sought. Unfortunately, it is almost impossible to search for subfossil timber of a specific age since all bog oaks look alike and most are unstratified. While more and more timber was processed in the hope of bridging the gaps, it became increasingly obvious that the solution might lie in widening the area from which material was drawn—in the same way that timbers from Tudor St. London had resolved the problem of the first millennium AD.

As a direct result of (a) processing more Irish material and (b) concentrating on the acquisition of suitable groups of English oak, a complete series of overlaps had been established across all of the remaining gaps by 1982. First, GBX had been joined to the existing Long Chronology to form a unit over 4,000 years in length, using Irish material. Second, the Long Chronology had been linked to GB2 by an 800-year chronology from Swan Carr, south of Durham (Baillie et al., 1983). Interestingly, the gap between the Long Chronology and GB2 turned out to be a single year! Third, a tentative link from GB2 via timbers from the archaeological sites at Navan and Dorsey, Co. Armagh, via timbers from the Roman excavations at Carlisle in northern England, and via a combined chronology from Roman London, known to end in AD 252, had dated the end of the entire Irish prehistoric complex as the year 229 BC (Baillie, 1983b).

It is important to realize that while this work was continuing in Ireland, similar progress was being made independently in Germany, with chronologies back to 700 BC by 1980 (Hollstein, 1980), back to 963 BC by 1981 (Schmidt, 1981), and back to 2061 BC by 1982 (Schmidt and Schwabedissen, 1982). So from 1982, increased efforts were made to compare the chronologies between the two regions. The result of collaboration between workers at Belfast, Stuttgart, and Cologne has been confirmation of the Belfast dating for the prehistoric complex and the resolution of a problem in the German complex at 550 BC. Thus a complete European complex, replicated for at least 5,000 years, is now in existence (Pilcher et al., 1984).

Gaps

It is interesting to note that while the chronology is complete for over 7 millennia there are still real gaps in the Irish material between 13 and 116 BC and between 947 and 949 BC. This implies to some extent that, given the random sampling used, the gaps are real and represent depletions in the population being sampled rather than an artifact of our sampling procedure. As discussed above, the depletions in the AD period appear to be anthropogenic in origin. It seems likely that the prehistoric depletion periods [and there are other points of depletion which stop short of being gaps, e.g., the "1546" link where the chronology relies on only one timber (Baillie *et al.*, 1983)] are due to natural phenomena. This might be particularly true of low-lying bogs where episodes of flooding could have eliminated most of the oaks.

At Garry Bog, Co. Antrim, the large number of random samples collected and cross-matched allowed us to plot the number of trees represented on the bog per 20-year period. The trees were not equally represented on the bog at all periods, in fact there are obvious periods of abundance and depletion. It is likely that this is a reflection of the ability of the bog to support oak growth at some periods and not at others. Flooding levels within Garry Bog, which can now be seen as clay layers or lenses within the peat, suggest that periodic enrichment of the peat with minerals from flood water may have led to renewed growth of oaks. Such considerations of oak growth at a single site lead naturally to questions about the reasons for oak growth and preservation in bogs in general. There are now no modern analogues of the bog oak forests in Ireland. In England, there are a few examples of fenland which support oak on an organic substrate, and oak is supposed to form part of the normal succession from open water through fenland to raised bog. Certainly large areas of the East Anglian Fenland and of the fenlands to the south of Lough Neagh supported an oak/alder woodland in the past.

Such considerations lead to further speculations on the size of the population of oak trees that gave rise to the subfossil oaks we see today. It is normal in an archaeological context to think of organic preservation as a rare process, one that preserved only a tiny fraction of the original population. However, there is at least a possibility that the oak stumps and trunks within the matrix of a bog represent all the oak trees that have ever grown on that bog. This is an area of research where cooperation with commerical peat exploitation might enable population studies to be made on the subfossil oaks within a bog. If this could be done we could follow the evolution and decline of a bog woodland and its eventual death and inundation by peat growth.

Applications

We have already looked at the possibility of using subfossil oaks to unravel some aspects of bog forest history. There have been two other major areas of exploitation of the chronologies, apart from the dating of buildings and archaeological sites. As already mentioned, the primary reason for attempting to construct a long chronology was to provide precisely dated samples for radiocarbon calibration. The large size of the bog oak trunks, as well as their relative proximity to roads and to our laboratory, made it possible to harvest large samples, to collect large amounts of timber, and to supply samples of 200 g per 20 years for radiocarbon measurement. These large samples combined with a highly refined measurement procedure have permitted more accurate measurements to be obtained than was possible with the bristlecone pine wood (Pearson, 1980). So far, the work on radio-carbon calibration using European oak has confirmed the general trend of the calibration demonstrated using bristlecone pine wood and has also con-firmed that short-term variations of magnitudes as great as 2% can occur over a short time. These short-term variations or "wiggles" have serious implications for the interpretation of archaeological events dated by radio-carbon methods (Baillie and Pilcher, 1983) and for the interpretation of some radiocarbon-dated vegetation changes observed in pollen-analytical studies. It is possible that the apparent synchroneity of dated vegetation changes could be the result of the ^{14}C variations rather than the realities of vegetation change.

The basic premise of dendrochronology is that climate controls the ring widths of trees. Since cross-dating can be demonstrated, it is logical to assume that the ring patterns contain a record of some aspects of past climate. Using modern trees, the relationship between tree growth and recent instrumental climate measurements can be demonstrated (Pilcher and Gray, 1982). We know that the relationship between tree growth and cli-mate is not a simple one, particularly in temperate regions. Different aspects of climate limit tree growth at different times of the year. Oak growth, for example, is favoured by warm summers but also by cold in the preceding winter. Thus, no simple predictive equation can be written that will enable us to decode the climate signal in a single oak tree ring chronology. Howev-er, by taking a number of sites on different soils in different areas, it is possible to extract a useful record of both temperature and rainfall (Pilcher and Hughes, 1982; Briffa et al., 1983; Hughes et al., 1984). The scope of this work in the prehistoric period is obviously limited by the small number of possible chronologies.

From the present standpoint, it is unlikely that tree rings will provide more than a rather generalized reflection of climate for the postglacial in Europe. One of the severe problems in using bog oaks for any study of climate would be the lack of modern analogues. Having said this, there is information in tree rings other than ring widths (Hughes et al., 1982). Some recent climate reconstructions have been based on wood density rather than ring width

(Hughes *et al.*, 1984). The information on past radiocarbon variations mentioned above may of itself provide information on past climate variations. Furthermore, variations in the stable isotope ratios ($^{12}C/^{13}C$ and $^{16}O/^{18}O$) in the wood may be related to changes in environmental conditions and, as such, may prove useful in climatic reconstruction. There is thus a considerable amount of accurately dated information waiting to be extracted from postglacial tree rings.

References

Baillie, M. G. L. (1973). *Tree-Ring Bull.* **33,** 15–28.

Baillie, M. G. L. (1974). *Ulster Folklife* **20,** 1–23.

Baillie, M. G. L. (1975). *Ulster J. Archaeol.* **38,** 25–32.

Baillie, M. G. L. (1977a). *Tree-Ring Bull.* **37,** 1–12.

Baillie, M. G. L. (1977b). *Tree-Ring Bull.* **37,** 13–20.

Baillie, M. G. L. (1979). *Proc. Symp. Archaeol. Sci. (Jan. 1978),* pp. 19–32. Univ. of Bradford.

Baillie, M. G. L. (1980). *Curr. Archaeol.* **7,** 61–63.

Baillie, M. G. L. (1981). *In* "Irish Antiquity" (D. Ó Corráin, ed.), pp. 3–22. Tower Books, Cork.

Baillie, M. G. L. (1982). "Tree-ring Dating and Archaeology." Croom Helm, London.

Baillie, M. G. L. (1983a). *Proc. 22nd Symp. Archaeometry* Univ. of Bradford, 73–82.

Baillie, M. G. L. (1983b). *In* "Archaeology, Dendrochronology and the Radiocarbon Calibration Curve" (B. S. Ottaway, ed.), Occasional Paper No. 9, pp. 51–24. Univ. of Edinburgh, Edinburgh.

Baillie, M. G. L., and Pilcher, J. R. (1983). *In* "Archaeology, Dendrochronology and the Radiocarbon Calibration Curve" (B. S. Ottaway, ed.), Occasional Paper No. 9, pp. 51–63. Univ. of Edinburgh, Edinburgh.

Baillie, M. G. L., Pilcher, J. R., and Pearson, G. W. (1983). *Radiocarbon* **25,** 171–178.

Becker, B., and Schirmer, W. (1977). *Boreas* **6,** 303–321.

Briffa, K. R., Jones, P. D., Wigley, T. M. L., Pilcher, J. R., and Baillie, M. G. L. (1983). *J. Climatol.* **3,** 233–242.

De Jong, A. F. M., Mook, W. G., and Becker, B. (1979). *Nature (London)* **280,** 48–49.

Ferguson, C. W., and Graybill, D. A. (1983). *Radiocarbon* **25,** 287–288.

Hollstein, E. (1980). "Mittel Europaische Eichenchronologie." Mainz am Rhein.

Hughes, M., Kelly, M., Pilcher, J., and LaMarche, V. (1982). "Climate from Tree-Rings." Cambridge Univ. Press, Cambridge.

Hughes, M. K., Schweingruber, F. H., Cartwright, D., and Kelly, P. M. (1984). *Nature (London)* **308,** 341–344.

McCracken, E. (1971). "The Irish Woods since Tudor Times." David and Charles, Newton Abbot.

Pearson, G. W. (1980). *Radiocarbon* **22,** 337–347.

Pearson, G. W., and Baillie, M. G. L. (1983). *Radiocarbon* **25,** 187–196.

Pearson, G. W., Pilcher, J. R., and Baillie, M. G. L. (1983). *Radiocarbon* **25,** 179–186.

Pearson, G. W., Pilcher, J. R., Baillie, M. G. L., and Hillam, J. (1977). *Nature (London)* **270,** 25–28.

Pilcher, J. R., and Baillie, M. G. L. (1980). *Tree-Ring Bull.* **40**, 23–34.
Pilcher, J. R., Baillie, M. G. L., Schmidt, B., and Becker, B. (1984). *Nature (London)* **312**, 150–152.
Pilcher, J. R., and Gray, B. (1982). *J. Ecol.* **70**, 297–304.
Pilcher, J. R., Hillam, J., Baillie, M. G. L., and Pearson, G. W. (1977). *New Phytol.* **79**, 713–29.
Pilcher, J. R., and Hughes, M. (1982). *In* "Climate Change in Later Prehistory" (A. F. Harding, ed.), pp. 75–84. Edinburgh Univ. Press, Edinburgh.
Schmidt, B. (1981). *Archaologisches Korrespondenzblatt* **11**, 361–363.
Schmidt, B., and Schwabedissen, H. (1982). *Archaologisches Korrespondenzblatt* **12**, 107–108.
Smith, A. G., Baillie, M. G. L., Hillam, J., Pilcher, J. R., and Pearson, G. R. (1971). *Proc. 8th Int. Conf. Radiocarbon Dating* **I**, A92–A95.
Smith, A. G., and Pilcher, J. R. (1973). *New Phytol.* **72**, 903–914.

PALAEOMAGNETISM*

Introduction

The widespread occurrence of ferrimagnetic minerals in Quaternary sediments and their propensity, under certain circumstances, to maintain a stable record of past changes in the earth's magnetic field provide the potential for magnetostratigraphic correlation on a wide range of time scales. This has resulted in the growing application of palaeomagnetic studies in Ireland, especially in the north.

Three parameters are used to define the geomagnetic field at any one point on the earth's surface: two are related to direction (declination and inclination, respectively the horizontal and vertical components) and one related to field strength (the intensity) (Thompson, 1974). In some rocks, sediments, and baked archaeological artifacts, a magnetic record known as the natural remanent magnetization (NRM), may become fixed in such a way that careful measurement can reconstruct the characteristics of the past geomagnetic field. Work on a wide range of datable materials such as lava flows, archaeological artifacts, and lacustrine deposits has shown changes in these geomagnetic parameters such as to provide chronostratigraphic marker horizons. These dated sequences constitute master chronologies and, on comparison with sample measurements, provide a transferred chronology.

Changes in the past direction of the geomagnetic field are classified into three temporal categories (McElhinney, 1973): (1) reversals of polarity, representing 180° changes in local field direction; (2) excursions, or medium-term deviations of field direction of some 45° or more; and (3) superimposed secular variations representing short-term deviations of the local magnetic vector.

*This section was written by Drs. Hirons, Edwards, and Thompson.

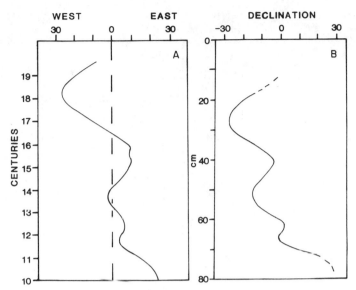

Fig. 6. Comparison of observatory and archaeomagnetic declination data plotted on a time scale (A) with whole-core declination of natural remanent magnetization from a short core from the North Basin of Lough Gall, Co. Armagh (B) (redrawn after Molyneux *et al.*, 1972).

The Pleistocene of Europe contains parts of three polarity chrons and several polarity subchrons. These describe subdivisions of time based on geomagnetic polarity with durations of approximately 10^5-10^6 and 10^4-10^5 years, respectively. The identification of geomagnetic excursions is problematic, particularly in latitudes where recent glaciation has interrupted sedimentation as in Ireland. Secular variations in declination and inclination of the geomagnetic field commonly encompass swings of 20–30° in amplitude on time scales of 2,000–3,000 years. Observatory records of secular change are available for the last 400 years in Europe (Malin and Bullard, 1981), and palaeomagnetic measurements from archaeological materials extend the record of European secular change to 1,000 years (Aitken, 1970, and Fig. 6). Beyond this the most continuous sequence of recorded secular changes is found in the detrital or chemical remanent records of lake sediments.

The stable record of past changes in secular geomagnetic parameters was first confirmed from repeated profiles from Lake Windermere (Mackereth, 1971; Creer *et al.*, 1972). Subsequent extension of this work has shown that similar patterns of both declination and inclination are to be found in the sediments of many European lakes in, for example, Finland (Stober and Thompson, 1977), Switzerland (Thompson and Kelts, 1974), France, and Greece (Creer *et al.*, 1977).

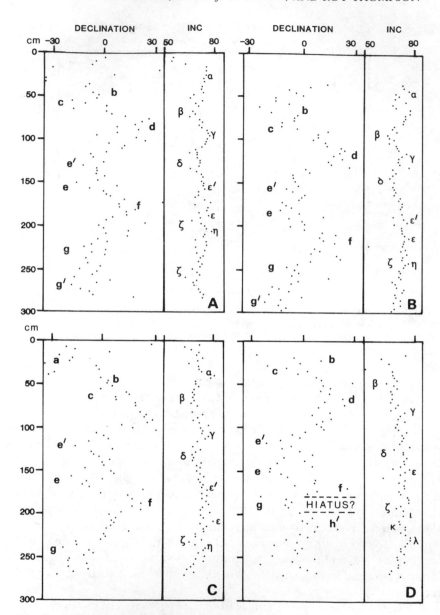

Fig. 7. A–D, Records of secular variation of declination and inclination from four Lough Neagh sediment cores. Turning poings a–h and α–λ are identified from Turner and Thompson (1981). Note postulated hiatus in D (redrawn after Thompson, 1973).

Ireland

Early work at Lough Neagh suggested that Ireland falls within the general European secular variation area. Most palaeomagnetic work carried out in Ireland has been concerned with secular variations as recorded in postglacial lake sediments (Molyneux et al., 1972; Thompson, 1973, 1975; Thompson and Edwards, 1982). The declination and inclination records observed in four 3-m cores taken from Lough Neagh are shown in Fig. 7. The turning points in the magnetic profiles are identified by reference to the recent British master curve of Turner and Thompson (1981) and labelled accordingly. These data show the palaeomagnetic record to be reproducible from different areas within Lough Neagh back to 6,000 bp, providing evidence for core-to-core correlation and also showing how the palaeomagnetic profiles can be used to identify anomalies in the sedimentary record. The postulated hiatus in core LN 11 (Fig. 7, D) was later confirmed by pollen analysis (O'Sullivan et al., 1973).

Comparison of observatory and archaeomagnetic records with short, 1-m Mackereth cores (Mackereth, 1969) from Lough Gall, Co. Armagh (Fig. 6), shows a good correspondence over the most recent time period. These measurements were obtained by the whole-core method whereby the unextruded sediments are subject to magnetic examination. This provides a very rapid, nondestructive dating and stratigraphic technique (Molyneux et al., 1972). The magnetic dating technique is particularly appropriate in sediments such as those at Lough Gall where the calcareous nature of the deposits makes the use of ^{14}C dating impractical. Similarly, in situations where contamination by old carbon due to mobilization of soils or peats has caused so-called radiocarbon reversals, palaeomagnetism can provide an alternative dating technique (Tolonen et al., 1975). For example, at Lough Neagh, palaeomagnetic measurements have been used to elucidate anomalies in the time–depth curve where ^{14}C dating of surficial sediments suggested incorporation of old organic residues from eroded deposits (O'Sullivan et al., 1973).

For longer time scales, ^{14}C techniques have been used to date secular variation in declination and inclination in lake sediments back to 10,000 bp (Mackereth, 1971; Thompson and Turner, 1979). A combination of multiple sediment core records from several British lakes has enabled problems related to the magnetic recording abilities of different sediments and to ^{14}C dating problems of limnic deposits to be minimized and a geomagnetic master curve produced (Turner and Thompson, 1981).

An Irish Palaeomagnetic Master Curve

Figure 8 presents an independent Irish palaeomagnetic master curve suggesting a dating scheme for the sequence of postglacial oscillations in decli-

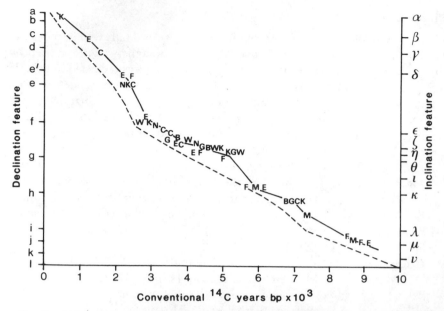

Fig. 8. Irish palaeomagnetic master curve. Broken lines show the position of the British master curve. Sites used in the construction of the curve are as follows: **B**,Ballynagilly; **C**, Lough Catherine (CATH6); **E**, Lough Catherine (CATH1); **F**, Fairy Water; **G**, Beaghmore; **K**, Killymaddy Lough; **M**, Meenadoan Bog; **N**, Lough Neagh; **W**, Weir's Lough.

nation and inclination. Declination features are plotted against the ordinate axis at equal intervals, and ¹⁴C estimates compared along the abscissa. In order to overcome the problems of the radiocarbon dating of lacustrine sediments, evidence from four independently dated magnetic profiles has been integrated with one further lake and four peat profiles, by means of cross-matched pollen sequences. The technique used was to match pollen zone boundaries of closely spaced sites and in this way to transfer their ¹⁴C dating to the master chronology. Thus 16 radiocarbon age determinations are included from Lough Catherine (Thompson and Edwards, 1982) along with dated pollen zone boundaries from nearby Meenadoan (Pearson, 1979; Pilcher and Larmour, 1982) and Fairy Water (K. J. Edwards and R. Larmour, unpublished data) peat sites. Three dates are included from Lough Neagh, but the upper and 2 lower Killymaddy Lough dates were considered too old due to incorporation of ¹⁴C-deficient carbon (Hirons, 1984) and only the remaining 6 were included. Dates from nearby Weir's Lough (Hirons, 1984) and fen peat sites at Ballynagilly (Pilcher and Smith, 1979) and Beaghmore (Pilcher, 1969) are used to supplement the Killymaddy data. Figure 8 shows the resulting 41 radiocarbon age determinations which

Table III   Summary Time Scale
for Declination and Inclination
Features from the Irish Geomagnetic
Master Curve (in years bp)

Declination		Inclination	
b	625	β	1150
c	1125	γ	1600
d	1400	δ	2100
é	2050	ε	3400
e	2350	ζ	4250
f	3000	η	4950
g	5250	θ	5300
h	6375	ι	5650
i	8225	κ	6350
		λ	8150
		μ	9050

could be confidently assigned to the palaeomagnetic oscillations, compared with the British master curve (broken) line which has been added. These data suggest that features of the Irish geomagnetic record are apparently several hundred years older than corresponding features in the British record. A summary time scale for declination and inclination features, which will prove particularly useful for dating palaeomagnetic profiles from the many areas of calcareous substrates in Ireland, is presented in Table III.

Conclusions

The rapidity of magnetic investigation allows many cores to be dated and examined for anomalous deposition rates, enabling the palaeoecologist to choose the most appropriate samples for further study. Other magnetic mineral properties (e.g., susceptibility, saturation isothermal remanent magnetization) which are dependent upon the sediment mineralogy rather than the past geomagnetic field can be of great value in palaeolimnological inference and sediment stratigraphic studies (Thompson et al., 1980). These include the tracing of sedimentary evidence for past catchment events (Thompson et al., 1975; Edwards and Rowntree, 1980) and the construction of total materials influx and, ultimately, of sediment budgets (Bloemendal et al., 1980).

References

Aitken, M. J. (1970). *Philos. Trans. R. Soc. London, Ser. A* **269**, 77–79.
Bloemendal, J., Oldfield, F., and Thompson, R. (1980). *Nature (London)* **280**, 50–53.

Creer, R. M., Thompson, R., Molyneux, L., and Mackereth, F. J. H. (1972). *Earth Planet. Sci. Lett.* **14,** 115–127.

Creer, K. M., Readman, P. W., Thompson, R., Hogg, T. E., Papaminopous, S., Stober, J. C., and Turner, G. (1977). *E.O.S.* **58,** 709.

Edwards, K. J., and Rowntree, K. M. (1980). *In* "Timescales in Geomorphology" (R. A. Cullingford, D. A. Davidson, and J. Lewin, eds.), pp. 207–223. Wiley, Chichester.

Hirons, K. R. (1984). "Palaeoenvironmental Investigations in East County Tyrone, Northern Ireland." Unpublished Ph.D. thesis, The Queen's Univ. of Belfast, Belfast.

Mackereth, F. J. H. (1969). *Limnol. Oceanogr.* **14,** 145–151.

Mackereth, F. J. H. (1971). *Earth Planet. Sci. Lett.* **12,** 332–338.

Malin, S. R. C., and Bullard, Sir E. (1981). *Philos. Trans. R. Soc. London, Ser. A* **299,** 357–425.

McElhinney, M. W. (1973). "Palaeomagnetism and Plate Tectonics." Cambridge Univ. Press, London.

Molyneux, L., Thompson, R., Oldfield, F., and McCallan, M. E. (1972). *Nature (London) Phys. Sci.* **237,** 42–43.

O'Sullivan, P. E., Oldfield, F., and Battarbee, R. W. (1973). *In* "Quaternary Plant Ecology" (H. J. B. Birks and R. G. West, eds.), pp. 267–278. Blackwell, Oxford.

Pearson, G. W. (1979). *Radiocarbon* **21,** 274–290.

Pilcher, J. R. (1969). *Ulster J. Archaeol.* **32,** 73–90.

Pilcher, J. R., and Larmour, R. (1982). *Proc. R. Ir. Acad., Sect. B* **82,** 277–295.

Pilcher, J. R., and Smith, A. G. (1979). *Philos. Trans. R. Soc. London, Ser. B* **286,** 345–369.

Stober, J. C., and Thompson, R. (1977). *Earth Planet. Sci. Lett.* **37,** 139–149.

Thompson, R. (1973). *Nature (London)* **242,** 182–184.

Thompson, R. (1974). *Sci. Prog. (Oxford)* **62,** 349–373.

Thompson, R. (1975). *Geophys. J. R. Astron. Soc.* **43,** 847–859.

Thompson, R., and Edwards, K. J. (1982). *Boreas* **11,** 335–349.

Thompson, R., and Kelts, K. (1974). *Sedimentology* **21,** 577–596.

Thompson, R., and Turner, G. M. (1979). *Geophys. Res. Lett.* **6,** 249–252.

Thompson, R., Battarbee, R. W., O'Sullivan, P. E., and Oldfield, F. (1975). *Limnol. Oceanogr.* **20,** 687–698.

Thompson, R., Bloemendal, J., Dearing, J. H., Oldfield, F., Rummery, T. A., Stober, J. C., and Turner, G. M. (1980). *Science* **207,** 481–486.

Tolonen, K., Siiriainen, A., and Thompson, R. (1975). *Ann. Bot. Fenn.* **12,** 161–164.

Turner, G., and Thompson, R. (1981). *Geophys. J. R. Astron. Soc.* **65,** 703–725.

CHAPTER 13

ECONOMIC ASPECTS OF THE QUATERNARY

William P. Warren*
Michael O'Meara
Eugene P. Daly†

*Quaternary and Geotechnical Section**
Groundwater Section†
Geological Survey of Ireland
Dublin, Ireland

Michael J. Gardiner and Edward B. Culleton

An Foras Talúntais
Dublin, Ireland

SANDS, GRAVELS, AND CLAYS‡

Introduction

Although Quaternary deposits occur almost everywhere in Ireland, often to considerable depths, their chaotic nature placed limits on their exploitation for industrial applications, as a large deposit of uniform material best suits the needs of modern industry. Despite the effectiveness of modern geophysical techniques, a thick cover of glacial sediment adds greatly to the difficulties of evaluation and subsequent exploitation of any valuable deposit that is concealed by this drift. However, a contrary view might be taken, namely, that these concealed deposits have been preserved because they could not be detected by shallow probings in the past.

‡This section was written by Michael O'Meara.

THE QUATERNARY HISTORY OF IRELAND

309

As the nature of the unconsolidated drifts could be readily seen in stream cuttings, they could easily be dug. Thus, since he first arrived in the country about a thousand years after the ice had finally disappeared, man has exploited these drifts to the greatest possible advantage; nevertheless, their uses have steadily declined until now they are almost confined to the provisions of sand and gravel for the, albeit important, construction industry. Before dealing with the very limited range of Quaternary deposits that now have economic potential, the wider role that they have played since man first ventured into this outpost of Europe will be outlined.

Much of the unattributed material in the following pages has been derived from the old unpublished records of the Geological Survey. The remainder is based on recollection of field observation and orally gathered data accumulated during the author's years as an officer of the Geological Survey.

Historical

The first immigrants of which there is definite evidence were a simple Mesolithic food-gathering people who fashioned their primitive durable tools from flint and chert nodules that could be picked up from beach and river gravels derived from Cretaceous and Carboniferous bedrock (see Woodman, Chap. 11, this volume). The later Neolithic farmers gradually penetrated to most parts of the island, still using flint and also chert which was widely available from the expanse of Carboniferous limestone in areas where flint was not present. These people also used tough igneous and metamorphic rocks for the heavier implements required for forest clearance. Their most spectacular relics are the huge megalithic burial chambers and ceremonial monuments, which they constructed from large blocks of stone left stranded by the ice sheets. For their ceremonial monuments more striking exotic stones such as granite and quartz rocks were used, perhaps for aesthetic or mystical reasons. It would seem that the site and even the design of their circles and alignments were dictated by the supply of the desired stones that had been transported by the ice to within a convenient distance. The Neolithic people also introduced the art of pottery. We can only surmize that the potters of those days found by trial a local clay that best suited their purpose.

The wealth of Bronze Age gold ornaments that have been discovered in Ireland indicates that gold must have been won in Ireland extensively. The most likely source of this gold is placer deposits in the river gravels of south Wicklow (Reeves, 1971). As no "mother lode" has yet been discovered in the region, despite extensive trenching, it is widely supposed that the gold was picked up by the ice from the gossans of the ores in the Avoca Mineral

Belt and concentrated by glaciofluvial action. Repeated working has skimmed off the major content of gold in these gravels, which are now no longer economic to work.

Raths, built after the arrival of the Celts, were for the most part sited where there was a sufficient depth of till to allow easy digging of the surrounding ditches, and the excavated material was used to build the inner ramparts. The surprising freshness of these raths (that still dot many parts of the countryside despite our adverse climate) indicates the stability of till which allowed the tunnelling out of underground passages and chambers without initial support. The stone-built versions known as cahers or cashels are confined to driftless areas. In contrast, the early Norman farmers preferred to take over the well-drained fertile soils in the lowlands, and they frequently built their mottes on isolated kames so as to have early warning of attack.

Perhaps the earliest references in literature to a distinctive glacial feature are those in the ancient Irish annals dating from the seventh century AD where the Esker Riadha, or the king's highway, was stated to be the route of one of the five great roads of Erin which were opened by Con of the Hundred Battles to give easy access to all parts of the country (Sollas, 1896). This feature, consisting of eskers and gravel mounds stretching from Dublin to Galway, gave a ready-made road across the impassable bogs in the Midlands. Even to this day, many roads follow winding courses along the crests of eskers. The Esker Riadha, which divided the island in two served as the boundary between the northern kingdom of Conn, Leath Conn, and the southern kingdom of Leath Mogha.

Glaciofluvial sands seem to have been first exploited to provide mortar for the early Christian churches, monasteries, and the round towers, but the secular population continued to live in temporary houses protected by the raths and to carry on a type of farming strongly oriented to stock rearing, particularly cattle. So there would be ample dung and lea to maintain the fertility of the limited land under cultivation. Although the use of chalk as a fertilizer by the Celts in Europe is recorded by Varro in *Rerum Rusticarum*, there is no mention of the use of lime in the Brehon Laws that, *inter alia*, minutely prescribed the code of farming which should be followed. Marling the land, as had been long practised in Britain, seems to have become widespread in about the seventeenth century in Ireland. For this purpose a convenient supply was widely available in the limestone tills (the "limestone gravel" of the early Geological Survey and variously known to the farmers as "corn gravel" in the Midlands and as "marl" in the east). Very heavy dressings were used, and the profusion of "marl pits" that pock mark the Southern Irish End-Moraine of the Midlandian Glaciation eloquently demonstrates the extent to which it was used, even on soils derived from the limestone drifts. In the supposed widely unglaciated Namurian terrain north of Killarney, large blocks of limestone have long been quarried to meet a local need for lime, but the introduction of subsidized ground limestone during the past 35 years has obviated the need for such stratagems.

Sand and Gravel

Distribution

Sands and gravels are widespread in all lowland regions of the country, with the notable exception of the drumlin belts (Fig. 1). Charlesworth (1939) postulated that drumlins and glaciofluvial gravels are mutually exclusive, and there is certainly an acute shortage of building gravels in the main drumlin belt covering major parts of Counties Leitrim, Fermanagh, Cavan, Monaghan, Armagh, and Down (but see McCabe, Chap. 4, this volume).

Outside the drumlin concentrations, glaciofluvial gravels occur freely all over the lowlands in the form of moraines, eskers, and outwash plains up to the limit of Midlandian glaciation. Even outside the limits of the ice, long trains of outwash gravels occur along the main river valleys that drained away the excess meltwater. Accordingly, there are deposits within easy access to meet most modern gravel requirements. In the Cork area, however, the gravels of the lower Lee Valley are being rapidly exhausted, and many alternative sources have been, or are in danger of being, sterilized by suburban sprawl. This is a problem that will have to be faced by many local authorities, particularly those south of the Southern Irish End-Moraine.

Suitability

Gravels that have been waterborne for some distance tend to be cleaner and better sorted, and so eskers, kames, deltas of former glacial lakes, and outwash terraces are the most suitable for exploitation. The upstanding esker ridges, with a thin cover of ablation material and of little agricultural value, are ideally suited to easy exploitation. The eskers are mostly composed of beds of fairly clean, fine sands with some coarser gravels. While in the past individual beds best suited for the particular purpose were dug separately, mechanical extraction is now used in most pits so that the entire pit face is extracted simultaneously, and the various grades required are obtained by mechanical crushing, washing, and screening. In most Irish gravel deposits there is a high proportion of fines, so it is usually necessary to add crushed stone aggregates to meet the particular grading that is now specified for concrete used in construction and in the manufacture of various concrete products.

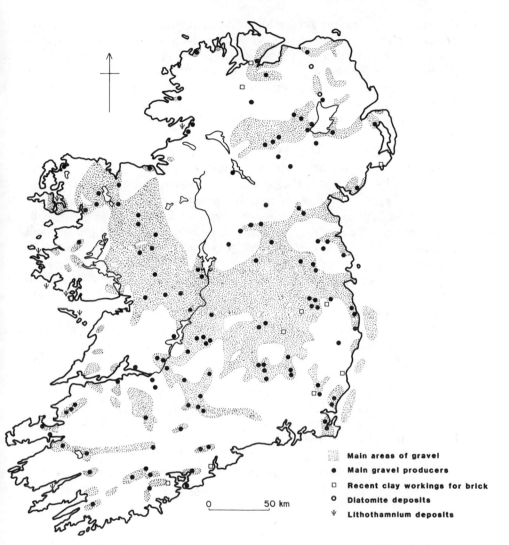

Fig. 1.   Distribution of Quaternary deposits with economic potential in Ireland.

Extraction and Use

Most of the sand and gravel requirements are met by a few large companies, each of which may produce from about 15 separate pits strategically
situated countrywide so as to reduce haulage. Many county councils operate
one or several pits to supply their own needs, while dozens of small oper-
ators may sell run of the pit gravel or turn out a particular product such as

concrete blocks. It is doubtful if all the irregular outputs are included in the official production for sand and gravel. The latest returns available are for 1976, and these together with those for the previous 5 years indicate a steady production for the entire country of about 9 million tons per year with an annual variation less than 1 million tons. The figures given for sand and gravel used to produce ready-mixed concrete are about 5 million tons per year (Central Statistics Offices, Dublin and Belfast, personal communication).

The ownership of sand and gravel is usually vested in the owner of the surface who may part only with what is often the most fertile part of his property as his own financial exigency may dictate, and this has resulted in many piecemeal workings scarring the countryside. The horrific moonscape about Blessington, 30 km southwest of Dublin, arising from the craters of sporadic pits and pinnacles of unworked calcreted gravels, is a glaring example of the damage to the environment that can be caused by uncontrolled gravel exploitation. More recent laws enable local authorities to prevent such despoilation. The healing of scars caused by gravel pits calls for some ingenuity by the developers, as a considerable lowering of the surface is inevitable.

The acquisition of a site for major gravel operation requires careful consideration by the would-be developer, as the deposit must be large enough to give a reasonable life, have a supply of water for washing, and allow for progressive restoration as the workings advance. A price of about three times the agricultural value is normal. First the soil and then the overburden are bulldozed and stored separately for rehabilitation of the surface. A working face is then opened to a depth of up to 20 m, and the gravel is brought down by chains of scraper buckets into front loaders which convey it to a hopper where the oversize material (>40 mm) is screened out for crushing. The annual output of an individual pit may vary from 100,000 to 1 million tons, and a quarter of the output may be sold as dry screened, run of the pit gravel; the rest goes to the treatment plant where it is washed and screened into the following grades:

Coarse gravel	20–40 mm
Medium gravel	10–20 mm
Fine gravel	5–10 mm
Graded sand	5 mm

The graded sand goes to a sand treatment plant, where it is further separated into fine concrete sand and mortar sand, and the tailings go to the settling pond and the sediment preserved to serve as the final layer before the soil is restored.

As gravels are so readily available, no serious assessment of their economic potential has yet been attempted; this applies even more to the extensive deposits of gravel offshore, because it is not considered that these could be extracted as cheaply as onshore deposits. However, as environmental constraints become more stringent and the demand for gravel in Britain increases, the exploitation of the extensive banks of gravel lying at depths of 5–20 m all along the southeast coast is likely in the near future. Large gravel banks also occur off the west coast but at depths of over 100 m, so these have no foreseeable potential.

Limestone is the predominant rock component of most Irish gravels; it also enhances their quality, as limestone absorbs less cement and bitumen than more siliceous rocks. The presence of deleterious substances such as pyrite and mica is very rare. Mica may be present in gravels with a high schist component and occurs freely in the gravels derived from granite, but these are mostly used for horticulture where it has a beneficial effect. A sand derived from the decay of a Carboniferous sandstone at Carrickleck, Co. Meath, is taken to Drogheda for treatment, where a variety of products such as glass sand, moulding sand, and filter sands are turned out. Production is about 30,000 tons per year. Dune sands along the coast are also used for amber glass. Beach sands at Arklow, Co. Wicklow, and at Achill Island have thin layers of heavy minerals, principally magnetite, that has been winnowed but by the currents, but the possibilities of these as a source of metal have not been investigated. At Massbrook on the west shore of Lough Conn, Co. Mayo, a naturally graded granitic sand is extensively dug for direct use in sand filters for water purification. Although they have caused problems in other countries, alkali aggregate reactions in concrete have not been reported in Ireland. The most common of these is the alkali–silica reaction which in certain circumstances results from reaction between the cement and cherts in the aggregate.

Calcareous Sands as Manures

Shell sands have been used as a fertilizer in all the maritime counties of Ireland, except in the east, from as far back as the twelfth century, being dredged at depths of 3–10 m in numerous inlets and estuaries. Analyses of these sands showed a content of 50–80% calcium carbonate in a readily assimilable form, and the manurial value was further enhanced by the animal content of many of the shells. Some reliable estimates of the quantity taken in individual bays show that several million tons of shell sand must have been taken for fertilizers in the early part of the last century. A gross yearly take of 1 million tons from Cork Harbour, Kinsale, and Oyster Bay was estimated by Kane (1844, pp. 285–291) with a further 300,000 tons being taken in Youghal Harbour, while Portlock (1843) estimated that

600,000 tons of shell sand per year were taken from the banks exposed at low tide in Lough Foyle. The soils on the tombolo at the Magharee Islands in Tralee Bay, which are now extensively used for growing onions, derive their fertility from the continual use of shell sand. The taking of these sands is now greatly reduced since subsidized ground limestone is cheaply available and since restrictions have been imposed on the taking of these sands lest it should accelerate coastal erosion.

Lithothamnion ("Coral Sand")

Amazing claims are made as to the beneficial effect that *Lithothamnion* or "coral sand," has on soils with regard to both the quantity and quality of the resulting produce, and these are well supported by practical experience of its use in France and Britain. However, soil scientists remain sceptical, as, apart from its high carbonate content, they cannot identify the presence of other fertilizing ingredients that would support these claims. Without doubt it has a remarkable content of trace elements that could offset all known deficiency diseases of both plants and animals, but this may be only part of its beneficial effect as the main benefit seems to be in improving the texture of the soils. The sand is claimed to rectify the slurrying of soil caused by constant cropping and excessive use of artificial manures. If it could facilitate the assimilation of the considerable amounts of cattle and pig slurries now being spread on the land, it would be of immense benefit in reducing the pollution of our rivers and lakes.

It seems that *Lithothamnion* (a calcified seaweed) flourishes in certain currents of the Gulf Stream and is deposited freely in places where these currents impinge on the coast of Northern Europe: Brittany, Cornwall, and the west coasts of Ireland and Scotland. Until relatively recently it was dredged in Bantry Bay and Roundstone Bay for local use. Surveys (Keary, 1967) have shown that, while *Lithothamnion* in Bantry Bay is rather limited, it is widespread in the deep inlets of the sea in Connemara and on the west coast of Co. Mayo. The largest and most easily exploitable deposits are in Kilkieran Bay, where a factory is contemplated for the processing of this abundant source.

Clay

The most abundant and widespread of all Quaternary deposits is till, but its chaotic boulder content has made it largely unsuited to many practical uses. Nevertheless, its role in the provision of the parent material of nearly all arable soils in Ireland must rank it as the most valuable of all our geological deposits. The rapid expansion of towns and cities that took place starting at the end of the eighteenth century found it pressed into the provision of a convenient, but indifferent, brick clay.

Bricks for the Cities

In Belfast, tills derived from the Triassic marls and sandstones along the Lagan Valley produced an excellent brick even though it was inclined to "sweat" owing to the presence of salt. Subsequently, the marls on the perimeter of the city were used until the city overran the pits. Similarly, about Dublin, the local "Irish Sea Till"—a clayey, though highly calcareous, till with few large stones—was, when weathered to some depth, used to provide an indifferent brick until it, too, was built over. While brick continued to be used for internal building, two pits, at Dolphins Barn and Kill o'the Grange, survived until the early 1930s when concrete blocks displaced brick for internal building (these bricks were not acceptable for outer facings which is almost the sole use of brick in modern buildings). Another brick factory at Courtown, Co. Wexford, using "Irish Sea Till," survived until 1976 mainly manufacturing land drainage pipes for which a porous product was not only acceptable but was preferable. Now for land drainage the lighter and quicker laid plastic pipes are widely used.

Local Brick Industries

The rapid expansion of better class buildings during the last century, when the transport of bulky material was both expensive and time consuming, favoured the setting up of small local brick industries. This brick industry followed easily from the craft of pottery, which had turned out crocks for dairying from the earliest times. Some idea of the extent of these brickmaking operations may be gathered from the statement by Kane (1844) that bricks were made in every parish in Co. Tyrone. In his *Economic Geology of Ireland* (1889), Kinahan listed about 120 locations spread over every county, and his list is by no means exhaustive. Although there is rarely any mention of the material used, it would seem that this was mostly Quaternary clays, principally alluvium. Kinahan mentioned that till was used in Co. Tyrone, where the till was puddled to allow the stones to sink and the suspended clay allowed to settle in ponds. Leached stone-free ablation till, occurring in pockets of the ground moraine of the Midlandian ice, was dug to a depth of about 1 m, and when the surface soil was returned the unleached calcareous till was more productive than before. This practice produced an excellent brick and excellent soil free from ugly scars. Such brick was in keen demand and was produced in scattered locations in Counties Kildare and Laois. Alluvial clays in the callows all along the River Shannon were used extensively although their high content of lime resulted in a poor to indifferent brick. The glacial lakes that were dammed by the ice for considerable periods usually left wide spreads of fairly uniform clay, but, however, their laminated character tends to cause warping on burning. These clays have been used at Coalisland, Co. Tyrone, and Blessington, Co. Wicklow, but at the latter place the clay was so calcareous that only the leached top could be used after being seasoned for several months.

Modern Clay Industry

The principal demand for brick now is for facings in prestige and pseudo-Georgian buildings, for which a high quality brick of uniform colour is demanded. To produce the required article by modern mechanized methods, a thick uniform deposit is essential. Thick beds of slightly sandy shales in the Upper Carboniferous and Permo–Triassic formations are the usual sources of raw material, while the Slane brick factory in Co. Meath uses Ordovician shale. The number of brick industries based on Quaternary deposits has declined steadily since 1900 when there were 14, falling to 11 in 1930 and to 3 in 1950.

The demand for land drainage pipes that arose from postwar land reclamation projects, brought a short-lived revival, but the last Quaternary-based operation using alluvial clay at Clarecastle, Co. Clare, closed in 1976. The cement factory at Limerick uses nearby estuarine muds for the clay component. But, apart from this, the industrial application of Quaternary clays seems to be decidedly limited.

References

Charlesworth, J. K. (1939). *Proc. R. Ir. Acad., Sect.* B **45**, 255–295.
Kane, R. (1844). "Industrial Resources of Ireland." Hodges and Smith, Dublin.
Keary, R. (1967). *Sci. Proc. R. Dublin Soc.* **3**, 75–83.
Kinahan, G. H. (1889). "Economic geology of Ireland." *J. Geol. Soc. Irel.* **8**, 1–514.
Portlock, J. E. (1843). "Report on the Geology of Londonderry and of Parts of Tyrone and Fermanagh." Williken, Hodges and Smith, Dublin.
Reeves, T. J. (1971). *Geol. Surv. Irel. Bull.* **1**, 75–58.
Sollas, W. T. (1896). *Trans. R. Dublin Soc.* **5**, 785–822.

SOIL AND PEAT PRODUCTIVITY*

Introduction

Although the role of agriculture in the economy of Ireland is steadily declining, it is still an integral and pervasive element, a generator of finance

*This section was written by Drs. Gardiner and Culleton.

to stimulate other activities, and a supplier of food and raw materials for home consumption and export. In 1980, it provided 18% (£1.17 billion) of the island's gross domestic product; 60% of this was exported. Soil productivity is a major determinant of agricultural output. This productivity, in turn, is greatly influenced by soil characteristics which are largely determined by Quaternary processes.

Various systems of land classification have been used to categorize the potential of Irish soils for crop production. These are invariably based on soil survey data (Culleton and Gardiner, Chap. 7, this volume). The General Soil Map of Ireland (Gardiner and Radford, 1980) is a product of various sources of information, but it provides a common legend and a framework for land classification.

The early land classification systems were mostly qualitative with about eight use range classes [e.g., Symons (1963) and Gardiner and Ryan (1969)]. Quantitative criteria have been rather limited. The most notable exceptions were the grazing potential study of Lee and Diamond (1972), the determination of the suitability of soils for cereals (Lee and Spillane, 1970) and for sugar beet (Lee and Ryan 1966), and, more recently, the determination of wheat and barley yields on different soil types (Conry, 1980). Tree growth in forests on various soil types has also been determined by researchers such as Bulfin et al. (1973) and Cruickshank (1982).

The Land Resource Base

In land resource appraisal for crop production, the significance of soil resources may be viewed in two different ways depending on the probable cost/benefit ratio of investment (Walsh and Gardiner, 1976). In one, the positive element is foremost and emphasizes the potential for intensification and enhanced returns. In the other, either potential output is low or investment costs are high due to certain soil limitations; such land is at a competitive disadvantage. Although the farming system has a significant influence on net returns, these two approaches largely coincide with farming on arable and nonarable land, respectively.

Nonarable Land

Nonarable land has permanent limitations which currently make it uneconomical for cultivation. Taking into account Ireland's relatively high rainfall, its diverse topography, and Quaternary history, it is not surprising that nonarable land constitutes a relatively high proportion (52%) of the total land resource base by comparison with other countries, namely: United Kingdom, 41%; Italy, 41%; France, 35%; West Germany, 30%; Belgium, 20%; and Netherlands, 1% (J. J. Scully, personal communication).

Table I   Physiographic Divisions and Extent
of Major Kinds of Nonarable Land
in Ireland

Category	Hectares (millions)	Percentage[a] (all land)
Mountain and hill land	1.07	12.8[b]
Peatland		
Blanket peat	0.90	10.5
Basin peat	0.37	5.1
Wet mineral lowland	1.95	23.7
Total	4.29	52.1

[a] Based on the principal soil within each mapping unit.

[b] High-level blanket peat is included under blanket peat.

Table I gives the breakdown of nonarable land into a number of major categories, and their distribution is shown on Fig. 2.

Mountain (>365 m OD) and Hill Land (150–365 m OD).   While this category is extensive (12.8%), especially in the southwest, west, and north, it is not so important in discussing soil productivity in relation to the Quaternary to which its major altitude, slope, and climatic limitations cannot be attributed. Here the farming system is based on extensive grazing with dry cattle and sheep. Livestock densities of different major soils within this category vary from 25 to 75 livestock units (lu) per 100 hectares (ha) (Walsh and Lee, 1977), depending largely on altitude and rainfall. Overall, it has a relatively low agricultural output. Its capacity is less than one-third that of low-level pasture, and it can sustain only 8% of the total livestock population. The shallow stony landscape typified by Connemara is not mountain and hill land *sensu stricto,* but is also included (1.7%) in this category because of close similarity in terms of soil type and agricultural potential. This condition is largely a product of Quaternary glaciation.

Peatland.   Peatland covers 15.6% (1.27 million ha) of the island's land surface (Hammond, 1981). Its evolution during the Quaternary has been traced by Culleton and Gardiner (Chap. 7, this volume). Although many areas occupied by peat have a low population density, the economic significance arises from the fact that large tracts of reclaimed peat may be suitable for agricultural development, while surrounding farms may be improved through development of contiguous bog areas.

Wet Mineral Lowland.   This category (23.7%) occurs most extensively in north–central and western Ireland. The soils consist mainly of imperme-

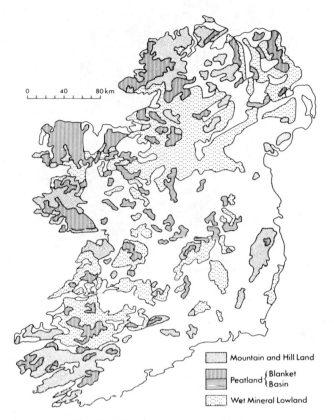

0   40   80 km

Mountain and Hill Land

Peatland { Blanket / Basin

Wet Mineral Lowland

Fig. 2. Distribution of nonarable land.

able surface-water gleys, and the land use options are mainly extensive grazing or forestry. Grazing capacity indices (with low nitrogen levels) for individual soil units within this category range from 137 to 202 lu/100 ha. Using a mean value of 160 lu/100 ha, it is capable of carrying 3.17 million livestock units. Hence its importance in agricultural development. Field drainage can bring about major improvements in its land use and productivity.

Arable Land

Arable soils in Ireland are almost entirely confined to lowland areas and are predominantly derived from glacial deposits. Parent materials consist of drift composed mainly of lower Palaeozoic shale and Lower Carboniferous limestone, which usually give rise to soils characterized by medium to coarse texture, free drainage, and strong structure. Some 4 million ha or 48% of land in Ireland is arable. Based on the kind and degree of permanent

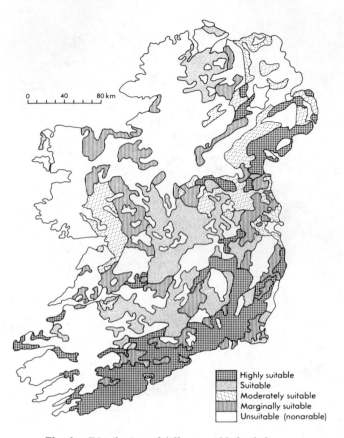

Fig. 3. Distribution of different arable land classes.

soil characteristics and on climatic limitations this land may be divided into four suitability classes (Table II). The occurrence in the south and southeast (Fig. 3) of the highest proportions of the highly suitable and suitable classes is a function of better climatic conditions as well as of suitable soils. This is also reflected in the pattern of tillage crop distribution, for example, wheat and sugar beet crop concentration (Gillmor, 1979), and in the distribution of beet crop yields (Lee and Comerford, 1970).

Effects of Textural Differences for Tillage Crops.   Texturally, soils of classes 1 and 2 are clustered in the sandy loam to loam categories, while those of classes 3 and 4 tend to occur outside this range, being either very coarse or fine textured. Under Irish climatic conditions clay loams, silty clay loams, or clays are usually less suitable for cultivation, although recent trends (Conry, 1980) indicate that winter cereals may yield better on heavier

Table II Extent of Arable Land in Ireland

Class	Tillage suitability	Hectares (millions)	Percentage (all soils)
1	Highly suitable	1.2	14.6
2	Suitable	1.2	15.0
3	Moderately suitable	0.6	7.5
4	Marginally suitable	0.9	10.4
	Total	3.9	47.5

soils. The importance of texture in assessing suitability of soils for cultivation is confirmed by their preferential use for certain crops. Malting barley, for example, has been favoured on light soils (mostly sandy loam and loam) in areas of relatively low rainfall.

Quaternary processes have played a major role in dictating the distribution of arable soils. For example, the Clonroche Series (Gardiner and Ryan, 1964), which is extensive in the southeast and is formed from till of lower Palaeozoic shale composition, is a highly suitable tillage soil. But, where these formations were replaced by till of Irish Sea provenance, impermeable heavy textured soils (Macamore Series, Gardiner and Ryan, 1964) which are unsuitable for cultivation are found. Similarly, soils of the Dunboyne Series in Co. Meath (Finch *et al.*, 1983) consist largely of shales and limestone but, because of heavy texture due to the incorporation of till of Irish Sea provenance, are only marginally suited to cultivation.

Base Status and Parent Material Composition. The impact of Quaternary glaciation on agriculture is reflected in the varying lime requirements of different soils. For best yields most crops, including pasture, require a pH between 6.5 and 7.0. At these pH levels, plant nutrition, both from major and trace elements, is improved, and biological activity and soil structure are enhanced. For these reasons liming has been intensively practised in Ireland, going back to the mid-nineteenth century.

Over 1 million tons of ground limestone are necessary to counteract annual leaching losses, but soils differ considerably in their natural base status and therefore in their individual lime requirements. This depends on the nature of their parent materials. For example, three widely occurring soils which have a calcareous influence in their parent material (Macamore, Baggotstown, and Patrickswell series) are compared in Fig. 4 with three which have little or no calcareous influence (Kells, Borris, and Clonroche series). Even though a lowering of surface pH levels may take place in any of these soils through leaching, those formed from calcareous materials have greater base reserves and significantly lower lime requirements.

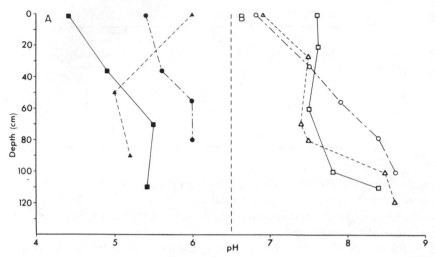

Fig. 4.   The influence of noncalcareous (A) and calcareous (B) parent materials on pH and lime requirements of different soils. ■, Acid Brown Earth on Silurian shale till (Kells); ▲, Acid Brown Earth on granite till (Borris); ●, Acid Brown Earth on Ordovician shale till (Clonroche); □, Gley on glaciomarine sediments (Macamore); △, Grey Brown Podzolic on Carboniferous limestone gravels (Baggotstown); ○, Grey Brown Podzolic on Carboniferous limestone till (Patrickswell).

Production Potential

Soils may be grouped according to the potential uses to which they are most adaptable, based principally on the significance of their more permanent characteristics. For a quantitative assessment, reliable yield data for each soil are required; these can only be provided by detailed field experimentation and yield observations over a number of years.

Pasture and Livestock

Ireland has very favourable climatic conditions for grass growth. Average annual rainfall and frequency are high, and soil temperatures are above the grass–growth threshold (5.5°C) except for a short winter period. Not surprisingly, then, the pattern of land use is dominated by grassland (>90%), and cattle production and dairying account for over 65% (1978) of gross agricultural output (Sheehy *et al.*, 1981). In 1982 the cattle population was 8,124,300.

Obviously, in seeking scope for increased output, dairying and cattle provide the most important sector. But this potential varies with different soils. For example, estimates of grazing capacity per 100 ha range from 50 livestock units (lu) for the mountain and hill land category, to 162 lu (mean)

Table III Grazing Capacity of Representative Soil Series

County	Soil series	Parent material	Great Soil Group	Grazing capacity (lu/100 ha)	
				Using 48 kg N/ha	Using 231 kg N/ha
Wexford[a]	Clonroche	Till, mainly Lower Palaeozoic shale and slate	Brown Earth	227	282
Wexford[b]	Macamore	Till, Irish Sea provenance	Gley	161	198
Carlow[a]	Kellistown	Till, mainly Carboniferous limestone	Grey Brown Podzolic	220	274
Carlow[a]	Kiltealy	Till, mainly granite	Brown Podzolic	205	254
Limerick[a]	Elton	Drift, mainly Carboniferous limestone	Grey Brown Podzolic	220	274
Limerick[a]	Howardstown	Till, mainly Carboniferous limestone	Gley	173	212
Clare[a]	Burren	Mainly Carboniferous limestone bedrock	Rendzina	37–74	—
Clare[a]	Kilrush	Till, Upper Carboniferous shales and sandstone	Gley	148	—
Leitrim[b]	Ballinamore	Till, mainly siliceous Carboniferous limestone	Gley	136	—
Leitrim[b]	Banagher	Reclaimed fen peat	Peat	124	—
Westmeath[c]	Patrickswell	Till, mainly Carboniferous limestone	Grey Brown Podzolic	215	265
Westmeath[c]	Banagher	Reclaimed fen peat	Peat	168	207
Meath[d]	Dunboyne	Till, mixed Irish Sea muds, limestone, and shale	Grey Brown Podzolic	215	269
Meath[d]	Ashbourne	Till, mixed Irish Sea muds, limestone, and shale	Gley	185	227

[a] Lee and Diamond (1972).
[b] Lee and Walsh (1973).
[c] Lee (1977).
[d] Lee (1983).

for the wet mineral lowland category, and up to 227 lu for certain soils in the arable land category (Lee and Diamond, 1972). More detailed statistics show the potential grazing capacity of representative soil series throughout the country (Table III). Considering that the national average stocking rate for lowland soils is approximately 148 lu/100 ha (Lee, 1977), the scope for increased production is obvious. The figures for Co. Meath illustrate the influence of drainage on soil productivity: the moderately well-drained Dunboyne Series has a much greater grazing capacity than the poorly drained Ashbourne Series, although both are derived from similar parent material. Lee and Diamond (1972) showed that between 1957 and 1970 livestock units increased overall by 78,000. But growth rates on arable soils were three to four times greater than on the wet mineral lowland soils. Thus, soil quality seems highly significant in determining the potential and the willingness of farmers to increase livestock numbers. By assigning potential grazing capacity indices to the soils of Ireland, it was concluded that livestock numbers could be almost doubled, given optimum management and proper financial incentives.

Arable Crops Including Horticulture

Although the extent of arable land in Ireland is about 48%, the amount under cultivation has seldom exceeded 10%. A significant decline in recent times is related to commodity price relativities and to a greater degree of specialization. The result is a shift of cultivated crop production away from smaller holdings.

There has been no major systematic attempt to quantify the potential of individual soils for arable crop production. Therefore, the amount of information is limited. Lee and Ryan (1966) examined sugar yields on four contrasting soil series in Co. Wexford, and Lee and Spillane (1970) studied wheat yields on the same soils (Table IV). The influence of soil type and drainage is reflected by the lower yields on the excessively drained and poorly drained soils which are broadly within the same climatic zone. Conry (1980) compared the yield potential for winter and spring barley on four different tillage soils. Perhaps the most important finding here was the significantly increasing yield trend with heavier soils for winter barley.

Considerable scope exists to expand the acreage devoted to arable crops. Even omitting the moderately and marginally suitable categories there are over 2.4 million ha of suitable arable land. This represents approximately four times the present cultivated area (594,000 ha). Climate is obviously a limiting factor for some crops in certain areas, but, overall, significant expansion in crop production is feasible, given favourable economic incentives.

Table IV   Wheat and Sugar Yields on Different Soil Series

Soil series	Drainage	Mean yield (tons/ha) (1964–1965) Wheat	Sugar
Clonroche	Well drained	4.2	7.0
Screen	Excessively drained	3.8	5.6
Macamore	Poorly drained	3.8	—
Rathangan	Imperfectly drained	3.5	5.2
Broadway	Well drained	—	8.2

Forestry As an Alternative Land-Using Enterprise

Where land is marginal for agriculture, forestry may be considered as an alternative. The main criteria for judgment are economic returns and employment potential. Much attention has been devoted to these aspects, especially on the more difficult, wet mineral lowland soils in the drumlin region. Here yields of up to 26 m^3/ha per year (Yield Class 26) have been measured for Sitka spruce (Bulfin et al., 1973), although the country-wide average is Yield Class 14. Bulfin and Hickey (1978) indicated that on these soils both net annual revenue and employment potential are higher for forestry than for agriculture and that consequently there should be a considerable expansion in afforestation. There has been a very active forestry planting programme in Ireland over the past 60 years, with targets in excess of 10,000 ha per year. On the wet mineral lowland in particular, returns to investment in forestry greatly exceeded the average return to management and investment income in farming, even for the year 1978 when the average returns to agriculture per hectare were high (Convery, 1981). However, these were average figures, and the forest yields on some of these soils could be much lower. The deferred nature of returns to forest investment is a serious problem.

Potential of the Peatlands

Peatlands cover 15.6% of the country (Hammond, 1981). They may be divided into two broad categories: (1) blanket peat and (2) basin peat. Development of blanket peat, which is located mainly in mountainous and western regions, is very difficult due to adverse climatic conditions and the impermeable nature of the peat itself. Only when less than 1 m deep does it have any real potential for agricultural development. Basin peat however, is located mainly in the low-lying Midlands where the climate is more favourable and the peat itself is less impermeable. Developing the Midland peats has been the responsibility of Bord na Móna since 1946, and up to 1977 a total of 75 million tons of peat for burning had been produced. In 1976 alone output was 5 million tons, plus over 1 million tons of moss peat, giving a total sales revenue of £27 million (Healy, 1978) and employment to over 5,000 people at 22 locations.

Research has shown that grass is the most suitable crop for shallow basin peat, but for arable cropping greater depths are required. MacNaeidhe (1978) reported that stocking rates can be as high as 3.4 store cattle per ha, with a seasonal live-weight gain of 0.8 kg/day over a 184-day grazing period. Research results also indicate that high yields of arable and horticultural crops can be obtained provided proper management techniques are used after the peat has been reclaimed.

Physical Constraints to Increased Productivity

Drainage Impedance

Collins (1981) summarized the adverse effects of Quaternary glaciation on Irish soils as follows: (a) formation of large, mainly impermeable, drumlin fields; (b) a sluggish arterial drainage pattern; (c) the deposition of Irish Sea muds and their intermixing with tills along the east coast; and (d) the scouring clean of any drift covering over the solid rock in a number of places. The combination of high rainfall, sluggish rivers, and slowly permeable tills has led to water surplus in many Irish soils. Central, therefore, to increasing their agricultural potential has been the improvement of arterial and field drainage. A total of 1.865 million ha have been drained or reclaimed since the late 1940s (Bruton and Convery, 1982; Wilcock, 1979). While arterial drainage is necessary, mainly because of flooding, its success in terms of increased agricultural potential is largely determined by the range of soils in the catchment. Field drainage is required on most soils (1.95 million ha) included in the wet mineral lowland category (Fig. 2). These areas are extensively covered by glacial sediments derived mainly, but not always, from the underlying bedrock. Their structure, compaction properties, clay mineral composition, and textural characteristics have a large bearing on their drainage characteristics.

The fine-textured Upper Carboniferous shales give rise to extensive areas of poorly drained soils in north-central Ireland, in west Co. Limerick, Co. Clare, and on the Castlecomer Plateau. These very impermeable soils are often referred to as surface-water gleys, suggesting that impedance occurs irrespective of landscape position. An indication of their fine-textured nature by comparison with soils from other sources is seen in Fig. 5.

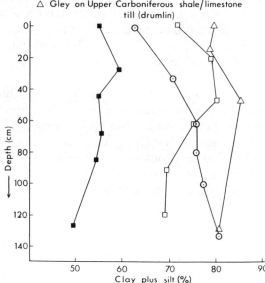

Fig. 5. Poor drainage in relation to texture and parent material.

Poorly drained land also occurs in certain situations on Old Red Sandstone, but the exact nature of the drainage problem is difficult to identify. Mulqueen and Gleeson (1981) suggest that seepage is largely responsible. This arises not only from strongly jointed bedrock; most seepage of agricultural importance comes through layered tills and meltwater deposits. Before successful drainage can be accomplished, the stratification of these Quaternary deposits must be established. A comprehensive survey (Galvin, 1969) has shown that seepage comprises over one-third of drainage problems encountered in Irish soils.

Glacial sedimentary processes have also profoundly affected soil drainage in north-central Ireland. Moulding and compaction of the till in this extensive drumlin field has given rise to large areas of very poorly drained soil. In places, these drumlins themselves impede drainage, leading to the formation of interdrumlin lakes and peat deposits. The drainage problem is compounded by frequent sharp drumlin slopes which accentuate land-use problems and encourage seepage and surface run-off.

MICHAEL J. GARDINER AND EDWARD B. CULLETON

Conclusions

Most soils in Ireland are derived from Quaternary deposits, hence there is a direct link between Quaternary geology and agricultural productivity. Agriculture plays a major role in the economy, and much of its potential as well as its limitations stem from the land resource base. Although there is a high proportion of nonarable land (52%), availability of suitable soils is not a constraint to a significant expansion of both livestock numbers and arable crops.

The extent of soils suitable for cultivation is about four times the present arable crop acreage. Many soil characteristics such as impermeability can be traced to parent material composition and Quaternary origin. This is nowhere better illustrated than on the Upper Carboniferous shale soils where there are major field drainage problems and along parts of the east coast where glacial drift of Irish Sea provenance has virtually obliterated the influence of the underlying rock formations.

References

Bruton, R., and Convery, F. (1982). "Land Drainage Policy in Ireland." Economic and Social Research Institute, Dublin.
Bulfin, M., and Hickey, B. (1978). In "Leitrim Resource Survey," Part IIIA, pp. 1–58. An Foras Talúntais, Dublin.
Bulfin, M., Gallagher, G., and Dillon, J. (1973). In "Leitrim Resource Survey," Part I, pp. 49–56. An Foras Talúntais, Dublin.
Collins, J. F. (1981). Agric. Rec. Special Anniversary Issue, pp. 3–8.
Conry, M. J. (1980). In "Plant Sciences and Crop Husbandry Research Report," pp. 7–8. An Foras Talúntais, Dublin.
Convery, F. J. (1981). Agric. Record Sept. 1981, 36–40.
Cruickshank, J. G. (1982). In "Northern Ireland Environment and Natural Resources" (J. G. Cruickshank and D. N. Wilcock, eds.), pp. 165–184. The Queen's Univ. of Belfast, and The New Univ. of Ulster, Belfast and Coleraine.
Finch, T. F., Gardiner, M. J., Radford, T., and Comey, A. (1983). "Soils of Co. Meath." An Foras Talúntais, Dublin.
Galvin, L. (1969). Ir. J. Agric. Res. 8, 1–18.
Gardiner, M. J., and Radford, T. (1980). "Soil Associations of Ireland and Their Land Use Potential." An Foras Talúntais, Dublin.
Gardiner, M. J., and Ryan, P. (1964). "Soils of Co. Wexford." An Foras Talúntais, Dublin.
Gardiner, M. J., and Ryan, P. (1969). Ir. J. Agric. Res. 8, 95–109.
Gillmor, D. (1979). In "Atlas of Ireland" (J. P. Haughton, ed.), p. 63. Royal Irish Academy, Dublin.
Hammond, R. F. (1981). "The Peatlands of Ireland," 2nd Edn. An Foras Talúntais, Dublin.
Healy, G. (1978). In "Peat Symposium" (M. J. Conry, ed.), pp. 39–49. Irish Society, Agronomy and Land Use, Tullamore.
Lee, J. (1977). In "Soils of Co. Westmeath," pp. 46–50. An Foras Talúntais, Dublin.
Lee, J. (1983). In "Soils of Co. Meath," pp. 71–77. An Foras Talúntais, Dublin.

Lee, J., and Comerford, C. K. (1970). *J. Int. Sugar Beet* **5**, 32–41.
Lee, J., and Diamond, S. (1972). "The Potential of Irish Land for Livestock Production." An Foras Talúntais, Dublin.
Lee, J., and Ryan, P. (1966). *Ir. J. Agric. Res.* **5**, 237–248.
Lee, J., and Spillane, P. (1970). *Ir. J. Agric. Res.* **9**, 239–250.
Lee, J., and Walsh, M. (1973). *In* "County Leitrim Resource Survey," Part I, pp. 44–48. An Foras Talúntais, Dublin.
MacNaeidhe, F. (1978). *In* "Peat Symposium" (M. J. Conry, ed.), pp. 4–7. Irish Society, Agronomy and Land Use, Tullamore.
Mulqueen, J., and Gleeson, T. (1981). *In* "Land Drainage" (M. J. Gardiner, ed.), pp. 11–33. Balkema, Rotterdam.
Sheehy, S. J., O'Brien, J. T., and McClelland, S. D. (1981). *In* "Co-operation in Ireland," pp. 1–70. Cooperation North, Dublin.
Symons, L. J. (ed.) (1963). "Land Use in Northern Ireland," Univ. London Press, London.
Walsh, T., and Gardiner, M. J. (1976). "Land Resource Appraisal for Economic Development." An Foras Talúntais, Dublin.
Walsh, T., and Lee, J. (1977). "Hill Land Potential for Livestock—A Background Review." An Foras Talúntais, Dublin.
Wilcock, D. N. (1979). *J. Environ. Manage.* **8**, 137–149.

HYDROGEOLOGY*

Introduction

Hydrogeology is the study of groundwater, the underground part of the hydrological cycle. Quaternary deposits have an important influence on the hydrological cycle in Ireland as they form the uppermost layer and cover very large areas of the country. These deposits consist of numerous lithologies of variable thickness which are found in many different topographic settings. As a result of their extensive nature they also affect groundwater movement and chemistry in the rock aquifers which they overlie. The effect of the Quaternary deposits on groundwater movement is largely a function of their permeability and to a lesser extent of their thickness and area. Owing to their extensive nature and relatively high storage and in spite of their variable permeability, these Quaternary deposits are an important source of baseflow in the summer, particularly in areas with no rock aquifers.

In many areas the Quaternary deposits have not been developed for water supply, and the level of investigation of those which have been developed has, with some notable exceptions, been quite low. Hence there are not much data available, and what data there are are variable in quality.

*This section was written by Eugene P. Daly.

Deposits of Low Permeability

Quaternary deposits of low permeability such as clay, silt, till, and peat restrict the amount of recharge water reaching the underlying aquifers. Where such deposits are sufficiently thick and/or fine grained, they will cut off infiltration completely thereby confining the underlying aquifers and may, in certain hydraulic circumstances, produce artesian conditions. In this situation the potential of the aquifer is reduced. Bennett (1978) described this as a widespread condition in the north of the country and in the Triassic sandstone aquifer of the Lagan Valley (Bennett, 1979) in particular. A positive effect of the low permeability is that the movement of contaminants into an aquifer may be prevented or the concentration sharply reduced by physical, chemical, and biological processes active in these strata. For example, landfills on these materials have been classified as containment sites (Daly and Wright, 1982) because leachate is prevented from migrating underground.

The physical properties of peat are unlike those of any other geological material. At saturation it consists of about 5–15% dry matter. The D'Arcy equation is not considered to be applicable to flow, particularly in well-decomposed peat. Fissures can develop even in saturated peat, and the permeability varies with the degree of decomposition. Peat has a very high absorption capacity, which is an advantage when considering cut-over bogs as waste disposal sites. An important aspect of peat bogs is their response to drainage, namely, that an unsaturated zone is created and the resultant increase in storage capacity reduces and delays winter floods and contributes groundwater to baseflow in summer and autumn.

Deposits of High Permeability

The Quaternary deposits of high permeability are the various types of sand and gravel. Where these deposits are sufficiently thick, extensive, saturated, and clean they are considered to be aquifers in their own right. Most of this section is concerned with the sand and gravel aquifers which are mainly of glaciofluvial origin. Where these deposits are not sufficiently extensive or perhaps even unsaturated, they are still important as they will allow a high proportion of recharge water to enter an underlying rock aquifer with which they are in hydraulic continuity. Furthermore, where such Quaternary material is permanently saturated, it will be a valuable source of additional storage for the rock aquifer and make more water available for development (Wright, 1979). This is particularly important in Ireland where most of the rock aquifers have relatively low storage coefficients, less than 2%, compared to 5–15% in the sands and gravels. An example of this is provided by the Colligan River gravels which overlie the limestone aquifer at Dungarvan, Co. Waterford. Small lenses of sand and gravel frequently act as a conduit for the natural discharge of large aquifers, as for example, with large limestone springs or valley gravels connecting an underlying aquifer with a river (Misstear *et al.*, 1980).

Distribution and Provenance of Sand and Gravel Deposits

These deposits achieve their maximum thickness in association with various halt stages of deglaciation and may be accompanied by an end moraine. Sand and gravel aquifers of glaciofluvial origin generally take the form of various types of outwash (outwash plain, valley train, or delta) in front of the moraine, kames and eskers behind the moraine, or are part of the morainic complex itself. Over 50 such deposits have been developed in some form in Ireland as a source of water supply.

In Table V an attempt is made to relate individual aquifers to the principal glacial stages, to indicate the dominant type of deposit present, and to summarize the information available on some of the more important of these aquifers. Obviously the sands and gravels are shown related to the most recent glacial event although in most cases it is probable that the deposits are the result of more than one event. For example, at the Curragh and Kilmanagh (Table V) two sand and gravel units are separated by a till, whereas at Carrigtwohill a thin layer of peat (0.6 m) separates units of till and sand and gravel. An aquifer may also consist of more than one type of deposit, thus, at Kilmanagh there is a small esker situated in the middle of a long valley train deposit.

Most of the coarse postglacial alluvial deposits are not sufficiently extensive to be considered as aquifers in their own right. These deposits generally consist of material reworked locally from glaciofluvial deposits and may be considered with them. A number of such coastal aquifers as aeolian sand dunes, raised gravel beaches, and glaciomarine sands and gravels have been developed around the coast. Glaciofluvial deposits, which constitute some of the most productive aquifers in the country, are well distributed throughout Ireland.

Grain Size

The bulk of the material in these glaciofluvial aquifers is coarse and very poorly sorted. There are frequently thin beds or lenses of finer grained and/or better sorted material. From Fig. 6 it can be seen that the grading envelopes of the bulk of the material in a number of aquifers from different environments, locations, and modes of deposition are reasonably similar. Furthermore, the proportion of fines (material < 0.06 mm in diameter) is less than 3% (after sampling by shell and auger).

Table V A Summary of the Hydraulic Characteristics of a Number of Sand and Gravel Aquifers in Ireland and Their Relationships to the Principal Glacial Events

Stage	Substage	Name/location of related aquifer	Principal type of deposit present	Approximate thickness of Quaternary succession (m)	Well yields[a] (m³/day)	Specific capacity[a] (m³/day/m)	Transmissivity[a] (m²/day)	Permeability[a] (m/day)	Specific yield
MIDLANDIAN	Late to postglacial	Fanad, Co. Donegal, Belmullet, Co. Mayo	Aeolian sands	10–20	600–1,100	120	320	50	
		Carndonagh, Co. Donegal and Clogher Head, Co. Louth	Raised beach	12	400–1,100	40–200			
	Dunany	Dromiskin and Dundalk, Co. Louth	Glaciomarine gravels	18–28	100–2,000	50–800	20–1,600	5–80	
		Meath Hill and Cross Guns, Co. Meath	Moraine	15	1,600				
		Cooley Peninsula, Co. Louth, and E. Co. Galway	Kames and esker (Galway)	10–30	1,100–1,600	500–1,400	1,200	100	0.04
		N. of Ireland (Price and Foster, 1974)	Valley train	15–35	600–3,800	100–1,200	60–4,700	6–470	0.05–0.15
	Galtrim	Ballyfin, Co. Laois, and Galtrim, Co. Meath	Moraine	15–20	200–250	30–80	100–250		
		Rosemount and Ballymore, Co. W'Meath	Eskers	15	250	300–550	500	50	
		Curragh, Co. Kildare	Outwash plain	50–75	300–1,100	100–1,220	200–1,530	200	
	Colbinstown	S. Co. Laois and N. Co. Kilkenny	Moraine	20–25	300–1,500	200–300	380–1,200	100	0.09–0.11
		Tallaght, Co. Dublin	Valley train	40	2,000–2,800	103–126	250–300	25–60	0.09
		Kilcoole, Co. Wicklow	Moraine	30	450	42			
	Cork/Kerry Ice	{ Killorglin, Co. Kerry	Outwash delta	50					
		{ Carrigtwohill and Briny, Co. Cork	Valley train	10–45	650–1,632	36–286	250–700	12–50	
	Blessington/ Ballyglanders	SE Co. Limerick, SW Co. Tipperary	Moraines	20–40	up to 650	up to 200			
		Kilmanagh, Co. Kilkenny	Valley train	10–25	100–500	100–1,900	400–5,800	60–630	0.07–0.09
		Blessington, Co. Wicklow, and Aherlow, Co. Tipperary	Outwash delta	100					
	Irish Sea Ice	{ SE Wexford	Moraine	10–30	1,100–1,350	154–190	115–416	6–21	
		{ Screen Hills, Co. Wexford	Kames	15–50	925	74	50	16	
	Hacketstown	Slaney, Suir, and S. Nore rivers	Valley train	12–21	up to 2400	218–727	270–1,700	30–337	
Munsterian		Colligan River, Co. Waterford	Valley train	up to 15					

[a] These values are affected by a variable standard of well completion.

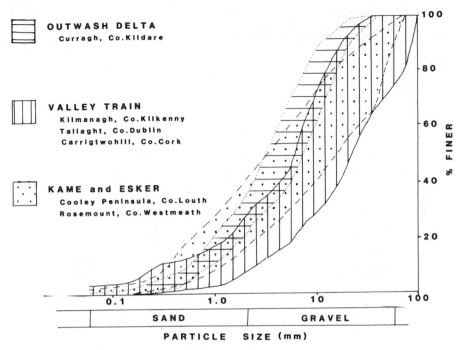

Fig. 6. Grading envelopes of three different types of sand and gravel aquifer.

The valley train deposits are coarser and less well sorted than the outwash plain material (Table VI), probably due to the ability of the more highly charged meltwater in narrow valleys to carry coarser material. The analyses of the kame and esker deposits show that they are finer grained but slightly less well sorted than the valley train material. In general, the outwash deposits are more laterally consistent and exhibit near classic sedimentological stratigraphy, whereas the morainic deposits are quite inconsistent. The eskers and kame deposits are intermediate between the two extremes.

The eskers and valley train aquifers are usually less than 30 m thick and 40 km² in area. They are long and narrow, normally 50–200 m wide in the case of the eskers and 200–1,000 m for the more productive parts of the valley train deposits. The rest of the aquifers are more uniform in shape; they are frequently up to 60 m thick and usually range in area from 40–200 km².

Hydrogeology

The rapid changes in lithology, thickness, and extent of these aquifers, which is mainly due to their mode of deposition, results in varying hydrogeological conditions over relatively short distances. These aquifers are unconsolidated, with intergranular permeability.

Table VI   Results of Grain Size Analyses Carried Out on Samples
from a Number of Deposits Throughout Ireland

Location	Type of deposit	Effective size (mm)	D_{50} size (mm)	Uniformity coefficient
Enler Valley, Co. Down (Price and Foster, 1974)	Valley train		7–40	4–30
Carrigtwohill, Co. Cork	Valley train	0.3–0.8	6–11	18.7–26
Tallaght, Co. Dublin	Valley train	0.7–1.4	9.3–23	18.6–19.3
Kilmanagh, Co. Kilkenny	Valley train	0.65–2.5	7.1–22	12.4–17
Curragh, Co. Kildare	Outwash delta	0.44–0.88	3–7	9.3–10.7
Rosemount, Co. Westmeath	Esker	0.35–0.65	3.2–8.8	16.3–23
Cooley Penninsula, Co. Louth	Kames	0.43–1.1	6–14	23.6–25.6
Dromiskin, Co. Louth	Glaciomarine gravels	1.25	9	11.2
Fanad, Co. Donegal, and Belmullet, Co. Mayo	Sand dunes	0.2–0.25	0.31–0.35	1.6
Edenderry, Co. Offaly[a]	Till	0.26–0.35	7.3–10	42.8–46
Kilmanagh, Co. Kilkenny[a]	Till	0.09	2.6	53.3

[a] Values for the two tills are included for comparison.

The aquifer coefficients of permeability and storage are a function mainly of grain size and shape, sorting, and clay content and are quite variable even within one aquifer. The related coefficient of transmissivity is a function of permeability and saturated thickness. The valley train aquifers are the most permeable and the morainic deposits the least. The unconfined storage in these aquifers ranges from 5–15% with the average being around 9%. These storage values are low for this type of aquifer and probably reflect the poor sorting and very large grain size of a substantial proportion of these deposits (Fig. 6).

From the data presented in Table V it is apparent that the valley train aquifers are the most productive of the Quaternary deposits and have been successfully developed at a number of locations. When comparing the values in Table V it should be borne in mind that well completion in these deposits is difficult and is normally not adequately carried out. Hence, large well losses result and the yields reported are frequently less, and more variable, than they might be.

The hydraulic conditions in these aquifers are normally a function of the structure of the particular deposit. In some valley train deposits the low-lying part of the aquifer in the river floodplain may be confined, whereas the river terraces along the valley sides are unconfined, as, for example, in the Enler Valley, at Comber, Co. Down (Price and Foster, 1974). Other valley train aquifers may be in hydraulic continuity, with a river or with an underlying aquifer for all or part of their length.

Individual esker ridges are generally unconfined, but the areas between a number of closely spaced eskers in the same system normally are confined. In the case of kames and morainic gravels, the confined and unconfined areas reflect the structure and are scattered irregularly throughout the deposit.

Owing to the relatively high permeability of these deposits the water table gradient is usually quite low. At the Curragh it is 1:500 (Daly, 1981). In the narrower valley train aquifers it is usually a little higher. The annual water table fluctuation in these aquifers is normally less than 1.0 m. As a result of the limited confined areas, short flow paths, and shallow water table/piezometric gradients, artesian flows, where they occur, will normally be small.

Another consequence of the high permeability of sands and gravels is that a large proportion of annual rainfall becomes direct recharge. The highest proportion of direct recharge occurs in the unconfined areas, hence it follows that the recharge pattern for the different types of deposit will be similar to that described above for the hydraulic conditions. Some of these aquifers and particularly the valley train deposits receive indirect recharge from the passage of influent rivers over the deposit or from the interception of shallow groundwater flow in aquitards with which they are in contact.

Hydrochemistry

The groundwater types found in the sand and gravel aquifers are quite varied and primarily depend on the material from which they are derived. Analyses of representative samples taken from Quaternary aquifers throughout Ireland are shown in Table VII. These aquifers of limited area and high permeability result in short flow paths and rapid throughput with, consequently, relatively low dissolved solids. The bulk of the groundwater in these aquifers is normal calcium/magnesium bicarbonate water (analyses 1, 2, 8, 9, and 11) which reflects the widespread distribution of limestone and its extensive glacial and glaciofluvial dispersal. These limestone waters are hard (total hardness > 250 mg/l) with neutral to slightly alkaline pH (7.0–7.5). The groundwaters of sand and gravel aquifers derived from non-calcareous rocks such as Ordovician (analysis 3), Devonian (4), and metamorphic strata (5) are soft (total hardness < 150 mg/l) and less mineralized, with a neutral to acid pH (6.0–7.0).

The chloride concentrations (25–50 mg/l) in coastal aquifers (analyses 3, 4, 5, and 9) are slightly higher than those inland (15–20 mg/l), reflecting the marine influence and slightly higher evapotranspiration. The high levels of sodium, chloride, magnesium, and the Mg:Ca ratio in the sample from Dromiskin in Co. Louth (analysis 6) is consistent with the marine origin of

Table VII Analyses of Representative Samples from a Number of Quaternary Deposits Throughout Ireland

Borehole number and location / Parameter	KNY 18/93 Oldtown, Co. Kilkenny (1)	No. 2 Rosemount, Co. W'Meath (2)	WEX 26/11 Enniscorthy, Co. Wexford (3)	No. 2 Kilcloyne, Co. Cork (4)	Dug Well Geesala Belmullet, Co. Mayo (5)	Co. Co. B/H Drumiskin, Co. Louth (6)	LS 16/58 Camaross, Co. Laois (7)	LS 2/3 Clonaslee, Co. Laois (8)	No. 5 Ardtully Beg, Co. Louth (9)	Glarryford B/H Ballymena, Co. Antrim (10)	Seven Springs Pollardstown, Co. Kildare (11)
Date sampled	Aug. 1979	May 1982	Feb. 1980	Oct. 1970	Feb. 1982	Feb. 1982	Nov. 1976	Aug. 1979	Dec. 1982	May 1977	March 1980
Laboratory	State Lab	State Lab	State Lab	IIRS	State Lab	State Lab	State Lab	State Lab	IIRS	GSNI	State Lab
pH	7.4	7.1	6.80	6.4	6.05	8.1	6.4	7.25	7.0	7.7	7.16
Total dissolved solids (mg/l)	336	474	157	160	148	518	162	280	~338	~300	389
Total hardness (as $CaCO_3$) (mg/l)	298	354	110	90	25	191	112	272	320	162	372
Total alkalinity (as $CaCO_3$) (mg/l)	278	354	60	66	18	300	103	273	233	118	352
Calcium (mg/l)	103	131	20	24	5	4	34	98	100	31	120
Magnesium (mg/l)	9.7	6.6	14.6	7.3	2.9	44	6.6	6.3	17	21	17
Sodium (mg/l)	6.4	8.0	14.0	—	23.0	107.9	10	14.5	10	—	7.8
Potassium (mg/l)	1.6	1.0	3.9	—	2.6	7.2	1.0	2.1	1.9	—	1.1
Bicarbonate (mg/l)	339	432	73	81	22	366	126	333	284	144	429
Chloride (mg/l)	21	16	25	38	38	90	18	16	30	24	17
Sulphate (mg/l)	neg[a]	neg	26	10	3	neg	nil[b]	6	22.4	—	4
Nitrate (as N) (mg/l)	4.2	0.65	5.5	3	1.1	2.1	2	1.4	5.5	6.95	3.1
Free + saline ammonia (as N) (mg/l)	0.018	0.51	0.033	nil	0.3	0.14	0.042	0.018	0.1	0.07	—
Manganese (mg/l)	nil	0.09	nil	—	0.01	nil	nil	nil	—	<0.02	—
Iron (mg/l)	0.08	0.45	neg	nil	<0.1	<0.15	0.46	0.08	0.1	<0.08	—
Principal source of material from which deposit is derived	Carboniferous limestones	Carboniferous limestones (argillaceous)	Ordovician and Silurian strata	Devonian rocks	Metamorphic strata	Carboniferous limestone and Silurian strata	Devonian sandstone rock aquifer overlain by Silurian and Devonian drift	Carboniferous limestone drift	Carboniferous limestone (Cullen and O'Dwyer, 1983)	Tertiary basalt	Carboniferous limestone

[a] neg, negligible.
[b] nil, zero.

this deposit. The excess alkalinity of 109 mg/l (as $CaCO_3$) and slightly higher mineralization suggest restricted flow conditions and greater residence time.

Most of the rainfall recharging rock aquifers in Ireland moves through Quaternary sediments. The material in these sediments has an important influence on the hydrochemistry of the underlying aquifer. Two analyses from boreholes in the Devonian sandstones illustrate this process. In one borehole (analysis 7) the cover is derived from Silurian and Devonian strata and the water is soft with low pH, and in the other (analysis 8) the cover is derived from limestone and the water is hard with an alkaline pH.

The quality of groundwater in Quaternary aquifers is invariably satisfactory and suitable for a wide range of uses unless subject to local pollution. In shallow aquifers, groundwater temperature will normally approximate the annual mean daily air temperature of 9–11°C and will not vary by more than ±1°C throughout the year.

Exploration Methods

The sequence of hydrogeological investigation required to prove the existence of sand and gravel aquifers is generally similar to those methods used in establishing other geological resources. Initial compilation work is followed by field investigations, analysis, drilling, and overall assessment and prediction.

Useful information on Quaternary deposits is available in the Geological Surveys, for instance, on geological maps, air photographs, well records, and Quaternary maps. Other sources of information are the National Soil Survey, as well as mining companies and university departments working in the areas of interest. The data available can be expected to be variable both in quantity and quality for different locations. Armed with this information, an investigator should be able to decide whether there are any thick Quaternary deposits in the areas of interest, and to identify target zones for further investigation. A review of the literature is then necessary to put any deposits in their wider Quaternary framework and to provide clues as to the types(s) of deposit which may be present.

The next step is carrying out detailed well surveys, water quality sampling, and some mapping of the Quaternary deposits over the target areas. These investigations should confirm the presence of thick surficial sediments and whether or not permeable material is present. Maps of water table/piezometric surface, depth to bedrock, and hydrochemistry can then be constructed if sufficient data are available. Geophysical surveys will then normally be required to provide detailed data on the depth, aerial extent, and stratigraphy of the aquifer, to locate the more permeable areas and to provide sufficient information to enable sites to be selected for exploration boreholes. To date electrical resistivity surveys have been the most suc-

Table VIII   Range of Resistivity Values
for Quaternary Deposits in Ireland[a]

Type of material	Resistivity (m)
Soil	100–300
Peat	20–50
Alluvium	10–100
Clay	60–100
Boulder clay	80–150
Sand	115–140
Dry sands and gravels	400–1200
Saturated sands and gravels	150–400
Bedrock	Normally greater than 500

[a] Sources: Geological Survey, Reynolds (1980), B. S.
Williams (personal communication).

cessful of the geophysical methods used in groundwater exploration in Ireland. These surveys usually consist of electrical soundings which are used for depth investigations and horizontal profiling (the process whereby lateral variations in resistivity are detected). The range of resistivity values for the principal Quaternary deposits is given in Table VIII.

The final stage of most investigations will consist of the drilling of a number of narrow diameter boreholes to confirm the presence of permeable material, to recover samples of the deposit for sieve analysis, and to calibrate the resistivity soundings.

Well Construction

The construction of high-yielding water wells in sand and gravel aquifers is a little more complicated than in consolidated formations. The essential elements of any well in sands and gravels are the support of the sediments during drilling and construction of the well in a manner that will permit sand-free water to be pumped on completion.

Normally, shallow wells are drilled by percussion; the borehole walls are supported by casing. Deeper ones are drilled by rotary using direct or reverse circulation with the support being provided by the circulating fluid. The well is completed by installing a permanent casing and well screen string (Fig. 7) to support the formation but also to let the water through. Accurate sampling and often geophysical logging (natural gamma, Fig. 7) are necessary to determine the screen slot size and to place the screen opposite the correct (normally the coarsest) horizon. All aspects of constructing wells in these sediments are discussed in E. Johnson Inc. (1966).

In Ireland sand and gravel aquifers have frequently been developed by wide diameter wells (up to 2 m) dug by mechanical digger to depths of up

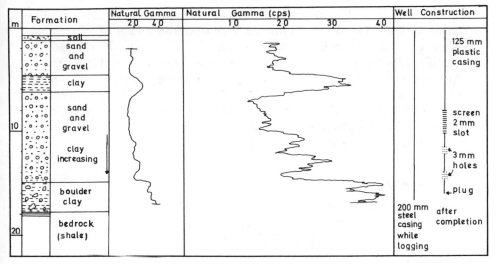

Fig. 7. Geological and geophysical logs and well construction in borehole KNY 18/92 at Oldtown, Co. Kilkenny.

to 7 m. This is a quick and inexpensive method of developing medium-sized supplies (up to 500 m^3/day), but it is only feasible in limited parts of most sand and gravel aquifers.

Groundwater Resources

Geological formations considered to be aquifers underlie some 30% of the land surface of Ireland. From data shown by Wright *et al.* (1979) and Bennett (1979), it is estimated that the Quaternary aquifers account for about 20% of this area. These aquifers contain about 25% of the groundwater resources, making them the second most important aquifer group after the Carboniferous limestones.

The groundwater resources are largely undeveloped, particularly those of the Quaternary aquifers. The principal reasons for this situation, in the case of the sands and gravels, are the lack of detailed Quaternary mapping over much of the country, the absence of the necessary hydrogeological and drilling expertise in the past, and a general lack of awareness of the potential of these aquifers.

Where practical, the aquifers should be developed where they are thickest and coarsest. There are particular problems with esker and valley train aquifers because of their long and narrow shape. In the case of valley train aquifers where it is intended to utilize induced infiltration from nearby rivers, the boreholes will normally be located close to the rivers, particularly where the aquifer and river are in continuity. Where it is considered undesirable to induce infiltration, for instance, where a stream is polluted, or where it is wished to avoid reducing low flows at critical periods, boreholes should be situated some distance from the stream and preferably where it is not in continuity with the aquifer. Owing to their shape the number of suitable drilling locations is quite limited for esker aquifers. Another problem with narrow aquifers is the difficulty in evaluating pumping test data owing to the complex boundary conditions which are likely to exist at the sides of eskers and valley train deposits. The remainder of the deposits, deltas, kames, and moraines, are likely to be most productive close to the aquifer's natural discharge areas.

Conclusions

A high proportion of the lowland area of Ireland is covered with significant thicknesses of Quaternary material. This has an important impact on the underground part of the hydrological cycle, either as aquifers or in the way they affect water moving into underlying rock aquifers.

The Quaternary deposits are generally thickest in those areas close to the major halt stages of the various ice sheets. It is in these regions that most of the important sand and gravel aquifers have been located and are likely to be discovered in the future. Hence a knowledge of glacial history and morphology is crucial in locating the aquifers and in determining the type of sand and gravel deposit. The grain size distribution, hydraulic characteristics, and suggested locations for development vary with the type of deposit. The groundwater from these aquifers is generally of potable quality and suitable for most uses. The chemistry of the water is primarily a function of the material from which the deposits are derived. In the future it is probable that many waste disposal sites will be situated in areas of thick Quaternary deposits of low permeability.

Quaternary aquifers contain substantial groundwater resources and are of considerable importance on a national scale. The deposits can be developed quickly and cheaply provided sufficient expertise is applied to the task.

Acknowledgments

The author acknowledges the assistance of Geoff Wright and Donal Daly of the Geological Survey of Ireland, Peter Bennett of the Geological Survey of Northern Ireland, and Kevin Cullen, consultant.

References

Bennett, J. R. P. (1978). *Open File Rep.* **59,** Geological Survey of Northern Ireland, Belfast.
Bennett, J. R. P. (1979). *In* "Hydrogeology in Ireland." Irish National Committee of the International Hydrological Programme, Dublin.

Cullen, K. T., and O'Dwyer, K. (1983). "Report on a Water Well Drilling Project at Ardtully Beg, Co. Louth." Unpublished Consultant's Report, Dublin.

Daly, D. (1981). "Pollardstown Fen. A hydrogeological assessment of the effects of drainage on the water supply to the Grand Canal." Unpublished Geological Survey internal report, Dublin.

Daly, D., and Wright, G. R. (1982). *Geol. Surv. Irel. Circ.* 82/1.

E. Johnson, Inc. (1966). "Groundwater and Wells." Johnson Division, U.O.P. Inc., St. Paul, Minnesota.

Misstear, B. D. R., Daly, E. P., Daly, D., and Lloyd, J. W. (1980). *Geol. Surv. Irel. Rep. Ser.* RS 80/3.

Price, M., and Foster, S. S. D. (1974). *Proc. Inst. Civ. Eng.* **57,** 451–466.

Reynolds, G. A. (1980). *Trans. Inst. Min. Metall. Sect. B* **89,** 44–49.

Wright, G. R. (1979). *In* "Hydrogeology in Ireland." Irish National Committee of the International Hydrological Programme, Dublin.

Wright, G. R., Aldwell, C. R. A., Daly, D., and Daly, E. P. (1979). *In* "European Community Atlas of Groundwater Resources." Commission of the European Communities, Brussels.

MINERAL PROSPECTING*

Current Approach

Although Ireland has a long history of small-scale mining notably in copper, lead, iron, zinc, and gold (Jackson, 1968; Cole, 1922), our received wisdom, until recently, has been that Ireland had no mineral wealth of any consequence. This view probably had its source in the fact that the first detailed survey of the geology of Ireland on a large scale, carried out by the Geological Survey in the nineteenth century, identified no new major resource nor area of high mineral potential. Thus the discovery of a large lead/zinc ore body at Tynagh in Co. Galway in 1961 was surprising and introduced to prospecting companies the possibility of further such deposits. The discovery of the Tynagh ore body had resulted from a number of factors, the most important being the identification by Geological Survey personnel of a well-defined fan of glacially dispersed mineralized boulders and a known history of small-scale mining activity in the area. It is fair to conclude that the find at Tynagh that was to herald a prospecting and mining boom in the 1960s and 1970s, was initiated at least in part by a careful consideration of the influence of glaciation in dispersing mineralized float from its bedrock source.

Later, Donovan (1965) traced a distinct geochemical anomaly that extended in till at least 3 km eastward from the Tynagh ore body. In the realization that other ore bodies might lie covered by Quaternary sediments, prospecting companies began prospecting in this medium using

*This section was written by Dr. Warren.

geochemical techniques which rapidly became standard but which at the time of the Tynagh discovery were in their infancy in Ireland. This involved both stream sediment and soil sampling. Stream sediment sampling was generally used in the reconnaissance phase and soil sampling (at depths of 20–30 cm) in the follow-up phase where targets had been identified. Soil sampling was done on a grid basis, narrowing the grid as the target area became smaller. Rarely were the Quaternary sediments analyzed to determine the direction of carriage or dispersal of mineralization.

In the decade following the discovery of Tynagh, further important ore bodies were discovered at Silvermines (lead/zinc), Gortdrum (copper/silver), and Navan (lead/zinc). A number of subeconomic mineral deposits (notably at Keel, Aherlow, Ballinalack, and Mallow) was also identified. This awakened the public to the fact that Ireland had, after all, considerable mineral resources, and the prospecting effort seeking further deposits increased accordingly (Gardiner et al., 1982). Thus stimulated, Irish mineral exploration entered a phase of intense activity which has continued for a period of about 15 years and has been concentrated largely on the area underlain by the Carboniferous limestone of the Midlands where all the important discoveries were made. The Midlands also contain some of the thickest, most extensive, and most complex suites of Quaternary sediments.

All the important discoveries to date have been in the "easy" areas of relatively thin Quaternary cover where geological targeting was to some extent possible and mineralized bedrock outcropped or subcropped close to the surface. In most cases little or no account was taken of the nature of the Quaternary sediments that overlay the mineralized bedrock, but conditions were such that prospecting was possible despite this handicap. However, now that the easy-to-find deposits have been found and attention through the 1970s and early 1980s has been focused largely in the very difficult area of thickest glaciofluvial, glaciolacustrine, and glacial sediments in an attempt to prospect the Carboniferous limestone of the Midlands (Fig. 8), the nature and distribution of the Quaternary sediments cannot be ignored nor their analysis treated as a luxury. The same applies to areas underlain by Lower Palaeozoic or Devonian rocks where complex till sequences are common and may not necessarily relate to surface indications of direction of ice transport, as, for example, in the drumlins of the northeast.

Although the outline morphology of the Quaternary sediments of the country, including the drumlin belts, the midland area of glaciofluvial/glaciolacustrine features, and the area of little or no constructional Quaternary features in the south, is well known, there is little known of the thickness of sediments or the vertical stratigraphy of most of the inland counties. Thus, prospecting geologists often find themselves sampling for geochemistry or

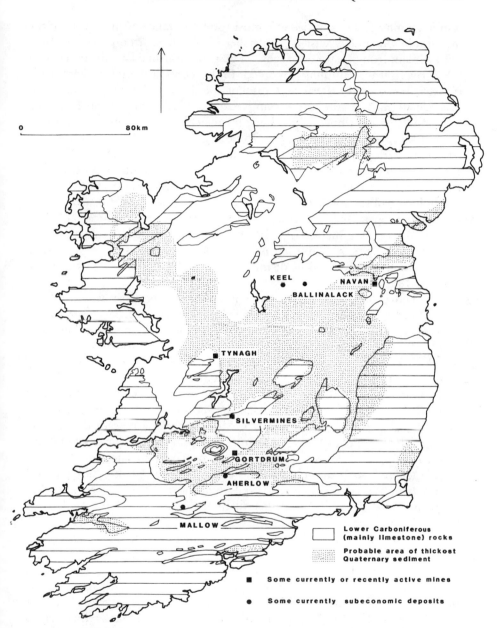

Fig. 8. Some recently, and currently, active mines and some currently subeconomic mineralization in relation to the distribution of Lower Carboniferous, mainly limestone, rocks and the probable area of thickest Quaternary sediments. All of these deposits were discovered in the 1960s and early 1970s.

testing geophysical properties in almost total ignorance of the nature of the medium in which they wish to apply their skills. In an attempt to overcome this problem from the point of view of geochemical analysis, the practice of "deep overburden sampling" evolved. In this context the term overburden is restricted to the Quaternary sediments, and the technique amounted to deep Quaternary sediment sampling. During the decade of the 1970s the technique was developed using both hand-held and tractor-mounted percussion samplers, and it has come to be regarded as "the main change in Irish exploration geochemical procedures since 1971" (Cazalet, 1982, p. 156). Yet there were, in 1980, fewer geochemical samples processed in Ireland from deep sampling than from soil sampling sources. This is considerably less than the total from all surface sampling sources (Cazalet, 1982). Although programmes are conducted from the reconnaissance stage onward using deep sediment sampling, it appears that in perhaps a majority of cases the technique is used in checking shallow anomalies or confirmatory tests of other positive indices of mineralization (Steiger and Poustie, 1979; Cazalet, 1982).

The theory behind deep sampling is to sample as close as possible to bedrock on the principle that identified anomalies will be closer to the bedrock source. As the effective penetration of both the hand-held and tractor-mounted light instruments used is in the region of 5–6 m (Steiger and Poustie, 1979; Cazalet, 1982), the aim of getting close to bedrock may be totally frustrated, particularly where large boulders are encountered. The main shortcoming of the approach, however, is that such samples are taken blind. Thus, the relationship between the sediment unit sampled and that occurring at the surface is unknown. Similarly, variations in the stratigraphy of the sediment through which the sampling tool has passed will not be apparent. And, even where close interval profiling is done, the size of the sample taken is such that identification of the nature of the sediment and differentiation between sediment types will be difficult. There are other shortcomings relating to the ability of the normal sampler used (Holman through flow type) to sample effectively in all sedimentary media.

Even when surface sampling is the chief mode of geochemical data collection in a prospecting programme, and this is still the most common method of sampling (Cazalet, 1982), sediment fabrics are rarely if ever analyzed to determine direction of sediment transport. In a surface or near-surface sampling exercise the possibility of fabric analysis is there, whereas at a depth of 5–6 m where the sample is retrieved in a percussion sampler there is no possibility of checking the fabric.

The Prospecting Medium

In prospecting in areas covered by Quaternary sediments in Ireland, the following broad categories of sediments are encountered: till, glaciofluvial deposits, glaciolacustrine deposits, fluvial deposits, lacustrine deposits, and peat, all of which may be found in either mountain or lowland situations. Of these, till is generally the most common and the most useful medium to the prospector. Till is particuarly useful as it generally is a first-order derivative of bedrock and contains a strong fabric indicative of transport direction.

Glaciofluvial and glaciolacustrine deposits are usually second-order derivatives of bedrock, and although they may present some difficulties in prospecting they can be useful as a tool. Fluvial and lacustrine postglacial sediments may be third- or even fourth-order derivatives of bedrock. They are rarely first order and therefore usually give no more than a regional overview when sampled for geochemistry. Peat is a very difficult medium, for, although subpeat geochemistry may with difficulty be interpreted at times from the lower peat layers, this is usually only a reflection of an already dispersed geochemical pattern in glacial or associated deposits.

Till

In conducting a geochemical survey in till, it is important to identify as closely as possible the depositional mechanism, whether it is a basal lodgement or melt-out till or a supraglacial melt-out or flow till, where this is possible, since this factor will have implications as to the distance of transport and the vertical distribution of the geochemical characteristics of the till. Also, it is clear that a geochemical pattern derived from a sampling network that is restricted to a single sedimentary unit is statistically more reliable than one that encompases a number of different sediment types, the boundaries of which are unknown. For reasons of sampling consistency a sampling grid should not include both weathered and unweathered tills, for it is probable that leaching, particularly of sulphide minerals, will have occurred in the weathered till (Shilts, 1971).

The displacement distance of bedrock debris in the form of lodgement till is unlikely to be as great as for melt-out till, particularly supraglacial melt-out till (Shilts, 1971). A carry-over of 3–4 km is the normal experience in Ireland, although in cases this can be as much as 20 km, probably where the debris has been carried in the englacial or supraglacial mode. In addition, there is likely to be a greater consistency in the geochemical characteristics of a lodgment till owing to the "plastering-on" mode of deposition, while basal melt-out tills will reflect the banding of debris as it was carried in the ice sheet or glacier, producing tills of patchy lithological composition at the surface and considerable vertical variation owing to its sheetlike composition, each sheet reflecting a dirt band in the former glacier. Upsheared bands of melt-out till deposited close to the ice front may cause very confusing anomalies if the nature of deposition is not understood. A hypothetical example of an extreme case is illustrated in Fig. 9, where drilling a surface or

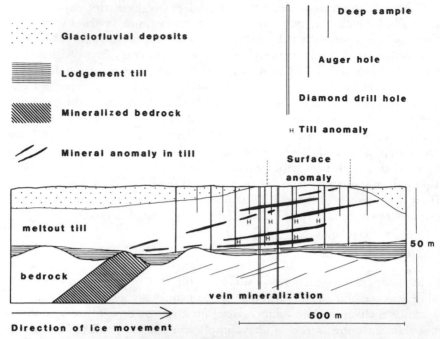

Fig. 9. Schematic diagram representing a hypothetical situation in which a surface geochemical anomaly is confirmed using deep sampling, then followed closer to bedrock by augering but misses the true mineral source owing to a lack of understanding of the "up-sheared" melt-out till sequence. In this case the vein mineralization is thought to be the source.

subsurface anomaly may intersect vein mineralization which is thought to be the source of the anomaly while missing the main deposit.

Where using boulder trains or fans of mineralized float to trace the bedrock source, it is essential to take fabrics in the associated till to construct a proper fan or train. Qualitative assessment of the shape of a fan or the size of boulders can be misleading. In such situations it is also important to know whether the boulders were transported subglacially or englacially/supraglacially, for it is likely that subglacially transported boulders will have travelled much less than those transported in either of the other modes. Furthermore, analysis of boulder size is relevant only if it is known the boulders were transported subglacially, that is, if they are associated with lodgement till. Boulders carried englacially or supraglacially may not reduce significantly in size in a down-ice direction (see Shilts, 1971). On the other hand, large mineralized boulders are more likely to be identified than small ones, and a number of separate fans set close together may merge and seem like a single fan. This highlights the need for fabric analyses of tills associated with mineralized float. In practice, each mineralized boulder occurrence should be a site for fabric analysis.

As Shilts (1971) has pointed out, labile minerals, particularly sulphides, are likely to have been leached in large measure from weathered tills, but the scavenging properties of the phyllosilicates and secondary oxides that dominate the clay-size fraction in weathered tills are such that they gather cations from the leached minerals roughly in proportion to their former presence in the unweathered till. Thus, geochemical analyses of weathered tills should be restricted to the clay-size fraction.

Glaciofluvial and Glaciolacustrine Deposits

Parts of the Midlands are dominated by glaciofluvial and coarse glaciolacustrine sediments, which form a more difficult prospecting medium. In addition to these being second-order derivatives of bedrock, direction of transport on a regional scale is more difficult to determine. Having determined the direction of glaciofluvial transport, it is likely to be necessary to determine the direction of glacial transport as the sediments will, most likely, have been transported initially by glacial action.

Weathering is a further complicating factor. Sands and gravels are likely to be more easily weathered than less permeable tills, and, because the clay-size fraction is unlikely to be evenly distributed as in till, scavenging will be selective and depend on the occurrence of clay lenses or drapes. Thus, anomalously high values in such sediments may simply reflect a clay lens which in an unweathered deposit would yield only background values. Nevertheless the development of accurate models of deglaciation should enable mineralized float and true geochemical anomalies based on analyses of unweathered sediments to be traced to the bedrock source.

Postglacial Sediments

Prospecting in postglacial sediments, whether alluvial or lacustrine, is one remove further from the bedrock source than the material from which they are derived (this is rarely bedrock itself). Furthermore, stream sediments represent a very selective sample of a catchment area. Lake sediments have the same general shortcomings, but in addition they will tend to give only a very generalized view of the geochemistry of the catchment area. They may be useful as a reconnaissance tool in larger and less accessible terrains than Ireland, in providing a regional overview (see Davenport, 1982).

Peat in this context, and as pointed out above, is, from the point of view of geochemistry, a masking agent. Research is needed in this area into the effects of minerals in the underlying sediments on growth characteristics of bog vegetation, whether living or fossil.

An Approach to Prospecting

It is clear from the foregoing that any geochemically based mineral exploration venture in Ireland will benefit from a detailed examination of the Quaternary geology of the exploration area. It is clearly desirable that, before commencing a programme of geochemical sampling in Quaternary sediments, a prospecting company should acquire at least a general outline knowledge of the major stratigraphic units. Also, wherever possible, samples should be taken directly from till (lodgment till where this can be distinguished) as a first-order derivative of bedrock. Generally, surface or near-surface sampling will be in the weathered zone of the sediment. In such cases only the clay-size fraction should be analyzed. If unweathered tills are also sampled, in deep sediment sampling or profiling, these samples should always be distinguished from the weathered surface samples and should not be compared with them. Routine sampling in shallow pits allows clear identification of the sediment type. It also allows routine fabric analysis and sampling of the C horizon of the weathered till for geochemistry. In areas of thick sediment cover (5–6 m) reverse circulation drilling in selected localities should outline the basic stratigraphy. This is also important as a control for any geophysical tests that might be carried out.

If a reliable model of the pattern of glaciation is assembled along with the pattern of geochemical dispersion, then the simultaneous pursuit of the geochemical indicators and the directional indicators will bear fruit. In areas where there is a very large amount of glaciofluvial/glaciolacustrine sediment, so that restricting sampling to tills would leave large areas unexamined, the nature of the sediment should be clearly understood, direction of fluvial transport ascertained, and the relationship to adjacent/subjacent till determined. The pattern of glaciation and deglaciation must be identified if sense is to be made of the pattern of geochemical dispersion in such sediments. Weathering in these deposits will cause a much greater problem than weathering in tills and, if at all possible, sampling should be carried out in the unweathered horizon only. Otherwise, the clay fraction must be abstracted for analysis, but the sample should not be taken in small clay lenses or horizons, nor should the results be directly compared with the till geochemistry.

In geophysical prospecting it is not crucial to understand the precise glacial/deglacial pattern, but the stratigraphy, lithology, and thickness of sedimentary units must be understood if interpretation is to be valid. This is particularly true where airborne reconnaissance electromagnetic (EM) or radiometric surveys are carried out, but it is also the case where ground-based follow-up geoelectric surveys, particularly induced polarization (IP), are done (see Reynolds, 1980).

Although aware of the difficulties presented by thick glacial or glaciofluvial sediments in prospecting, many geologists and prospecting managers have yet come to realize that these sediments can be a valuable prospecting tool and that in areas of thick cover an understanding of the nature of the Quaternary deposits is indispensable to a good prospecting programme. Discussions with prospecting geologists have also revealed that even where there is a conviction of the doubtless value of systematic analysis of the Quaternary deposits to identify sedimentary units and direction of transport, pressure to "cover the ground" using conventional methods often prevents the use of this approach.

The fact remains that the easy targets have been found and the difficult terrain which has been the subject of intensive conventional (mainly geochemical) prospecting has yielded no new ore body since the discovery of Navan in 1970. The difficulty of this terrain is due almost exclusively to the thickness, complexity, and ubiquity of Quaternary sediments, yet there is no perceptible move on the part of prospecting companies to seek a detailed knowledge of the Quaternary geology of Ireland or to analyze the sediments in which they prospect in order to interpret geochemical anomalies, mineralized boulder trains, or geophysical reports. On the contrary, the recent trends in mineral exploration, in deep geochemical sampling of the Quaternary sediments, and in geologically oriented drilling (see Cazalet, 1982) are indicative of an attempt to "see through" these deposits.

This situation owes much to a general lack of expertise in, and familiarity with, Quaternary geology on the part of prospecting geologists in Ireland. This in turn is largely due to the fact that in both the Irish and British university systems, Quaternary sediments are usually not regarded as part of geology, and students are not taught Quaternary geology. In this situation the value of a thorough knowledge of the Quaternary geology to a prospecting geologist is not as self-evident as it might be to the Quaternary specialist. The efficiency of the method can really only be demonstrated in a prospecting programme in which there has been a strong Quaternary input from the outset. This is unlikely to be done in Ireland in the near future as it begs the question of conviction on the part of prospecting companies. However, in Scandinavia prospecting techniques involving thorough analysis of Quaternary sediments are becoming more common and can be expected to be the norm in the near future. In Canada, also, there is a growing tendency to accept the necessity of a deeper knowledge of Quaternary sediments in prospecting (Shilts, 1983). Thus, any change in emphasis in this area will probably enter Ireland by example from abroad.

Conclusions

There is an acute need for active research into prospecting methods in glacial sediments in the Irish context, particularly in the type of glaciofluvial complexes that typify the midland areas. It is also very apparent that geologists graduating from Irish universities have little or no knowledge of Quaternary geology and need training in this area. This points to a need for a research and teaching institute in Quaternary studies geared to training Irish geologists, either at the undergraduate or postgraduate level, in Quaternary geology. A research institute or department of Quaternary Studies within one of the universities, offering a basic postgraduate course in Quaternary Studies and postgraduate research facilities, would probably be the most efficient and cost-effective means of promoting a better understanding of what are the fundamentals of the nation's economy—soil parent materials, sand and gravel resources, mineral exploration (both industrial and metallic), and peat.

References

Cazalet, P. C. D. (1982). In "Mineral Exploration in Ireland Progress and Developments 1971–1981 Wexford Conference 1981" (A. G. Brown and J. Pyne, eds.), pp. 148–156. Irish Association for Economic Geology, Dublin.
Cole, G. A. J. (1922). "Memoir and Map of Localities of Minerals of Economic Importance and Metalliferous Mines in Ireland." Geological Survey of Ireland, Dublin.
Davenport, P. H. (1982). In "Prospecting in Areas of Glaciated Terrain—1982" (P. H. Davenport, ed.), pp. 57–81. Canadian Institute of Mining and Metallurgy, Montreal.
Donovan, P. R. (1965). "Geochemistry Dispersion in Relation to Base Metal Deposits in Glacial Terrain in West-Central Eire." Unpublished Ph.D. thesis, Imperial College, London.
Gardiner, P. R. R., Pyne, J. F., and McArdle, P. (1982). Min. Mag. 147, 366–369.
Jackson, J. S. (1968). Archaeol. Austriaca 43, 92–114.
Reynolds, G. A. (1980). Trans. Inst. Min. Metall. 89, 844–849.
Shilts, W. W. (1971). Can. Min. J. 92, 45–50.
Shilts, W. W. (1983). In "10th International Geochemical Exploration Symposium—3rd Symposium on Methods of Geochemical Prospecting," pp. 73–75. Espoo, Helsinki.
Steiger, R., and Poustie, A. (1979). In "Prospecting on Areas of Glaciated Terrain 1979," pp. 22–29. The Institution of Mining and Metallurgy, London.

INDEX

Italicized entries refer to figures or tables.

A

Aasleagh Falls, 36
Abies alba (fir), 57, *159*, 160, 162, 163
Abramis bramis (bream), 235, *236*
Accipiter gentilis (goshawk), *238, 240, 241*
Accipiter nisus (sparrow hawk), *241*
Accipitridae, *238*
Acer pseudoplatanus (sycamore), *190*, 215
Aceramic, 198, 269
Achill Island, 315
Acidification, 138
Aeolian deposits, *97*, 112, 334, *see also*
    Loess
Aeolian phenomena, *97*, 110, 112, 333
Aghamore, *128*
Aghavannagh, *22*, 26
Agher, *188*, 196
Agriculture, 318
    pre-elm decline, 196, 197, 199
    prehistoric, 10, 143, 144, 178, 179, 182,
        195–216, *197*, 200, 202, 203, *204*,
        205, 207, 208, 209, 259–275, 287–288
    slash-and-burn, 201
Aherlow, *334, 344, 345*
Ailsa Craig microgranite, 45
Alaska, 174
Alca impennis (great auk), *239, 241*
Alca torda (razorbill), *239*
Alces alces [elk (European) or moose
    (North American)], 227, 230, 244, 245,
    257, 258
Alces latifrons [elk (extinct)], 230

Alcidae, *239*
Alder, *see Alnus glutinosa*
Algae, 173
Allerød interstadial, *11*, 124, 129, 164, *165,
    166*, 173, 174, 176
Alluvium, 27, 28, 317, 318, 333, *340*, 349
Alnus glutinosa (alder), 143, *159*, 160, *168*,
    178, *181, 190*, 192, 193, *194*, 195, 200
Alopex (arctic fox), 232
Alosa fallax killarniensis (twaite shad), 235
Alpine glaciation, 89, 90
Altiplanation surfaces, 96, 110
Altnahinch, 286, *287, 288, 290*
Amino acid dating, 118
Amphibians, 236–237, 248
An Foras Talúntais (The Agricultural In-
    stitute), 134
Anas crecca (teal), *238*
Anas penelope (wigeon), *238*
Anas platyrhynchos (mallard), *238*
Anas querquedula (garganey), *238*
Anatidae, *238*
Andrews Wood, *80*
Anglian stage, *8*
Anglo–Norman, 214
Anguilla anguilla (catadromous eel), 235,
    257
Anguis fragilis (slow worm), 237
Animals, *see* Fauna; *see also* specific species
Animals, domestioc, 198, 259
Annagassan, *87*
Annagh Fault, *22*, 24, 25
Annagh Hill, 21, *22*, 24, 25